John Gwyn Jeffreys

British Conchology or, an Account of the Mollusca which now Inhabit the British Isles and the Surrounding Seas

Volume III: Marine Shells

John Gwyn Jeffreys

British Conchology or, an Account of the Mollusca which now Inhabit the British Isles and the Surrounding Seas
Volume III: Marine Shells

ISBN/EAN: 9783744729307

Printed in Europe, USA, Canada, Australia, Japan

Cover: Foto ©berggeist007 / pixelio.de

More available books at **www.hansebooks.com**

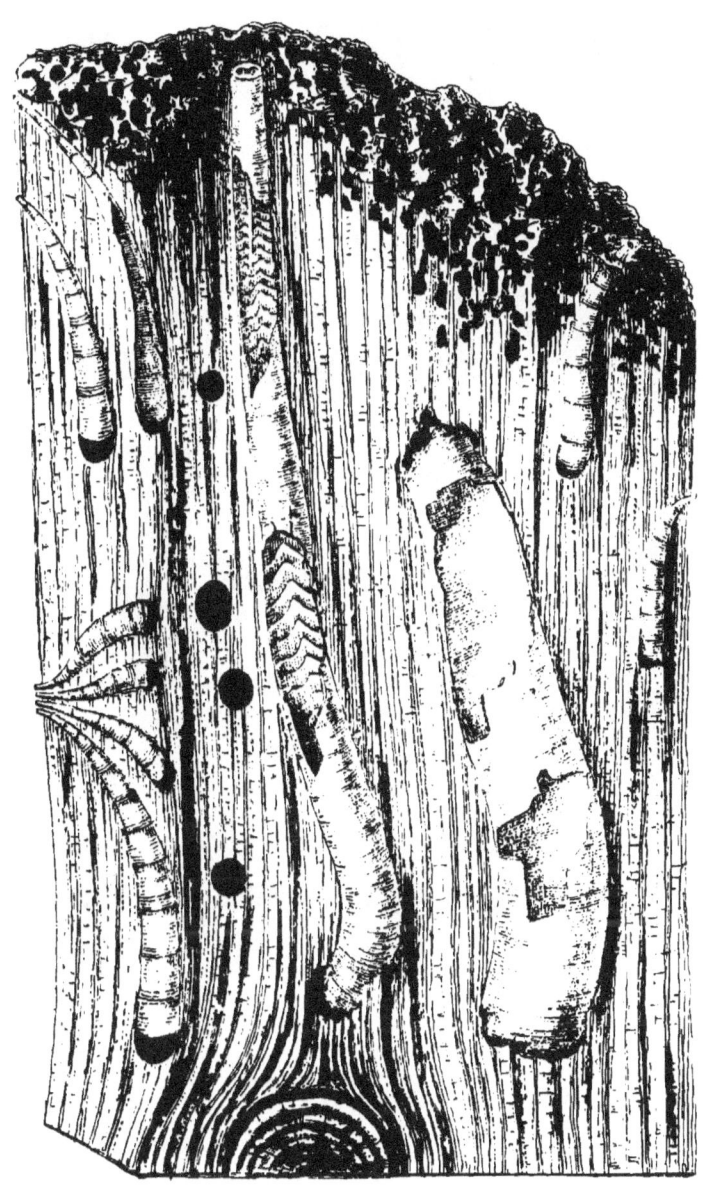

Teredo Norvegica

BRITISH CONCHOLOGY,

OR AN ACCOUNT OF

THE MOLLUSCA

WHICH NOW INHABIT THE BRITISH ISLES AND THE SURROUNDING SEAS.

VOLUME III.

MARINE SHELLS,

COMPRISING THE REMAINING CONCHIFERA, THE SOLENOCONCHIA, AND GASTEROPODA AS FAR AS LITTORINA.

By JOHN GWYN JEFFREYS, F.R.S., F.G.S., &c.

LONDON:
JOHN VAN VOORST, PATERNOSTER ROW.
MDCCCLXV.

[*The right of Translation is reserved.*]

PRINTED BY TAYLOR AND FRANCIS, RED LION COURT, FLEET STREET.

Family XVI. SOLENIDÆ, Latreille.

BODY elongated in a transverse direction: *mantle* closed in front, with its borders adhering together, open at the anterior end for the passage of a foot, and forming at the posterior end two conical tubes or siphons of different lengths, which are more or less enclosed in a common sheath: *gills* two on each side, long and narrow: *palps* corresponding with the gills in number and position, long, slender, and triangular: *foot* large and muscular, adapted for burrowing in sand.

SHELL shaped like the body, equivalve, open or gaping at both ends: *epidermis* strong and persistent, overlapping the front or ventral edges of the shell: *beaks* small: *ligament* external: *hinge* strengthened inside by a ridge: *teeth* consisting of one or two thorn-like cardinals in each valve, which are erect, curved, and interlock; laterals partly recumbent, in some cases rudimentary or wanting: *muscular scars* irregular: *pallial scar* sinuated.

This and the succeeding families of marine Conchifera differ from those described in the second volume in having the mantle more or less closed in front. Professor Oken imagined that, in a biological point of view, they typify the Nudibranchs and Salpæ; but he gave no reason for this fanciful analogy. Although the *Solenidæ* appear at first sight to constitute a natural and simple group, it will be found to comprise certain forms which connect it with other families. This resemblance has probably misled some systematists, and induced them to associate with *Solen* such very dissimilar genera as *Psammobia, Lutraria, Panopæa,* and *Mya*. The structure of the hinge, however, will always serve to distinguish any one of them from the rest.

Dr. Carpenter says that the external layer of the shells in the present family is composed of cells which form elongated prisms, with walls as straight and parallel as those of *Pinna*; but their axes are nearly conformable with the surface, cropping out somewhat obliquely upon the exterior, where their rounded terminations with distinct nuclear spots may sometimes be seen. The internal layer is very dense, and nearly homogeneous; but evident traces of cells are occasionally to be met with. Most of the *Solen* tribe are littoral, and live in sand which they penetrate for that purpose; a few are found at various and often considerable depths of water, and these prefer a more muddy habitat. None of their remains have been discovered in any geological formation older than the lower tertiaries.

Genus I. SOLECUR'TUS*, De Blainville. Pl. I. f. 1.

BODY oblong, compressed: *mantle* capable of being inflated in front: *tubes* partly separated, extended, and occasionally strangulated, issuing from a common sheath: *foot* tongue-shaped, of an enormous size.

SHELL resembling in shape a kidney bean, rather solid, nearly equilateral, concentrically striated or sculptured diagonally with imbricated ribs: *teeth*, two cardinals in the right valve, and one in the left; laterals short and rudimentary: *pallial scar* having a broad and shallow fold.

For the reasons which I have given in the last volume (pp. 327 and 434), with respect to the systematic value of characters derived from the separation or union of the pallial tubes, I prefer not placing this genus and *Ceratisolen* in one family, and *Solen* in another. The relations of *Ceratisolen* to *Solen*, through *S. pellucidus*, are too close to warrant their being assigned to different families, and the transition from the last-named species to *S. ensis* is very slight and gradual.

* A short *Solen*.

It is the genus *Macha* of Oken, who described it in the 'Allgemeine Naturgeschichte' for 1835, giving the *Solen strigilatus* of Linné as the type. This, however, was eleven years after De Blainville's publication of *Solecurtus*. Herrmannsen at first cited the date of Oken's publication as 1815, but corrected the mistake in his 'Supplement;' he disapproves of the word *Solecurtus*, as well as of *Solenocurtus* and *Solenocurtis* (emendations of Sowerby and Swainson), and suggests *Cyrtosolen*. But if we proceed in this way to rectify the nomenclature of Natural History, few of the modern names would be recognizable in their new dresses. A century ago Linné complained of a deterioration in this respect—one of his axioms being "Veterum nomina plerumque præstantissima, recentiorum pejora fuere." I fear that the lapse of time has not brought with it any improvement. The present name appears to have been compounded according to a grammatical rule called Syncope, and it has a precedent in the word lapicida (for lapidicida) used by Livy and Varro. Leach's name of *Azor* appeared in the second edition (1844) of Brown's work on British and Irish Conchology; it had *S. antiquatus* for its type.

1. SOLECURTUS CAN'DIDUS *, Renier.

Solen candidus, Renier, Tav. Conch. Adriat. p. 1. *Solecurtus candidus*, F. & H. i. p. 263, pl. xv. f. 1, 2.

BODY of a uniform bright orange-yellow colour: *mantle* somewhat paler towards the margin: *tubes* united at their bases, where the siphonal mass is large and thick, and separated at their extremities; orifices fringed: *foot* pale orange with a whitish sole.

SHELL elliptical, rather convex, but compressed in the middle, solid, opaque, somewhat glossy: *sculpture*, 40 to 50 oblique and imbricated longitudinal striæ or slight ribs, of which nearly

* White.

two-thirds cover the ventral or front part of the shell, and radiate from the beak, the rest occupying the whole of the posterior side, and diverging from an angle formed by a junction with the first-mentioned set of striæ; this angle varies from acute to obtuse, according to the number of striæ; the anterior side is not thus striated; the surface is also covered with minute and crowded longitudinal striæ resembling those observable in species of *Psammobia*: *colour* pale yellowish-white: *epidermis* like oil-skin, yellowish with a brown tint in aged specimens: *margins* slightly incurved in front, obliquely truncated with a rounded contour at each end, nearly straight behind and parallel with the front, except in the middle, where the beak forms a projection equal to the indentation on the opposite side: *beaks* pointed and nearly straight: *ligament* chrysalis-shaped, prominent, dark horncolour: *hinge-line* straight: *hinge-plate* thick and strong, reflected over the ligament and abruptly truncated at the posterior end: *hinge* supported by a strong oblique shelf-like rib: *teeth*, in the right valve two blunt cardinals curving upwards from below the beak, the posterior being much larger than the other; the left valve has a similar cardinal on the anterior side, besides a short, triangular and oblique lateral on the posterior side: *inside* chalky-white, with a slightly nacreous gloss in some parts, incipient pearls being occasionally formed on the inner edge of the mantle; margin blunt: *pallial scar* well defined; sinus oblong, and extending two-thirds across the transverse diameter of the shell: *muscular scars* distinct; anterior irregularly pear-shaped, posterior triangularly oval. L. 0·9. B. 1·9.

Var. *oblonga*. Shell narrower in proportion to its breadth.

HABITAT: In sand, between low-water mark at spring tides (Lukis), and various depths seawards, from 20 to 85 fathoms, on different parts of the coast from the Shetland to the Channel Isles, but local; more common in Bantry Bay than elsewhere. Var. 1. Guernsey (Lukis); Polkernow Cove, Cornwall (Miss Lavars); Shetland (Barlee). Believing it to be the *Solen multistriatus* of Scacchi, I find it recorded as a fossil from the neighbourhood of Antwerp, and from Gravina in Apulia. Lamarck and Brocchi appear to have mistaken a white variety of *S. strigilatus* for this shell. Its foreign distribution comprises the coasts of France, Spain, Portugal, Italy, Algeria, the Canary Isles, and Madeira. All that is known of the habits of this pretty species, we owe to the Rev. R. N. Dennis, who informs me that he was greatly surprised, when at Herm, with its activity—adding, "A couple of specimens which I had in a milk-pan of salt water were on the crawl whenever I looked at them, really travelling at a great rate for mollusks, and, without the least respect for their neighbours' comforts, walking over, and upsetting all the weaker shell-fish which were with them."

I have retained the specific name *candidus* because it is now generally accepted; but the *Solen candidus* of Renier may have been only a white variety of *Solecurtus strigilatus*, a rather common Mediterranean shell. His description is too scanty for identification, and he does not even give the last-named species as an Adriatic shell. This variety was noticed by Linné in his 'Mus. Ulr. Reg.' Olivi was at first inclined to consider it a distinct species, but after finding intermediate specimens he reduced it to the rank of a variety; it was enumerated by Chiereghini also from the Adriatic under the name of *Solen albicans*. The present species is the *Psammobia scopula* of Turton, and *Adasius Loscombeus* of Leach. I have been asked why I notice any of the bizarre names given in Leach's 'Synopsis of the Mollusca of Great Britain,' since they are quite disregarded by British naturalists. I cannot, however, forget that it is a published work, and has been circulated on the Continent. There I know that these volumes have also found a place; and if I were to ignore the works of Leach, Brown, and other authors on the subject of which I treat, I feel that I should stand justly accused of having neglected the writings of my own countrymen, and of having thus caused some confusion or inconvenience to those who study the European Mollusca. I do not regret the trouble I have taken in making this concordance, hoping and believing that it will save the labour of my fellow-workmen. Turton must have been mistaken in saying that *S. strigilatus* had been dredged in Torbay, and found by General Bingham in Cornwall, and by Mrs. Loscombe in the Scilly Isles. The collection of that lady was sold by auction about 25 years ago, when I purchased, through the late Mr. G. B. Sowerby, all the supposed British Shells contained in it. Among them were specimens of *S. strigilatus* and many other undoubtedly Mediterranean species, as well as a few from the Arctic seas. *S. strigilatus* is a much larger shell than *S. candidus,* and usually rose-coloured with two white rays.

2. S. ANTIQUA'TUS*, Pultcney.

Solen antiquatus, Pult. Cat. Dors. p. 28, pl. iv. f. 5.   *Solecurtus coarctatus*, F. & H. i. p. 259, pl. xv. f. 3, and (animal) pl. I. f. 5.

BODY rather compressed, entirely white: *mantle* having its edges fringed with short cirri: *tubes* capable of being inflated to three times their ordinary diameter, united for a considerable distance from their bases, and separate at their extremities; orifice of the branchial tube cirrous, that of the excretory one plain: *gills* partly lodged in the lower portion of the siphonal sheath, the upper pair much shorter than the other: *palps* distinctly pectinated within, and less so on the outside: *foot* thick and fleshy.

SHELL elliptical, with an oblique outline, compressed throughout, but especially in the middle, solid, opaque, slightly glossy: *sculpture,* numerous and irregular concentric striæ, and minute longitudinal lines like those in *S. candidus,* but much less distinct; the surface is also covered with equally minute and

* Decayed.

close-set oblique striæ, which appear to be impressed by the persistent epidermis: *colour* chalky-white: *epidermis* yellowish-brown, wrinkled at the sides and composed partly of delicate fibres, which are obliquely arranged: *margins* and all other characters as in *S. candidus*, except that the hinge-plate is not so much reflected, the principal or larger cardinal teeth are jagged at their crowns, and the pallial sinus is broader and not so long. L. 1. B. 2·25.

HABITAT : Sand in 4 to 50 f., on all our coasts, although sparingly. Fossil in the raised sea-bed at Belfast (Grainger), and in the Coralline Crag, as well as the Italian upper tertiaries. Bohuslän appears to be its most northern limit, and the Canary Isles the most southern. It also inhabits the intermediate district and both sides of the Mediterranean, the Adriatic, and Ægean, at various depths ranging from 4 to 40 f.

"We have seen it break up its tubes voluntarily into fragments, in the manner of the Mediterranean *Solecurtus strigilatus*" (F. & H.). Clark says that the animal, when in confinement, exserts the belly of the mantle, inflated by water, beyond the margin of the shell; but the instant it is irritated, it can place every organ *à l'abri*. The shell differs from that of *S. candidus* in being flatter and wanting the divaricating striæ or ridges.

The *Solen coarctatus* of Gmelin (from a figure in Chemnitz) is described as inhabiting the Nicobar Isles, and does not appear to be the present species. Our shell is the *S. cultellus* of Pennant, but not of Linné.

I do not believe that *Siliquaria bidens*, Chemnitz, is a native of our seas, the only testimony in favour of it being that of Pulteney, Boys, Laskey, and Turton. It is the *Solen fragilis* of the three first-named authorities, and *Psammobia tæniata* of the last, as well as the *Solen divisus* of Spengler. The locality given by Chemnitz is the Nicobar Isles. De Gerville included it in his list from Brittany under the name of *Solen pellucidus*; and I was informed by M. Cailliaud that it had been taken alive on that coast. Gould considered that it might be the *Solen centralis* of Say, a common North-American species; but this is very doubtful.

Solen gibbus, Spengler, was recorded by Dr. Turton as British, under the name of *S. declivis*, Mrs. Loscombe being supposed to have found a specimen in the Scilly Isles. It is a West-Indian shell, and known as *S. Guineensis*, Chemnitz, and *S. caribbæus*, Lamarck. This species is likewise described by Gould as North-American.

Genus II. CE′RATISO′LEN *, Forbes. Pl. I. f. 2.

Body oblong, flattened: *mantle* slightly projecting above and below on the anterior side: *tubes* for the most part separated, and considerably extended: *foot* conical, and capable of being expanded into a club-like form.

Shell resembling a bean-pod in shape, thin, nearly equilateral, sculptured in the middle with extremely fine striæ, which radiate from the beaks: *hinge* strengthened inside by a short rib, which diverges obliquely from the beak in each valve towards the front margin: *teeth*, one cardinal in the right, and two in the left valve, besides short but distinct laterals: *pallial scar* broad, with a shallow fold.

Ceratisolen is a connecting link between *Solecurtus* and *Solen*. Its shell has the shape and nearly central beaks of the former, and the texture and teeth of the latter; but it differs from both in the hinge being strengthened by internal cross ribs. The animal has its mantle-tubes separate and extended as in *Solecurtus*. No other species appears to be known except our own,

* A pod-shaped *Solen*.

although the *Pharella* of Gray is closely allied to *Ceratisolen*. The present genus is considered by some authors synonymous with *Pharus* of Leach, originally a manuscript name, and since made to a certain extent intelligible in consequence of Dr. Gray having cited, in the British Museum Catalogue of Mollusca, "*Solen legumen*" as the type or example. But, in a scientific point of view, it does not seem to matter much whether the name of any group is merely manuscript, or inadequately defined. In order to constitute a genus, it is not sufficient for any naturalist, even if he should be gifted with an eagle-eye, to pounce upon a certain species and say, "That's my genus so and so." Something more is wanting. He ought to describe its characteristics, or at all events point out in what respects it can be distinguished from other genera. I fully concur in the recommendation of the British Association Committee, "that new genera or species be *amply* defined;" and one of the grounds of this recommendation seems also to be reasonable, viz. that a large proportion of the complicated mass of synonyms, which has now become the opprobrium of zoology, has originated from the slovenly and imperfect manner in which species and higher groups have been defined. A name accompanied by a sufficient description or diagnosis, and adopted by naturalists of recognized authority, supersedes in my opinion a prior name which, from the want of such accompaniment, was in fact "vox et præterea nihil." *Ceratisolen* has also euphony on its side. "Ejusmodi vocabula Græca lingua pulcherrima sunt." (Linné's ' Philosophia Botanica,' § 222.)

CERATISOLEN LEGU'MEN *, Linné.

Solen legumen, Linn. S. N. p. 1114. *Ceratisolen legumen*, F. & H. i. p. 256, pl. xiii. f. 2 (as *Solen legumen*), and (animal) pl. I. f. 4.

BODY compressed, yellowish-white: *mantle* suffused with red: edges of the open part fringed in front, but not at the sides: *tubes* rather long, for the most part separate and diverging, of a reddish hue; orifices cirrous: *foot* reddish-purple; when contracted it is oblong and truncated, and when extended the extremity becomes club-shaped.

SHELL pod-shaped, smaller at the anterior than at the posterior extremity, semitransparent, glossy and partially iridescent: *sculpture*, numerous and fine but irregular striæ in the line of growth, and a few slight and minute longitudinal striæ in the middle, which radiate from the beaks to the front: *colour* pale yellowish-white: *epidermis* like oil-skin, yellowish-green or sometimes light orange, puckered by the radiating striæ: *margins* nearly straight in front and behind, curved obliquely upwards at the anterior end, and rounded at the other end; the dorsal compartment or area appears to be separated from the ventral part in consequence of the epidermis being thinner and of a paler hue: *beaks* blunt, inclining a little to the anterior side, which they approach within about two-fifths of the whole distance: *ligament* long, narrow at first, and expanding gradually outwards, dark horncolour or black: *hinge-line* nearly straight: *hinge-plate* thick, and short, strengthened in each valve by a slightly curved rib to support the ligament, by a long angular rib on the anterior side, and by a callous and short rib running nearly at a right angle with the beak: *teeth*, in the right valve an erect and wedge-like cardinal, and in the left two similar cardinals, which resemble a pair of nippers; each valve has a lateral at no great distance from the cardinal teeth, of an irregular shape and bent towards each other, that in the right valve being sometimes double: *inside* chalky-white, but nacreous in some parts and occasionally exhibiting minute pearls; margin rather sharp: *pallial scar* indistinct; the fold is withdrawn far into the interior: *muscular scars* irregular; the anterior elongated and extending to the central rib, the posterior trapezoidal. L. 0·9. B. 4.

* A bean-pod.

Habitat: Large sandy bays at low-water mark of spring tides in the under-mentioned localities: Christchurch, Hants (Da Costa); Exmouth; Bideford; North and South Wales; south, east, and west of Ireland. It is thrown up in the greatest profusion on the sands at Pendine in Carmarthenshire. Mr. Grainger found a single valve in the Belfast deposit; and Mr. James Smith has included it in the list of Argyleshire fossils. North of Great Britain it has only been recorded by Müller as Scandinavian; but its southern range extends from Brittany to Sicily and Algeria. Mr. M'Andrew dredged it on the coast of Portugal in 15 to 20 f., and off Malaga in 4 f.; and he obtained it on the shore at Mogador. According to Weinkauff it is common at Bona in brackish water.

This elegant shell was first recognized as English by Lister. I must venture to dissent from Linné and subsequent writers, who referred it to 'Le Molan' of Adanson. It is the *Hypogæa hirudo* of Poli.

Genus III. SOLEN *, Linné. Pl. I. f. 3.

Body narrow: *mantle* thickened in front: *tubes* for the most part united, nearly sessile, or extensile in a limited degree: *foot* flexible, when in action conical and pointed, but when at rest disk-like.

Shell cylindrical, very inequilateral, divided into diagonal compartments, sculptured only by the lines of growth: *teeth*, one cardinal in the right valve, and mostly two in the left: laterals partly erect, sometimes wanting: *pallial scar* having a narrow sinus at the posterior extremity.

This kind of shell-fish was well known to the ancients; and the estimation in which they held it as an article of food induced them to observe its habits with

* Razorfish; supposed to be the Σωλήν of Aristotle.

an accuracy at least equal to that which is shown in the accounts given by certain naturalists of our own time. According to Aristotle the Σωλῆνες were said to withdraw into their holes on a noise being made, and to sink deeper when they perceived the motion of the iron implements used for their capture. Athenæus in his learned gossip of the philosophers at supper (answering in some particulars to the 'Noctes Ambrosianæ' of our modern Athens) quotes some verses of Epicharmus, commemorative of Hebe's marriage, in which slender Solens were enumerated among the dainties at the nuptial feast. They are also mentioned by other Greek writers. Sophron says that widows were especially fond of them; it does not appear what sort of consolation they afforded. Diphilus pretended to distinguish the male from the female Solen by their shells: that of the former was striped, and the fish a good remedy for the stone and similar complaints; while the shell of the female was of a uniform hue, and its fish more savoury. They were eaten boiled or fried; but the best way of cooking them was to roast them on a wood fire until they gaped. In Pennant's time they were brought up to table fried in batter. The last-named author had a strange notion that the Solens, "when in want of food, elevate one end a little above the surface, and protrude their bodies far out of the shell"! This is repeated by Montagu and Wood.

The razorfishes (or "spoutfishes," as they were called by Grew and other naturalists of former days) usually burrow in sand at the verge of low-water mark, not perpendicularly, but in a slanting direction at an angle of about 60 degrees. On the retreat of spring tides they may be seen nearly half out of their holes, apparently taking in a supply of oxygen for their gills. They are evidently sensible of vibratory movements in the air, as well as on the ground, taking alarm at greater or less distances according to the state of the atmosphere and direction of the wind. When the Solen is disturbed it squirts out water in a strong jet; and having thus compressed the volume of its body, it lengthens and darts out its dibble-shaped foot, and rapidly disappears below the surface to a depth of two or three feet. A Solen-hunt requires considerable alertness; for if you cannot approach near enough to catch them when partly exposed to view—and this is not easy, their muscular strength being, in proportion to their size, far greater than that of a man—and you delve with your hands after them, they will probably beat you in the race. The stake is much more important to them than to you, and it calls for all their energies. Fishermen entice them out of their holes by a pinch of salt, making (as they say) the razorfish believe that the tide is coming in. Reaumur, however, considered that the salt irritates them, and causes a painful pricking sensation in the mantle, which induces them to rise to the surface and endeavour to get rid of the annoyance by expelling the salt backwards. He also noticed the blind instinct which the Solen has when taken out of its hole, and held between the fingers in the open air, suspended vertically: it protrudes its foot several times in succession, as if it were in the act of burrowing into its native sands. The account given by Poli of Solen-fishing at Naples is curious. We know that the flow and ebb of the tide there are very slight, and different from what takes place on our own shores. He tells us that the lurking-place of the Solen is betrayed by a hole in the sand, agreeing in shape with the apertures of its tubes or siphons. Where the water is shallow the fisherman sprinkles some oil on the surface, in order to see these marks more clearly. He then steadies himself by leaning on a staff with his left hand, and feels for the Solen with his naked right foot. This he catches, and holds between his big toe and the next; but although his toes are protected by linen bands, the struggles of the Solen to escape are so violent, and the edges of the shell so sharp, that very often a severe wound is inflicted by it. Where the sea is five or six feet deep, another mode of fishing is adopted. It consists in the fisherman diving or swimming under water with his eyes open, and, after having found the holes, digging with his hands for the razorfish. Sometimes the Solen so forcibly resists being taken, that it will suffer its own foot to be torn away, or will even die rather than surrender. Their power of locomotion is not limited to burrowing; they can dart from place to place in the water as quickly as a scallop, and apparently in the same way. Pliny instances the razorfish as a luminous mollusk; but this has not been confirmed by any recent observation. The breadth of the shell is very remarkable in comparison with that of any other bivalve. In the west of France they are called "couteaux" or "couteliers." Another name ("seringues") was suggested by Reaumur as more appropriate.

A. Shell somewhat curved, flattened and thin; hinge near one end, and furnished with cardinal and lateral teeth. *Cultellus*, Schumacher.

1. Solen pellu'cidus[*], Pennant.

S. pellucidus, Penn. Br. Zool. iv. p. 84, pl. lxvi. f. 23; F. & H. i. p. 252, pl. xiii. f. 3, and (animal) pl. I. f. 2.

Body compressed, varying in colour from pale yellowish-

[*] Transparent.

white to brownish-yellow: *mantle* thick, protruded a little beyond the valves of the shell; edges plain: *tubes* contiguous and nearly sessile; orifice of each fringed with tentacles or cirri of different lengths, which are spotted with yellow or flake-white: *gills* unequal in size, the upper pair not being half the depth of the lower pair: *palps* small, smooth outside and pectinated within: *foot* tongue-shaped and flexible, lying when at rest across the shell on the anterior side: *liver* green.

SHELL usually sabre-shaped, but of various degrees of curvature, tapering to each extremity, scarcely transparent except in young specimens, glossy and partially iridescent: *sculpture*, numerous and fine but irregular striæ in the line of growth, and a few slight and minute longitudinal striæ in the middle, which radiate from the beaks to the front: *colour* yellowish-white, with sometimes faint transverse streaks of salmon-colour: *epidermis* like oil-skin, yellowish-green or light orange, puckered by the radiating striæ: *margins* gently curved in front and almost straight behind, rounded at the anterior end, and obliquely truncated at the other end; dorsal area apparently separated from the ventral part, in consequence of the epidermis being thinner and of a paler hue behind: *beaks* inconspicuous, inclining a little to the anterior side, which they approach within one-fifth of the whole distance: *ligament* lanceolate, yellowish-brown: *hinge-line* straight: *hinge-plate* short, strengthened in each valve by a slight rib to support the ligament, and by a short and thicker rib at the other end, which diverges inside towards the front anterior margin: *teeth* somewhat irregular; in the right valve an erect and wedge-like cardinal, and in the left two similar cardinals, which resemble a pair of nippers; the posterior cardinal in the left valve is often branched or forked; each valve has a lateral at no great distance from the cardinal teeth, of an irregular shape, and bent towards each other, that in the right valve being occasionally double: *inside* polished; margin sharp: *scars* indistinct; pallial sinus short. L. 0·4. B. 1·5.

HABITAT: Gregarious in various parts of the British seas, in sand (often mixed with mud), at depths of from 4 to 85 f. In a fossil state it occurs at Belfast and in the Coralline Crag; and Philippi appears to have recorded it from Palermo*, under the name of *S. tenuis*.

* In the preceding volume of this work I inadvertently mentioned

All writers on Scandinavian mollusca have enumerated the present species in their lists, from the Loffoden Isles to Kiel Bay, in 3-50 f.; Collard des Cherres, Cailliaud, and Taslé have found it in Brittany; M'Andrew dredged it in the south of Portugal, off Gibraltar, in the Gulf of Tunis, and Sicily, in 15-40 f.; and Weinkauff procured it by the same means at Algiers in 20 f.

This pretty shell was discovered by the Rev. Hugh Davies about the year 1770 on the Carnarvonshire coast. Clark says that on both the mantle-tubes "are a few, large, rather long, white filaments, springing from the body of the common sheath, just below the siphonal orifices." I did not observe them in any of the specimens that I examined. The foot is sometimes red or pink of various shades. The shells are not unfrequently taken from the stomachs of haddocks. They are occasionally distorted.

It is the *S. pygmæus* of Lamarck. *S. pellucidus*, Spengler (from Chemnitz) is a tropical species, from Nicobar.

B. Shell more or less curved, tubular, and rather solid; hinge at one end, and furnished with cardinal and lateral teeth. *Ensis*, Schumacher.

2. S. ENSIS*, Linné.

S. *ensis*, Linn. S. N. p. 1114; F. & H. i. p. 250, pl. xiv. f. 2.

BODY somewhat compressed, pale drab: *mantle* having a narrow fringed slit in the middle of the anterior side: *tubes*

Palermo and Panormi as two places, being misled by Philippi using both names, in his work on the Sicilian Testacea, sometimes as different habitats of the same species. Panormus or Panormum is the ancient name of Palermo.

* Scimitar.

very short, enclosed in a sheath, speckled with brown, and encircled near the orifices by two rows of irregular cirri: *gills* narrow, nearly of equal size, and adhering throughout: *palps* pale brown, thin and delicate, smooth outside and striated within: *foot* of a dull reddish hue, obliquely sloping at the extremity, which is studded with very minute papillæ, and covered with meandering red-brown lines in the interstices: *liver* brown and granular.

SHELL resembling in shape a French bean with the ends cut off, of nearly equal diameter, opaque, glossy and partially iridescent: *sculpture*, slight and irregular striæ in the line of growth, set at two different angles; those in front are parallel with the curve of the shell, while the striæ on the dorsal area or diagonal compartment run in an opposite direction: *colour* yellowish-white, with numerous reddish-brown longitudinal streaks crossing the dorsal area: *epidermis* membranous, yellowish-green, thicker in front than at the back: *margins* equally curved before and behind, truncated at each side, but more rounded at the anterior end, which is slightly constricted; dorsal area nearly equal in size to the rest of the shell: *beaks* inconspicuous, placed close to the anterior side: *ligament* very long and narrow, yellowish-brown: *hinge-line* straight: *hinge-plate* long, strengthened by a rib in each valve to support the ligament, and thickened at the anterior end: *teeth*, in the right valve an erect and wedge-like cardinal, enclosed in the left by two much stronger and nipper-like cardinals; laterals one in either valve, long, rib-like, erect at its extremity, and somewhat bent, that of the left valve overlapping the other: *inside* nacreous; edges thin: *pallial scar* distinct, with a shallow sinus at the posterior end: *muscular scars* of unequal size; the anterior linear, posterior oval. L. 0·5. B. 3·75.

HABITAT: Sandy bays from 3 to 20 f. Fossil in all our upper tertiaries, as well as in Norway and Italy. Its European distribution in a living state extends from the Faroe Isles (Landt) to Sicily (Maravigna) and the Black Sea (Eichwald); Algeria (Deshayes and Weinkauff); Canada and the United States (Bell, Gould, and others). The range of depth is from 2 to 20 f. in the north, and from 4 to 40 f. in the south of Europe.

The locomotion of this species is the same as that of *S. pellucidus*. Its foot is permeated by a series of aquiferous ducts or canals, causing a great expansibility of that organ. Gould says that the animal is "too long for the shell;" but its power of contraction equals that of its extension. A distorted specimen, found by Mr. Barlee, and now in the University Museum at Oxford, is bent in an extraordinary degree. Whether the curve of such a crooked generation might in course of ages be increased, so as to form a nearly complete circle, would be a curious speculation.

In the time of Aldrovandus it was called by the Venetians "cappa longa." Linné doubted whether it were not a variety of *S. siliqua*. The one certainly inhabits deeper water than the other, and they are closely related in form. The present species is Lister's *S. curvus* (accidentally binominal), the *Hypogæa falcata* of Poli, *Ensis magnus* of Schumacher, *S. ensiformis* of S. Wood, and *Ensatella Europæa* of Swainson.

3. S. si′liqua*, Linné.

S. siliqua, Linn. S. N. p. 1113; F. & H. i. p. 246, pl. xiv. f. 3, and (animal) pl. I. f. 1.

Body similar to that of *S. ensis*, except in being rather less compressed, and in the foot being yellowish-white, with its extremity abruptly truncated, and marked with extremely fine close-set and very pale lead-coloured lines.

Shell so closely resembling that of *S. ensis*, except in being of a much larger size, that it is sufficient to mention the few particulars in which they differ. This is almost straight instead of curved, much deeper in proportion to its breadth, and more solid; the margins at both ends are abruptly truncated; the cardinal teeth in the left valve are blunter, and sometimes cloven; and the lateral tooth in this valve is often double. L. 1. B. 8.

* A pod.

Var. *arcuata.* Shell usually smaller, more or less curved, but equally deep relatively to the breadth.

HABITAT: Common on all our sandy shores which are uncovered at spring tides; seldom beyond that limit, although in the Dredging Report of the British Association in 1850 it is stated to have been taken in the Orkneys at a depth of 12 f. The variety is found on many parts of our coasts, especially those of Ireland and Scotland—I have a specimen from Burra Firth in Unst, of unusual dimensions, viz. nearly $1\frac{1}{4}$ inch long or deep by 7 inches in breadth; Norway (Sars). This variety was noticed by Turton in his 'Conchological Dictionary;' it is referred by Forbes and Hanley to *S. ensis*. The late Dr. Lukis found it living with that species in Belgrave Bay, Guernsey, and sent me specimens of both for comparison. The typical form occurs in many of the newer or postpliocene deposits, as well as in the Norwich and Red Crag; Uddevalla (Malm); Sicily (Philippi). Its foreign range comprises Behring's Straits, the North-east coast of America, Faroe Isles, and all the intermediate shores to the Ægean, including the African side of the Mediterranean.

In Lister's days it was called in Yorkshire "Hose-fish," and caught (if the tide was out at night) by candle-light. He adds that they make a delicious sauce, and have the flavour of shrimps. "In Ireland it is much eaten in Lent" (Da Costa). Fleming says that when a little stale they are a tempting bait for cod and haddock. The teeth are liable to vary. In a specimen from Oxwich Bay near Swansea the laterals are placed closer than usual to the hinge; and that of the left valve is branched, as in *S. pellucidus*, and divided into three.

S. novacula of Montagu and *S. ligula* of Turton

(judging from a comparison of authentic specimens with the descriptions of those authors) are varieties of *S. siliqua*, and only distinguishable by the absence or size of some of their teeth. The present species is the *Hypogœa crinita* of Poli, *S. gladius* of Bolten, and *S. gladiolus* of Gray.

C. Shell straight, tubular, and rather solid: hinge at one end, and only furnished with a single lateral tooth in each valve.

4. S. VAGI'NA*, Linné.

S. vagina, Linn. S. N. p. 1113. *S. marginatus*, F. & H. i. p. 242, pl. xiv. f. 1, and (animal, siphon only) pl. I. f. 3.

Body cylindrical, pale yellowish-brown: *mantle* thickened: *tubes* wrinkled across; each is encircled by several rows of brownish spots, and near the extremity by a row of very short tentacular cirri; orifice of the lower tube distinctly scalloped, that of the upper one plain: *gills* long, linear, orange-brown: *palps* large, sharp-pointed: *foot* oblong, yellowish-white.

Shell exactly cylindrical, and of equal size throughout, somewhat glossy, opaque: *sculpture* as in the preceding two species: *colour* pale yellowish-brown (with an orange tint in aged specimens), and marked with streaks of a darker hue in the transverse line of growth: *epidermis* membranous, yellowish-brown: *margins* equally straight before and behind, obliquely truncated at the anterior end, which is deeply constricted (as if it had been tied while in a soft and plastic state with a string), and transversely truncated or very slightly curved at the posterior extremity; dorsal area or compartment not so distinct as in the other species: *beaks* inconspicuous, separated by the constriction from the anterior end: *ligament* very long and narrow, dark horncolour: *hinge-line* straight: *hinge-plate* long, strengthened by a rib in each valve to support the ligament, and thickened at the anterior end: *teeth*, in each valve a single wedge-like cardinal, resembling in shape the leaf of a water-lily, and attached to the hinge by an obliquely twisted stalk; the tooth in the right valve is outermost: *inside*

---

* A sheath.

chalky-white; edges thin: *pallial scar* well marked, placed far within; sinus deep, but narrow, defined by a broad line on each side, like the prongs of a steel fork: *muscular scars* deep; anterior linear, and parallel with the hinge-line; posterior oblong. L. 0·85. B. 5.

HABITAT: With *S. siliqua*, but more local. Guernsey and Jersey (Hanley); Weymouth (Pulteney); Exmouth (Clark); Kingsbridge (Montagu); Falmouth (R. L. King); Laugharne in Carmarthenshire and the adjacent coasts (Montagu and others); Anglesea (Pennant); north, east, and south of Ireland (Thompson and others). "Alluvial deposits," Belfast (Hyndman and Grainger); Italian tertiaries (Ménard de la Groye, Brocchi, and Philippi). Its exotic range comprises Norway (Lovén, and Asbjörnsen); north coast of Holland (Waarden-burgh); Heligoland (Frey and Leuckart); France (De Gerville and others); Portugal (M'Andrew); Italy, from Spezzia (Capellini) to Sicily (Maravigna); Adriatic (Chiereghini); Black Sea (Kutorga); Algeria (Deshayes and Weinkauff); St. Michael, Azores (Drouet); Red Sea (Philippi).

The ancient naturalists had some strange notions as to the sexes of the Mollusca. Aristotle, as is well known, believed in their spontaneous generation; but a different opinion prevailed about three centuries ago, when Belon and Rondelet described *S. siliqua* as the male, and *S. vagina* as the female of the same species. The reasons which they gave for this distinction were not altogether uncomplimentary to the fair sex, consisting in *S. vagina* being (although smaller) of a uniform complexion, and more sweet-savoured than the other. Reaumur and Deshayes have given accounts of the animal tolerably agreeing with my own. At Cherbourg, Lisbon, and Spezzia it is sold in the fish-markets, but not so much esteemed as *S. siliqua*. The flavour is said to be peculiar. Poli mentions its being so acrid, that none but the poorest would use this kind for food. The shell differs from that of the last-named species in being more regularly cylindrical, deeper in proportion to its breadth, and of an orange-grey instead of a purplish-green colour; the diagonal compartment is less marked; the sides are more truncated; the anterior end is constricted; and it has fewer teeth.

Linné's description of *S. vagina* in the 'Mus. Lud. Ulr. Reg.' is peculiarly appropriate to this species, although in the 12th edition of the 'Syst. Nat.' he appears to have united with it an Indian species which has been named *S. truncata* by W. Wood. Our species (as a mollusk) is the *Hypogæa tentaculata* of Poli, and (as a shell) the *S. marginatus* of Pulteney and Donovan. The last-mentioned author said it was not the *S. vagina* of Linné, but he gave no reason for saying so.

Family XVII. PANDORIDÆ, Gray.

Body oval or oblong : *mantle* having a slit on the anterior side for the passage of a foot, and forming on the other side a tubular sheath : *tubes* short, united nearly to their openings, which are fringed : *gills* two on either side, each pair being more or less united, long, narrow, and slightly curved : *palps* corresponding in number with the gills, and triangular : *foot* tongue-shaped.

Shell oval or oblong, inequivalve, pearly, gaping at the posterior side, which is flexuous and elongated, and projects upwards : *beaks* very small : *cartilage* at the posterior side, wholly internal, long and oblique : *hinge* strong : *teeth*, either a single cardinal in each valve, or an oblong plate, which is attached only to the cartilage and partly covers the hinge : *pallial scar* slight, and narrowly sinuated : *muscular scars* small.

The shape of the shell, its nacreous substance, and the absence of an external ligament are the chief characteristics that distinguish this small family. The two genera which compose it have a different hinge-structure, but are in other respects so closely allied, that it is more convenient to place them together. The *Pandoridæ* inhabit sand at various depths.

Genus I. PANDO'RA*, Hwass. Pl. I. f. 4.

BODY oval, compressed on one side and rather tumid on the other, thin, and gelatinous: *gills* free, except at their bases, where each pair is united, and terminating in the tubular sheath: *palps* short: *foot* small, thick, and swollen at the point.

SHELL oval, inequilateral, scaly and smooth; left valve flat and the other convex: *epidermis* membranous and thin: *teeth* consisting of a plate-like cardinal in each valve: *pallial scar* pitted at intervals: *muscular scars* well marked, roundish-oval.

The merit of instituting the genus *Pandora* is due to Hwass, a German justiciary, and not to Bruguière as is commonly supposed. Both gave the same species (*Tellina inæquivalvis*, Linné) as the type. This is clearly shown by the 11th volume of Chemnitz (p. 211), which was published between two and three years before the 'Encyclopédie Méthodique.' Carpenter has remarked the complete conformity that exists between the shells of the present genus and *Avicula*,—namely, in the regular prismatic arrangement of the cellular structure, the axes of the prisms being perpendicular to the surface; in the presence of distinct partitions between the cells, forming a persistent membrane, which is left after decalcification; and in the truly nacreous interior. The genus appears to be of comparatively recent origin; for (according to

* A mythological character.

Searles Wood) no well determined fossil species have been met with in any formation older than the Paris basin. The animal was included by Poli in his genus *Hypogœa*. For the shell Bolton proposed *Calopodium*, and Brown *Trutina*.

PANDORA INÆQUIVAL'VIS*, Linné.

Tellina inæquivalvis, Linn. S. N. p. 1118. *P. rostrata*, F. & H. i. p. 207, pl. viii. f. 1-4, and (animal, as *P. obtusa*) pl. G. f. 10.

BODY transparent, with flake-white specks; *mantle* thin, scarcely (if at all) protruded: *tubes* short, separate although nearly close together, issuing from a very slight, pellucid and membranous sheath, which extends beyond the shell at its posterior end, and is partly continued round the edges; orifices wide, plain but jagged: *gills* unequal-sized, the upper being twice the size of the lower pair, which are almost rudimentary; they are pectinated by the blood-vessels on both surfaces: *palps* very short, reddish-brown, striated transversely, and often overlapping each other: *foot* white: *liver* green: *ovary* red-brown.

SHELL irregularly triangular, right or convex valve considerably overlapping the other; it is variable in thickness and opacity, and somewhat glossy: *sculpture*, slight plait-like marks of growth, and sometimes a few imperfect longitudinal wrinkles on the flat valve, which are only perceptible in front and appear to radiate from the beak: *colour* pearl-white; epidermis filmy: *margins* rounded or obtusely angular on the anterior side, forming in front a nearly semicircular but oblique curve, which is prolonged at the posterior side to a blunt point; dorsal margin straight or slightly incurved, furnished in the right valve with a double furrow, and in the left with a double ridge, both of which extend from the beak to the posterior end or point: *beaks* extremely minute and tubercular; umbones not prominent: *cartilage* horncolour, running inwards on the posterior side at an acute angle with the dorsal margin, and occupying a groove in each valve, the sides of which are thickened: *hinge-line* straight, or more or less incurved: *hinge-plate* long, strengthened by a rib in the left valve, that fits into a slight furrow in the opposite valve: *teeth*, in the right valve

* Valves unequal in size.

an erect cardinal, set at a right angle with the hinge-line, and in the left valve a longer and somewhat horizontal cardinal, set at an acute angle with the upper margin of the anterior side; the teeth and cartilage are on opposite sides of the beak, and diverge from each other: *inside* highly polished and iridescent, slightly striated in a radiating direction; edges thin and sharp: *scars* more or less distinct, according to the thickness of the nacreous lining. L. 0·6. B. 1·25.

Var. 1. *tenuis.* Shell much smaller, and of a delicate texture, proportionally broader or more produced at each end, with an oblique and flexuous outline; dorsal margin straight.

Var. 2. *obtusa.* Shell smaller and thinner, longer in proportion to its breadth; the posterior side larger, and not so much produced or extended; dorsal margin also straight.

Monstr. Shell oval, with the sides shorter than usual; dorsal margin projecting a little outwards.

HABITAT: In sand, Channel Isles, at the recess of spring tides, and in shallow water; often among *Zostera marina.* Var. 1. Between 85 and 100 f. off Unst in Shetland. Var. 2. From 7 to 50 f. on all our coasts. The monstrosity is from the Hebrides and Shetland. In a fossil state the typical form occurs in the Coralline Crag, and the variety *obtusa* in the Red Crag; both are noticed by Philippi from different parts of the tertiary formation in Sicily. The first has only a southern range, from Guernsey to the Ægean; while the distribution of the other is wider, reaching to the Canaries in the same direction, and extending northward to Spitzbergen. In the Ægean, Forbes gave 4 f. for *P. inæquivalvis,* and 71-10 f. for *P. obtusa*; and at Mogador M'Andrew recorded the respective depths of 3 f. and 35-40 f. for the two forms. The observations made by M. Martin in the Gulf of Lyons showed similar results.

The animal is shy and easily alarmed. Lacaze-Duthiers, in his valuable essay on the development of the gills in Lamellibranchiate Conchifera (Ann. Sc. Nat. 4^e

sér. Zool. ii.), remarks with respect to this species, that the outer gill, which resembles a hood, might at first sight be taken for a single leaf, so disproportionately small is its size. He considers it a case of arrested development. Mr. Jordan says, "Whilst collecting specimens at Jersey, I noticed that they have a habit of squirting, like *Saxicava rugosa* and the *Pholades* when first touched; one individual ejected a fine stream, fully sixteen inches high." In Mr. Clark's description of the animal of var. *obtusa* the tubes are stated to be fringed at their orifices with fine white short cirri; the margin of the sheath, in some specimens, is marked by a fine orange line; and the base of the cirri and margins of the orifices are usually encircled by a dead-white narrow thread. The ovary is of a reddish-brown colour. I found it to contain in July an immense mass of vesicular ova in different states of growth; the more forward of them resembled in shape some species of *Cythere*. Adult specimens vary in their comparative length and proportions, as well as in the prominence of the ridges on the dorsal side. The difference between the typical shell and the variety *obtusa* apparently arises from the nature of their respective habitats—the one being sublittoral, and the other belonging to deeper water. An intermediate form has been taken by Cailliaud on the coast of Brittany, and by M'Andrew at Corunna. On a superficial view, indeed, it would seem as if a valid distinction existed in the length from the beak to the front margin being always greater in *P. inæquivalvis* (or *rostrata*), and on the posterior side in *P. obtusa*; but this only shows that varieties, as well as species, have some one character of their own. Such may be expected when the conditions of life vary. The extension of the posterior side in the typical form may be caused by the difference of locality. When the littoral zone is sandy, the surface is apt to be disturbed by waves and occasional storms, so that the stratum may be of a greater or less thickness at one time than at another: now it is covered by a deposit of material thrown up by the sea; in a few days this cover may be stript off. In order to prevent its tubes being choked by an accumulation of the imported material, the *Pandora* living between tide-marks gradually lengthens that end of its shell. The variety which inhabits deeper water is not exposed to fluctuations of this kind; it therefore does not require any such provision, and lies undisturbed in its level bed. This may explain the variation in the proportions of length and breadth which is exhibited by the two forms. The difference of thickness in the shells of *P. inæquivalvis* and its varieties also depends on habitation. I am inclined to think that, with regard to every species living both in the littoral and coralline zones, the shell is thicker in the former and thinner in the latter. Examples to illustrate this proposition occur in *Venus gallina* and its varieties *striatula* and *laminosa*, *Mactra solida* and its variety *elliptica*, *Trochus ziziphinus* and its small conical variety, *Buccinum undatum* and its variety *Zetlandica*, and in many other species. Experiments made by Dr. Davy, Forchhammer, and Bischoff have proved that the quantity of carbonate of lime held in solution by sea-water, and from which shells are produced, is greater on the coast than in the ocean; it is derived from the land, and brought down to the sea by rivers and streams, the washings of rain, and the action of waves. This fact ought not to be lost sight of in discriminating species from varieties of which the comparative solidity and size are the sole or chief criteria.

Lamarck at first named this species *P. margaritacea*,

and afterwards *P. rostrata*; the young is the *P. flexuosa* of Philippi, and the animal the *Hypogæa gibba* of Poli. It is also the *Trutina solenoides* of Brown. The variety *obtusa* was described by Meuschen as *Anomia tabacca*, by Montagu as *Solen pinna*, and by Leach as *P. glacialis*; the young is the *P. oblonga* of Philippi. Lamarck changed the specific name imposed by Linné, either from caprice (as seems to have been his custom), or on the ground that it denoted an essential character of the genus and therefore was superfluous. I am not satisfied with this reason, believing that all designations, whether generic or specific, are merely symbols of distinction, and that the law of priority in zoological nomenclature ought not to be disregarded because the name of one species represents a character that is common to others of the same genus. I have restored the original name, by which this species is well known throughout the greater part of Europe.

Genus II. LYON'SIA*, Turton. Pl. II. f. 1.

Body oblong, somewhat compressed, rather thick: *gills* forming apparently a single leaf on either side, in consequence of each pair being doubled upon itself: *palps* long and narrow: *foot* tongue-shaped, rather large, flattened, and provided with a byssal groove.

Shell oblong, nearly equilateral, finely striated lengthwise; right valve more convex than the left: *epidermis* fibrous: *hinge* furnished with a free plate or ossicle, which covers the cartilage: *muscular scars* slight; anterior oblong, posterior roundish.

A link connecting the *Pandoridæ* with the *Anatinidæ*, having the same shape and pearly nature as the former, and the peculiar hinge-process or ossicle of the latter

* Named after the late Mr. W. Lyons, an active British conchologist.

family. This relationship has also been remarked by Carpenter in his account of the microscopical structure of the shell. The mantle-tubes are united in the present genus, as well as in *Pandora*; they are separate in the *Anatinidæ*. Philippi considered *Lyonsia* to be closely allied to *Galeomma*; but I cannot see much resemblance between them. Mr. W. Wood was the first to notice the curious appendage which covers the hinge. It was conjectured by Clark that it acted like the check-tape of a trunk, to prevent its being opened too widely. This might be so if it were attached to the shell. I should be disposed to attribute to it quite a contrary action, and to believe that its use may be to strengthen the hinge, and to protect it from being squeezed too closely and broken, as is frequently the case with certain species of *Anatina* and *Thracia*. The ossicle of *Lyonsia* is of a different shape and position from that of the *Anatinidæ*. In those it is semiannular, and clasps the hinge *crosswise* with the two ends; in the present genus it is flat, and lies over the hinge *lengthwise*, with one end at the anterior and the other at the posterior side of it.

This genus has several synonyms, including *Magdala*, Leach, *Osteodesma*, Deshayes, and *Pandorina*, Scacchi.

Lyonsia Norve'gica*, Chemnitz.

Mya Norvegica, Chemn. Conch. Cab. x. p. 345,. t. 170. f. 1647, 8. *L. Norvegica*, F. & H. i. p. 214, pl. viii. f. 6-9, and (animal) pl. II. f. 3.

Body milk-white, sometimes with a tinge of yellow or pale brown : *mantle* thin ; edge studded at the anterior side with from 8 to 10 papillæ, which are of a darker hue in coloured individuals : *tubes* nearly sessile ; orifice of the lower tube fringed with a few short, thick, and close-set cirri ; upper tube having a plain bulbous orifice, but furnished with the

* Norwegian.

usual hyaline protrusile valve; this tube is speckled with minute sand-like points; each tube is encircled at its base by a few cylindrical filaments, which are somewhat longer than the tubes, and are occasionally speckled with flake-white: *gills* and *palps* pale brown: *foot* flexible, white, cloven at the heel, whence byssal filaments are produced.

SHELL irregularly rhomboidal, the left or convex valve somewhat overlapping the other, of a membranous consistency, opaque and lustreless: *sculpture*, numerous rows of fine granulated striæ, radiating from the beaks to the outer margins; between each of these striæ are five or six rows of minute and close-set tubercles or pores, which are connected with the tubular structure of the external layer of the shell; there are also occasional lines of growth: *colour* pale yellowish-white: *epidermis* light-brown, and having an agglutinating property, by means of which the surface becomes invested with a coat of sand and Foraminifera or other organic remains: *margins* broad and rounded on the anterior side, flexuous or somewhat indented in front, curved obliquely upwards to the posterior side, which is prolonged into a beak-like form and truncated at that end, with a double but indistinct ridge in the left valve, and a corresponding furrow in the right; dorsal margin incurved: *beaks* triangular, inclining to the anterior side; umbones rather prominent: *cartilage* golden-yellow, lying nearly parallel with the hinge-line, and contained in a groove in each valve, the sides of which are thickened: *hinge-line* obtusely angular: *hinge-plate* long, strengthened by a rib in the left valve, which fits into a slight furrow in the opposite valve: *ossicle* irregularly quadrangular, with the broader end towards the posterior side, where it is notched or forked: the narrower end is truncated and placed immediately under the beaks: *inside* highly polished and iridescent; edges thin, reflected or folded outwards in the right valve: *muscular scars* often double. L. 0·875. B. 1·7.

Var. *elongata*. Shell more slender, and transversely elongated: *Osteodesma elongata*, (Gray) Hanley, Rec. Sh. p. 25, pl. 13. f. 27.

HABITAT: All our coasts, in sand, from 4 to 86 f., but nowhere common. The variety has been found in the Hebrides and Shetland. *L. Norvegica* has not been noticed as a British fossil; but Philippi has recorded it from the newer tertiaries of Sicily. Its foreign range in a living state comprises the Sea of Ochotsk, and the coasts of Iceland, Scandinavia, France, Italy, Algeria, Greece, and Madeira, at depths varying from 10 to 70 f. in northern, and from 4 to 70 f. in southern latitudes.

Miss Hutchins is the reputed discoverer of this remarkable and interesting shell. According to Clark the gills consist of a single leaf on each side; the tubes appear to be enclosed in one sheath, which has the margin finely dentated; and each orifice is garnished with about eight white simple cirri, and as many black equidistant points at their external edges. He has not mentioned the filaments at the base of each tube. Such discrepancies are extremely perplexing; and if the identification of any species depended solely on characters afforded by the soft parts, the study of conchology would be almost impracticable. The faculty and means of observation, as well as the good faith possessed by Mr. Clark, were certainly not inferior to those which I have exercised; and yet how different is the result! The microscopic pustules covering the surface of the shell appear to be the ends of the tubes which compose the outer layer; these are open in the young, and closed in the adult. The inner layer is entirely nacreous. The shells are occasionally found in the stomach of the red gurnard. Dr. Lukis supposed that the young disguise themselves in their sandy coating more completely than the adult; but this is not always the case. The epidermis is of a gelatinous or viscous nature, and thus grains of sand as well as organic particles become attached to it.

It is the *Mya nitida* of Fabricius (but not of Müller), *M. striata* of Montagu, *Amphidesma corbuloides* of Lamarck, *Mya pellucida* and *Myatella Montagui* of Brown,

Tellina coruscans of Scacchi, and *Pandora? æquivalvis* of Philippi. It likewise appears to be the *Mya membranacea* of Gmelin, from Müller's 'Prodromus,' judging from the description; although Dr. Mörch informs me that the shell figured in Olafsen and Povelsen's Voyage to Iceland, and referred to by Müller for this species, represents *Astarte sulcata*, var. *elliptica*.

Family XVIII. ANATI'NIDÆ, D'Orbigny.

BODY oval or oblong: *mantle* very thin: *tubes* long; orifices fringed: *gills* one on each side: *palps* two on each side: *foot* lanceolate or tongue-shaped, small, and compressed.

SHELL oval or oblong, slightly inequivalve, gaping more or less on each side, and truncated at the posterior end: *beaks* small, inclining to the posterior side, mostly fissured: *epidermis* slight: *ligament* sometimes external and situate at the posterior side, besides invariably an internal *cartilage*, which is contained in a pit or receptacle under the beak in each valve: *hinge* furnished with a free crescentic ossicle, placed across the hinge-line at the anterior side of the beaks; otherwise toothless: *pallial scar* narrowly but deeply sinuated: *muscular scars* small and irregular.

The typical genus, *Anatina*, is a native of tropical seas. It may be distinguished from *Thracia* in the tubes being united, the pearly nature of the shell, and in having inside an oblique falciform rib, proceeding from the cartilage-pit towards the posterior side in each valve. This process is formed apparently in consequence of the beaks being fissured in that direction, and it serves as an upright girder to strengthen the shell. Something of the same kind, however, may be observed in most species of *Thracia*. The genus *Anatina* of Schumacher is different from that of Lamarck, and belongs to the *Mactridæ*. The *Anatinidæ* usually frequent a sandy or nullipore bottom at various depths; but a British species of *Thracia* (*T. distorta*) prefers a more secluded habitat, and occupies the deserted holes made by *Saxicava rugosa* in limestone, or other rock-cavities, as well as tufts of *Corallina officinalis*.

Genus THRA'CIA*, Leach. Pl. II. f. 2.

BODY oval: *tubes* separate.

SHELL oval, nearly equilateral, rather thin, having a tubercular or shagreen-like surface: *colour* sometimes tinged with yellow.

Montagu proposed his genus *Ligula* chiefly to receive the species which we now assign to *Thracia*; but, for the reasons which I have given in the second volume of this work (p. 433), it is inexpedient to retain that name in the Mollusca.

According to Dr. Carpenter the minute elevations or points, that roughen the surface of the shell, represent numerous isolated cells filled with calcareous matter, and forming a superficial coating superposed upon the ordinary external layer; the epidermis is extended over these points, and sinks down into their interspaces, just as the human epidermis covers the papillary surface of the true skin. The proper external layer is composed of polygonal cells, with sharply defined boundaries, having large nuclear spots strongly resembling some of those which are exhibited in *Mya arenaria*. The total quantity of animal matter or membrane contained in the substance of the shell is extremely small, although the cellular structure in all probability results from the calcification of animal tissue. The structure of the internal layer is scarcely distinguishable. The power of tension

* A Sea-Nymph.

continually exercised by the strong and elastic cartilage exceeds that of the shell; and the latter being the weaker body, gives way and is split in the conflict. Only one species (*T. distorta*), which is comparatively more solid than the others, resists the strain and remains uninjured.

The synonymy of the European species has been lamentably perplexed ever since the time of Pennant, notwithstanding the pains taken by Lovén and the authors of the 'British Mollusca' to unravel the tangled skein. This makes it extremely difficult to define with any certainty the geographical distribution of some of these species. Geologically *Thracia* appears to be an ancient genus. "Fossils of this form are found in the lower Oolites, and doubtfully so in the Carboniferous series" (S. Wood). It is the genus *Odoncincta* of Da Costa, and has received other equally barbarous names from modern authors. *Cochlodesma*, Couthouy, does not differ in any respect except in the absence of an ossicle: all the British species of *Thracia* possess this appendage.

A. Nearly equilateral.

1. THRACIA PRÆTE'NUIS*, Pulteney.

Mya prætenuis, Pult. Cat. Dors. p. 28. pl. iv. f. 7. *Cochlodesma prætenue*, F. & H. i. p. 235, pl. xv. f. 4.

BODY thin, clear white: *gills* strongly pectinated, each divided by an oblique furrow into two parts, the upper being less deep than the lower portion: *foot* white.

SHELL triangularly oval, compressed, opaque, somewhat glossy; right valve more convex than the left, and slightly overlapping it: *sculpture* (besides the usual marks of growth), close-set and microscopical transverse hair-like lines or scratches

* Very thin.

on every part except the posterior side, which is covered with numerous concentric rows of tubercles interspersed with fine striæ that appear to radiate from the tubercles: *colour* milk-white: *epidermis* membranous, creamcolour: *margins* semi-circular on the anterior side, moderately curved in front, with a slight indentation or flexuosity towards the posterior side, which is more rounded than truncated; posterior dorsal margin sloping and straight; anterior dorsal margin slightly curved: *beaks* projecting, with an abrupt excavation underneath, caused by the compression or fracture of the hinge; this part is defined by a sharp but irregular ridge in each valve: *ligament* exceedingly small (being only visible in fresh specimens), placed close to the hinge on the posterior side; it is dark horncolour: *cartilage* golden-yellow, contained in a triangular and shallow cup, which is solid, attached to the hinge-plate by a ledge, and projects inwards horizontally and at a right angle with the hinge; from the lower part of this cup or cartilage-pit in each valve runs an oblique and sharp ridge to the posterior adductor muscle, and the shell is considerably thickened in that part: *hinge-line* obtusely angular: *hinge-plate* narrow and thin: *ossicle* falciform, clasping the hinge close to the beak on the anterior side: *inside* chalky-white, except the muscular scars and below the cartilage-pit, where the surface is polished and nacreous; it is furnished with a slight rib in the line of fracture; edges sharp: *pallial* and *muscular scars* nearly marginal. L. 0·85. B. 1·3.

Var. *curta*. Shell more oval, or longer relatively to its breadth.

HABITAT: Land's End to Unst, from 4 to 60 f.; and at low water, spring tides, on the coasts of Kerry and Galway. The variety is from Shetland and Cork Harbour. Fossil in the Coralline Crag (S. Wood); Christiania (Sars); Palermo (Philippi). The extra-British distribution comprises Iceland, the Faroe Isles, Scandinavia (3–30 f.), the north of France, Adriatic, Naples, and Sicily.

The course of striation, or the arrangement of the microscopical granules, in this shell is the reverse of that in *Lyonsia Norvegica*, viz. transverse instead of longitudinal. Petiver called the present species *Chama prætenuis*, or the "small, white, thin *Spoon-hinge*." Some authors have referred it to Schumacher's genus *Periploma*; but his description and figure give a different and more complicated hinge-structure. Leach carved out of it two other genera, *Bontia* or *Bontæa* and *Galaxura*. It is the *Anatina truncata* of Lamarck, and *A. oblonga* of Philippi, the former having been identified by Collard des Cherres, and the latter by Sars, with typical specimens. Collard des Cherres enumerated it in his list as *Periploma myalis*, and Chicreghini as *Tellina fragilissima*; S. Wood described it as *Cochlodesma prætenerum*.

2. T. papyra′cea*, Poli.

Tellina papyracea, Poli, Test. Sic. i. p. 43, t. xv. f. 14, 18. *Thracia phaseolina*, F. & H. i. p. 221, pl. xviii. f. 5, 6, and (animal) pl. II. f. 4.

Body varying in colour from clear white to pale brown, covered with minute and numerous tubercles or papillæ, that give the surface a frosted appearance: *mantle* protruded considerably at each end; edges plain: *tubes* separate, cylindrical, but short and wide, capable of being much inflated and unequally distended, sometimes club-shaped at their extremities; the upper tube is marked with eight, and the lower with four faint longitudinal lines or streaks, which terminate at the orifices in the same relative number of short, thick, and blunt cirri: *gills* forming two large suboval plates, each divided in the middle by a deep and oblique furrow: they are smooth within and pectinated without: *palps* equal-sized, short, and triangular: *foot* flat and expansile, bluish-white.

Shell thinner than the last species, and more inequivalve, more convex (though compressed towards the front and sides), and more elongated transversely: *sculpture* similar, but more delicate: *epidermis* less persistent, and having usually a rusty tinge on the posterior side: *margins* not so much rounded on the anterior side, decidedly and more abruptly and obliquely

* Paper-like.

truncated at the posterior end, with a well-defined angle on that side; posterior dorsal margin somewhat recurved, instead of sloping; anterior dorsal margin longer: *beaks* less prominent, with a slighter and less distinct excavation below them: *ligament* rather large, but short, yellowish-white or pale brown, keeping the valves asunder on the posterior side, and when removed leaving a lanceolate gap: *cartilage* yellowish-brown; pits obliquely elongated sideways, and not projecting so far inwards as in *T. pratenuis*; connecting ridge at the bottom thicker and less distinct: *ossicle* semiannular, placed as in that species: other particulars the same, except that in the present species the beaks only (and not the hinge) are fissured, and the rib-like mark of repair in the interior is therefore wanting. L. 0·6. B. 1·1.

Var. 1. *gracilis*. Shell more slender, and approaching a cylindrical shape, thinner, more uniformly convex; posterior end shorter in proportion.

Var. 2. *villosiuscula*. Thicker, and less elongated transversely; posterior angle more rounded, but truncated. *Anatina villosiuscula*, Macgillivray in Edinb. New Phil. Journ. April 1827, p. 370, pl. i. f. 10, 11. *T. villosiuscula*, F. & H. i. p. 224, pl. xvii. f. 4, 7.

Monstr. Furrowed on the posterior slope, or having misshapen valves.

HABITAT: Sandy bays in the laminarian zone; rather common. Var. 1. Bantry Bay, and twenty miles north of Unst in 86 f. Var. 2. As widely diffused as the typical form, but usually in deeper water, or where the supply of calcareous material is more plentiful. Fossil at Belfast (Grainger); Kyles of Bute, and Lochgilphead (Geikie); Coralline Crag (S. Wood). Both the typical kind and the variety *villosiuscula* range from Iceland to the Ægean and Canaries, at depths of from 2 to 35 f.; an intermediate form has been taken by Steenstrup in Iceland, by Malm on the Swedish coast, and by M'Andrew in Vigo Bay. Mr. Malm's son found the variety subfossil at Uddevalla, and Sars in the newer part of the glacial formation near Christiania.

My finest specimen measures only two lines more in length and breadth than the average dimensions given in the description; but Lilljeborg obtained some in Finmark of a much larger size. I made an inexcusable blunder in stating (Ann. Nat. Hist. Sept. 1859) my belief that the present species was identical with *T. distorta*.

This is the *Tellina fragilis* of Pennant (but not of Linné); and that specific name ought perhaps to be retained. It is also the *Mya punctulata* of Renier, *Ligula pubescens* of Montagu ("smaller specimens"), *Amphidesma phaseolina* of Lamarck, and *Anatina truncata* of Macgillivray (Aberd. Moll.). Chiereghini enumerated it in his list of Adriatic shells as *Mya truncata*, according to his learned editor Dr. Nardo: this shows the difficulty of ascertaining the limits of geographical distribution, if we trust to local catalogues which have not been compiled by competent authorities, nor been subjected to such revision. The variety *villosiuscula* is the *T. ovata* of Brown, and *Anatina intermedia* of Clark.

3. T. PUBES'CENS*, Pulteney.

Mya pubescens, Pult. Cat. Dors. p. 27, pl. iv. f. 6. *T. pubescens*, F. & H. i. p. 226, pl. xvi. f. 2, 3.

SHELL of a gigantic size and considerable solidity compared with *T. papyracea*; it is also more of an oval shape, being proportionally longer from the beaks to the front margins, and shorter in a transverse direction; the smaller or left valve is flatter, and the inequality of the valves is more observable; the umbonal part is sculptured by rather strong, but obscure concentric ribs or folds; the whole surface is finely granulated, although more strongly at the sides; the colour is sandy, instead of white; it is of a dull hue; and the epidermis has more of a yellowish cast. L. 2. B. 3.

* Full-grown.

HABITAT: Cornwall, Devon, and Dorset; procured by trawling. The reputed Irish localities are doubtful: this species has been often mistaken for the adult of *T. papyracea*: the only specimen in Mr. J. D. Humphreys's extensive collection of shells from Dublin, Cork, and Bantry was marked by him "England." Mr. Grainger obtained it in a dead state at Belfast, where it is also found in a post-pliocene deposit, as well as in the Coralline Crag. The foreign localities of which I am assured are Morbihan (Macé and Taslé); Provence (Martin); Gibraltar, 8 f. (M'Andrew); and Ægean, 70 f. (Forbes). Philippi has recorded it as recent at Naples (on the authority of Scacchi) and fossil at Palermo.

My largest specimen is $2\frac{1}{3}$ inches long, and $3\frac{3}{4}$ broad. The young have the same characters as the adult, and are even more unlike *T. papyracea*.

The *Mya declivis* of Pennant, to which it was at one time referred, appears to have been the half-grown state of *M. truncata*. The present species is the *Anatina myalis* of Lamarck, and *T. Montagui* of Leach.

4. T. CONVEX'A*, W. Wood.

Mya convexa, W. Wood, Gen. Conch. i. p. 92, pl. 18. f. 1. *T. convexa*, F. & H. i. p. 229, pl. xvi. f. 1, 4.

SHELL nearly rectangular, extremely gibbous, except towards the front and posterior side, which are compressed to such an extent as to give a wedge-like aspect; it is thinner than *T. pubescens*, opaque, and somewhat glossy: *sculpture* much finer than that of the last species, and consisting of minute papillæ, which are equally disposed over the whole surface in transverse and undulating lines; the marks of growth are slight, but numerous: *colour* pale yellowish-brown: *epidermis* membranous and thin: *margins* rounded on the anterior side and in

* Convex.

front, with a slight indentation towards the posterior side, which is more or less obliquely truncated, and separated by a blunt angular ridge in each valve, with an obscure intermediate fold, making this side appear bicarinated; dorsal margins gently curved: *beaks* very prominent, obliquely inflected to the posterior side; the space below them on each side is deeply excavated: *ligament* short and cylindrical, greyish-horncolour, separating the valves by an elliptical gap: *cartilage* yellowish, contained in a narrow but solid receptacle, which lies parallel with the hinge-line, and does not project far within the shell; the receptacle is supported underneath by the ordinary rib-like process: *hinge-line* obtusely angular: *hinge-plate* narrow and slight: *ossicle* as in the other species: *inside* yellowish; edges blunt: *scars* nearly marginal. L. 2. B. 2·5.

HABITAT: 4–70 f. in suitable parts of the English, Irish, Scotch, and Shetland coasts; difficult to procure on account of its habit of burrowing rather deeply in muddy sand. Not uncommon in the "alluvial" deposit at Belfast (Hyndman and Grainger); Wexford (Sir Henry James); Coralline Crag (S. Wood); "glacial" formation near Drontheim, 400–500 feet above the present sea-level (Sars); Palermo (Philippi). It has been noticed as a Swedish and Norwegian species by Lovén, Sars, M'Andrew and Barrett, Danielssen, and Malm, at various depths between 8 and 100 f.; M'Andrew dredged it off Gibraltar in 45 f.; and Martin obtained it from fishermen on the coast of Provence, but smaller in size than northern examples.

This handsome shell may easily be recognized by its almost globular form. The young and fry correspond in shape with the adult; but they are white and not so convex, and their dorsal margins are quite straight. The ligament, as well as the epidermis are wonderfully preserved in fossil specimens dug out of the clay-bed at Belfast.

Montagu described it as a large form of *T. distorta*.

It is the *T. declivis* of Macgillivray, *T. ventricosa* of Philippi, and (apparently) the *T. Scheepmakeri* of Dunker.

B. Posterior side usually larger.

5. T. DISTOR'TA*, Montagu.

Mya distorta, Mont. Test. Brit. p. 42, t. i. f. 1. *T. distorta*, F. & H. i. p. 231, pl. xvii. f. 1, 2, 3, 8, and (animal) pl. II. f. 5.

BODY roundish-oval, white: *tubes* rather short, and somewhat more united towards their bases than in the other British *Thraciæ*; the branchial or lower tube is often extended more than half an inch, while the other remains quiescent; previously to the former being withdrawn, it is always globularly inflated at its extremity, which inflation increases until it extends near the margin of the shell, and the tube then suddenly collapses; during the inflation the terminal cirri disappear, and they only become visible when the tube is at rest: *gills* large and brown: *palps* nearly equal and pectinated: *foot* short and linguiform.

SHELL varying in shape from round to oval, more or less distorted and often sinuous, generally convex but sometimes flattened, more solid in proportion to its size than the other species, opaque and lustreless: *sculpture*, minute and crowded tubercles or granulations of equal size, arranged in concentric although irregular rows; marks of growth distinct: *colour* milk-white, with occasionally a yellowish tinge: *epidermis* membranous, abraded in front and only to be seen at the edges, dingy brown: *margins* rounded on the anterior side and in front, somewhat truncated or wedge-shaped on the posterior side, which is in most instances (but not invariably) larger or more elongated than the other side, and obscurely angulated; dorsal margins obtuse-angled: *beaks* sharp and entire, slightly inclined to the posterior side; umbones rather prominent: *ligament* short, of various shades of colour from yellowish to dark brown, separating the valves by an oval gap: *cartilage* strong, yellowish-brown or horncolour, contained in a thick triangular receptacle, which is set obliquely and projects considerably within each valve; fulcral

* Distorted.

rib indistinct: *hinge-line* obtusely angular: *hinge-plate* thick and strong: *ossicle* semilunar, slightly attached, and consequently often lost in dead specimens: *inside* creamcolour, somewhat glossy and nacreous; edges blunt: *scars* large and well defined. L. 0·6. B. 0·8.

Var. *truncata*. Shell oblong; front margin straight; posterior margin abruptly truncated. *Anatina truncata*, Turton. Dith. p. 46, t. 4. f. 6.

HABITAT: From Guernsey to Unst, in crevices of rocks and old oyster-shells, between 5 and 35 f., as well as occasionally buried in tufts of *Corallina officinalis* at low water; local, but widely diffused. The variety is from Exmouth, Tenby, and Cork. Fossil in the Coralline Crag (S. Wood); Palermo (Philippi). Foreign range: Finmark to the Cattegat, 3-40 f. (Lovén and others); north of France (De Gerville and others); Provence (Martin); Algeria (Deshayes and Weinkauff).

This is the smallest British *Thracia*; and its habitat is different from that of its congeners. It may also be known by its irregularly oval shape, its less angular outline, uniform granulation, and comparatively large cartilage-pit. The young are triangular, and somewhat resemble a *Montacuta*. A full-grown specimen taken from a narrow chink in a piece of limestone well exemplifies the mode in which shells are constructed. Part of the left valve had been crushed, apparently by accidental pressure; and in order to repair the damage, an inner layer was formed by exudation from the mantle, to which the broken fragments were cemented and still adhere. The distorted growth of this species shows that it does not excavate the holes in which it lives. It sometimes appropriates the labours of other animals, but never unjustly or consciously, like a plagiarist. The original and short-lived fabricators of the dwellings subsequently occupied by the *Thracia* are beyond the power of complaint; and all that can be said of them is

Sic vos non vobis saxa forate diu.

It constitutes the type of Fleurian de Bellevue's genus *Rupicola*. Pennant and Donovan described it as *Venus sinuosa*, Lamarck as *Anatina rupicola*, Philippi as *Erycina anodon*, *Anatina? pusilla*, *T. ovalis*, *T. fabula*, and *T. elongata*, Récluz as *Rupicola concentrica*, and Deshayes as *T. brevis*. Many other species have been made by Reeve from Mr. Cuming's specimens of this extremely variable shell. It appears to have been confounded by Kiener with *T. corbuloides*, Deshayes, on the supposition that it was a smaller form of that species.

Another species of *Thracia* (*Amphidesma truncata*, Brown, or *T. myopsis*, Beck) has been found in glacial beds, at Greenock by Mr. Stewart Kerr, and at Elie in Fifeshire by the Rev. Thomas Brown. To this species appears to have also belonged a shell named "*Cochlodesma*, n. s." by Professor King, which was lately brought up from the depth of 1000 f. or thereabouts, 100 miles west of Cape Clear, by Capt. Hoskyn in H.M.S. 'Porcupine'; and the fragments of which I have examined. *T. myopsis* now lives only in the Arctic seas.

Family XIX.
CORBU'LIDÆ, (CORBULADÆ) Fleming.

BODY oval or globular: *tubes* short and united: excretal tube furnished with a conspicuous valve: *foot* long and flexible.

SHELL oval, more or less inequivalve and open at the posterior end: *beaks* turned towards the posterior side: *cartilage* wholly internal, occupying a horizontal triangular cavity under the beak in each valve: *hinge* strong, furnished in some genera with a single erect cardinal tooth in one valve or both, besides a long lateral tooth on one or each side in either valve; in species of *Neæra* there is also a free calcareous ossicle: *pallial scar* slight, with a shallow sinus: *muscular scars* well marked.

These are of small size, and comprised in few genera: the species are numerous and prolific, characters which are probably correlative. The British genera are *Poromya*, *Neæra*, and *Corbula*. The first is a box studded with tiny pearls.

"No lesse praisworthie faire Neæra is."

Her shell resembles the body of a bird, without feet or wings, but having a stretched-out beak; and, although this age is not barren of artistic invention, it might serve as a graceful model for some work of fictile manufacture. The last has also an apposite name, and reminds one of a basket with a close-fitting lid. The hinge in each genus is constructed somewhat on the plan of the *Mactridæ*; but it does not possess an external ligament as well as an internal cartilage. The *Corbulidæ* live in mud and sand at various depths, but seldom between tide-marks. Lamarck called them "Corbulées," Latreille "Corbulæa," and Hinds "Corbulacea."

Genus I. POROMY'A*, Forbes. Pl. II. f. 3.

BODY roundish-oval, thin: *tubes* unequal in size, clothed with numerous long filaments: *foot* narrow and slender.

SHELL roundish-oval, slightly inequivalve and inequilateral, thin and pearly, with the outer layer composed of minute tubercles; posterior side angulated: *epidermis* membranous and thin: *teeth*, in the right valve a short but strong cardinal, and in the left a minute triangular cardinal and a ridge-like lateral on the posterior side.

* Passing into the genus *Mya*; or having, with the shape of that shell, a tubular structure.

The structure of the shell is very remarkable, although not differing much from that of *Thracia*. The external layer consists of crowded oblong cells having their ends outward, and the inner layer is nacreous; the cellular part is easily rubbed off. The mantle is said to be open in front, an unusual character in this group. Further particulars of the animal are desirable.

This genus is the *Embla* of Lovén, and (according to Chenu) the *Eucharis* of Récluz.

POROMYA GRANULA'TA*, Nyst and Westendorp.

Corbula granulata, Nyst & West. Coq. Foss. d'Anvers, p. 6, pl. 3. f. 3. *P. granulata*, F. & H. i. p. 204, pl. ix. f. 4–6, and (animal) pl. W. f. 2.

BODY creamcolour: *mantle* open in front: *tubes* encircled at their bases by a fringe of 18 or 20 tentacular filaments, which expand like the petals of a flower, and are sometimes folded back on the posterior side of the shell: *foot* very transparent.

SHELL somewhat quadrangular or rhomboidal (the right valve larger than the left and slightly overlapping it), moderately convex, fragile; externally it is opaque and of a dark hue, but when the superficial or granular coating is removed, it is semitransparent and glossy: *sculpture*, very minute and close-set tubercles of nearly equal size, arranged in longitudinal rows, and occasional but slight marks of growth: *colour* dusky outside, and whitish under the surface-layer: *epidermis* dark brown, visible only at the edges, and especially at the back (where it forms a kind of elongated ligament on both sides of the beak): *margins* rounded on the anterior side, slightly curved in front, indented near the posterior side, which is obliquely truncated and has a distinct ridge extending from the beak to the posterior angle, with a broad fold on either side of it; posterior dorsal margin longer and straighter than the other: *beaks* blunt and calyciform; umbones prominent: *cartilage* yellowish-brown, set rather obliquely in an obtusely angular receptacle, which does not project far within: *hinge-line* gradually curved: *hinge-plate* thickened on both sides of the beak: *teeth*, in the right valve an erect, blunt and tuber-

* Granulated.

cular cardinal; in the left valve a small, sunken and triangular cardinal, besides a long but slight laminar lateral on the posterior side: *inside* glossy and nacreous, closely but obscurely lineated lengthwise; edges sharp: *muscular scars* triangular, lying near the dorsal margins. L. 0·325. B. 0·375.

HABITAT: In mud among boulders, 40–45 f., close to Croulin Island, and in another part of the Sound of Skye; rare. Mr. Dawson found a worn and imperfect valve in shell-sand from Haroldswick Bay in the north of Shetland. Coralline Crag (S. Wood); newer tertiary beds near Antwerp (Nyst and Westendorp). Koren got it at Bergen; M'Andrew and Sars dredged it off the coasts of Finmark, the former in 45–90 f.; Deshayes obtained it from Sicily and Bona, and Tiberi at Naples; Forbes in the Ægean between 40 and 150 f.; and M'Andrew at Madeira in 20 f.

Clark conjectured that this might be the young of *Thracia convexa*, and he said that the present species has an ossicle in the hinge; but he did not see with my eyes. I have compared specimens of *P. granulata* and *T. convexa* of all sizes, from the fifteenth of an inch in length. Each exhibits a marked difference of outline: one is square, and the other triangular. I have also examined perfect examples of the *Poromya* from Scandinavia, Skye, and Naples; and in none of them could I detect an ossicle or any space for it. He also stated that the siphons of these two mollusca are equally short, and ornamented with cirri or filaments; but neither of these characters was noticed by him in his elaborate account of the only species of *Thracia* described in the 'History of the British Marine Testaceous Mollusca,' and he admitted that he had not seen the animal of *T. convexa* or of *P. granulata*.

Forbes described the recent shell as *P. anatinoides*,

Lovén as *Embla Korenii*, Deshayes as *Corbula vitrea*, and Tiberi as *Cumingia parthenopœa*.

Genus II. NEÆRA*, (NEARA) Gray. Pl. II. f. 4.

BODY globular, thin: *tubes* unequal in size, clothed with a few long filaments: *foot* lanceolate.

SHELL fig-shaped, inequilateral, thin; posterior end twisted and extended into a beak-like process: *epidermis* membranous: *teeth*, sometimes a small cardinal in each valve, or a crest-like laminar lateral on the posterior side of one valve or both; certain species have also a free calcareous ossicle.

The late Capt. Brown first suggested the generic separation of the present group of shells, which are distinguished no less by the singularity than by the elegance of their shape. His services in the cause of British conchology would have been greater if his attention had not been distracted by so many other branches of zoology. Good results, however, were produced by his publications, especially in promoting the faculty of observation in young persons. Clark repudiates the genus, and merges it in *Anatina*, on the ground that each has an ossicle. This leads to the consideration of the difficult question, what is a genus? Nor can I agree with him that we have but one species of *Neœra*. Perhaps in a few centuries hence, or sooner, his opinion on the last point may be found correct; or possibly the very notion of species may be classed among the vulgar errors of a half-enlightened age. What our Poet-laureate says is true, that

"Science moves, but slowly slowly.
Creeping on from point to point:"

or as Seneca puts it, "Multa hoc primum cognovimus

* A Sea-Nymph mentioned by Spenser.

sæculo, multa venientis ævi populus nobis ignota sciet;" but at present my opinion coincides with that of other naturalists, both as to the existence of species, and of those of *Neæra* in particular.

This genus is the *Cuspidaria* of Nardo. It contains many exotic species; the late Mr. Hinds described and enumerated seventeen in the 'Proceedings of the Zoological Society' for 1843, and Mr. A. Adams several more in the 'Annals and Magazine of Natural History' for March 1864. The name *Neæra* was originally used for a genus of Diptera; but no one is likely to be misled by the subsequent application of it to the Mollusca, unless perchance in consulting an index to any work on general zoology. Otherwise the name given by Nardo is more characteristic.

1. NEÆRA ABBREVIA'TA*, Forbes.

N. abbreviata, Forbes in Zool. Soc. Proc. 1843, p. 75: F. & H. i. p. 201, pl. vii. f. 7.

SHELL triangularly oval, obliquely twisted to the posterior side, nearly equivalve, extremely gibbous, fragile, semitransparent, slightly glossy and iridescent: *sculpture*, about a dozen concentric plaits or folds, besides numerous fine but irregular intermediate striæ; the surface is also marked by a few obscure longitudinal lines, and the posterior side by a sharp rib which runs outwards from behind the beak in a curved or flexuous direction: *colour* greyish-white: *epidermis* yellowish-brown, visible only at the edges and back: *margins* rounded on the anterior side and in front, indented or flexuous on the posterior side, which is short, wedge-like, and considerably compressed; dorsal margins nearly equal in length, and straight: *beaks* blunt, much inflected, somewhat inclined to the anterior side; umbones prominent; the dorsal area is deeply excavated: *cartilage* small, yellowish-brown, occupying an elliptical cavity in a parallel line with the hinge: *hinge-line* obtusely angular: *hinge-plate* narrow: *teeth*, a minute thorn-like cardinal in each

* Shortened.

valve, and a slight lateral on the posterior side of the right valve: *inside* glossy and nacreous; posterior side separated by a sharp rib: *scars* indistinct. L. 0·3. B. 0·4.

HABITAT: Loch Fyne (M'Andrew and Barlee); Skye and Shetland (Barlee); in 40-75 f., on a muddy ground. Fossil in the Belgian tertiaries (Nyst). Its known distribution elsewhere in a recent state is as follows:—Bohuslän (Lovén); Christiania, 40-100 f. (Asbjörnsen); Bergen, 40-50 f. (Danielssen); Finmark (Sars and Lilljeborg); dead valves in the Ægean, 75-105 f. (Forbes).

It is the *N. vitrea* of Lovén.

2. N. COSTELLA'TA*, Deshayes.

Corbula costellata, Desh. Exp. Scient. Mor. (Géologie) p. 86, t. vii. f. 1-3. *N. costellata*, F. & H. i. p. 199, pl. vii. f. 8, 9, and (animal) pl. G. f. 8, 9.

BODY gelatinous, clear white: *mantle* so transparent as to allow the pink gills and dark brown liver to be seen through it: *tubes* cylindrical, sometimes yellow with reddish or orange markings, and tinged with brown at their extremities; excretal tube much the smaller of the two; tentacular cirri white and plain, extending beyond the tubes: orifices fringed: *foot* narrow.

SHELL more slender than *N. abbreviata*, more inequivalve, much less ventricose and even somewhat compressed, equally fragile, semitransparent, glossy and iridescent: *sculpture*. 20-30 longitudinally radiating ribs, which are slighter and more like striæ on the anterior side and in front, but stronger and more distant towards the posterior side, especially the last two or three; these ribs vary in size and fineness; the prolonged part on the posterior side is also marked with two or three slight ribs, which are parallel with the dorsal line and extend to the rostral point: *colour* and *epidermis* as in the species last described: *margins* also similar, except behind, where the anterior dorsal margin is raised and appears high-shouldered, and the posterior dorsal margin is inflected and

* Fine-ribbed.

curved; rostral prolongation considerable, much more attenuated than in the other species: *beaks* small and mammillary; umbones by no means prominent; dorsal area narrowly excavated on the posterior side: *cartilage* orangecolour, contained in a triangular receptacle which shelves outwards: *hinge-line* straight: *hinge-plate* narrow and slight: *teeth*, an extremely minute tubercular cardinal in the left valve, and a strong erect and triangular lateral in the right valve on the posterior side: *inside* glossy, with a rib on the posterior side: *muscular scars* well marked; anterior irregularly oblong, posterior triangular. L. 0·25. B. 0·415.

Var. *lactea*. Shell milk-white, more glossy, transparent, and delicate, having only two ribs on the posterior angle, besides those on the rostral process.

HABITAT: Loch Fyne, 40–70 f., with the last species (M'Andrew and Barlee); Cumbrae, Firth of Clyde (Robertson); Skye and Shetland (Barlee and J. G. J.). The variety was dredged by me on a sandy bottom, in 78 f., from 40 to 50 miles east of the Whalsey Skerries, Shetland. Upper tertiaries of Greece (Deshayes); Antwerp (Nyst); Guise-Lamotte, France (De Koninck); Calabria (Philippi). It inhabits the coasts of Scandinavia at depths ranging between 10 and 100 f. (Lovén and others); Carthagena, in 30 f., and Gibraltar, in 45 f. (M'Andrew); Provence, in 60 f. (Martin); Gulf of Genoa, in 25 f. (J. G. J.); Adriatic (Chiereghini); Naples (v. Martens); Ægean, in 20–185 f. (Forbes); Malta, in 40 f., Gulf of Tunis, in 35 f., Madeira, in 18–24 f., and Teneriffe, in 20–35 f. (M'Andrew). Specimens dredged by the late Professor Barrett in deep water at Jamaica are scarcely distinguishable from those of the North Atlantic.

This exquisite shell cannot well be mistaken for *N. abbreviata*; their shape, sculpture, and dentition are very different.

Nyst seems to have been the earliest describer of it,

as *Corbula Waelii*; and the figures which he also gave are very exact. This was in 1843. The great French work on the expedition to the Morea was published eight years previously. Bory St. Vincent contributed the geological portion of this work, which contains a good representation of the shell; the only other notice of it appears in the index to the plates, where it is entered as "*C. costellata*, Deshayes." It is the *N. sulcata* of Lovén, *C. rostrato-costellata* of Acton, and *Tellina naticuta* of Chiereghini. The figures in Philippi's work on the Sicilian Testacea are not satisfactory; they were probably made up or "restored," for he says that all his specimens were "paullulum læsas."

3. N. ROSTRA'TA*, Spengler.

Mya rostrata, Spengl. in Skrivt. Selsk. iii. p. 42, t. 2. f. 16.

SHELL resembling a fig with a broad stalk, nearly equivalve except in the young, convex, more solid than the preceding species, opaque and almost lustreless: *sculpture*, numerous but slight concentric raised striæ or wrinkles, becoming more crowded and flexuous towards the posterior side; the upper angle on that side (which forms a long and diagonal crest or ridge, extending from behind the beak in each valve to the rostral point, and defined by an oblique rib) is crossed by close-set and somewhat curved striæ at a right angle to the transverse markings on the body of the shell: *colour* whitish: *epidermis* more persistent than in the other two species, pale yellowish-white: *margins* rounded on the anterior side and immediately in front, bending upwards and nearly in a straight course to the deep sinus or indentation caused by the extension of the posterior side; this part is remarkably twisted and elongated, being about two-fifths of the entire breadth of the shell; posterior dorsal margin curved inwards; anterior dorsal margin high-shouldered: *beaks* inflected: umbones rather prominent; dorsal excavation deep, wide on the anterior and narrow on the posterior side: *cartilage* small, golden-yellow,

* Beaked.

contained in an oval pit, which projects obliquely inwards: the cartilage is held together by a calcareous band or ossicle, placed as in *Lyonsia*, which is easily split and broken in two when the valves are separated; it then curls up, so that each half resembles the shelly appendage peculiar to *Thracia*: *hinge-line* straight: *hinge-plate* moderately broad: *teeth*, a lateral in each valve, which is triangular, erect, and rather long in the right valve, ridge-like and slight in the left: *inside* glossy and nacreous, obscurely striated lengthwise: *scars* indistinct. L. 0·45. B. 0·8.

Habitat: East coast of Shetland, 40 miles off the land, in 76 f., soft and muddy sand; a right valve only, with living specimens of the common kind, *N. cuspidata*. The foreign localities are, Bergen, among *Oculina prolifera* (Spengler); other parts of Norway, at various depths from 10 to 130 f. (Lovén, Asbjörnsen, Danielssen, and Sars); Sweden, 20–60 f. (Lovén and Malm); Gulf of Lyons, 80–100 f. (Martin); Toulon (Thorrent); Genoa (J. G. J.); Naples, 30–40 f. (Tiberi); Sicily (Philippi); and Ægean, 110–150 f. (Forbes). The *N. Chinensis* of Gray, from Mr. Hinds's explorations in the East Pacific, is closely allied to this species, if not identical with it.

This is a larger and stronger shell than *N. costellata*, much more elongated in proportion, and has a different kind of sculpture.

It is apparently the *Anatina longirostris* of Lamarck, and *Corbula cuspidata* of Brown, as it is certainly the *N. attenuata* of Forbes, and *N. renovata* of Tiberi. I have examined the types of these last two, as well as of Spengler's species. Of the two figures given by Philippi (vol. i. tab. i. f. 19) that on the left hand represents the present species, and the other (which is drawn partly from imagination) *N. cuspidata*.

4. N. CUSPIDA'TA*, Olivi.

Tellina cuspidata, Olivi, Zool. Adr. p. 101, tab. iv. f. 3. *N. cuspidata.* F. & H. i. p. 195, pl. vii. f. 4–6, and (animal) pl. G. f. 4-7.

BODY greyish or dirty white: *mantle* rather thin: *tubes* nearly sessile, sometimes mottled with pink; orifice of lower one fringed with 5 or 6 short cirri; the base of each tube is encircled by 6 rather long and slender filaments, which have cup-shaped extremities, like the polypidoms of many zoophytes: these filaments occasionally are knotted or studded at intervals with bulbs of an azure hue; the orifice of the upper or excretal tube is plain, but provided with the usual hyaline valve: *foot* long, flexible, and white.

SHELL obliquely triangular (left valve sensibly larger than the right), extremely gibbous and tumid, moderately solid, opaque, and almost lustreless: *sculpture*, numerous slight and irregular concentric striæ or wrinkles, becoming closer and flexuous towards the posterior side: the upper angle on that side is crest-like and striated as in *N. rostrata*, but it is not so distinctly defined, nor elongated to anything like the same extent: *colour* whitish under the *epidermis*, which is light chestnut or reddish-brown, thick (especially at the dorsal edges, where it has somewhat the appearance of a ligament), sometimes coated with sand or mud: *margins* rounded on the anterior side as well as in front, with an abrupt and deep sinus on the posterior side, which is somewhat twisted and comparatively short, being about one-half of the entire breadth of the shell; posterior dorsal margin incurved; anterior dorsal margin forming a rounded slope, but not projecting as in the last species: *beaks* inflected, and interlocking, or placed one on each side instead of opposite; umbones extremely prominent; dorsal excavation deep, heart-shaped on the anterior side, and trench-like on the posterior: *cartilage* and *ossicle* as in *N. rostrata*, but the former is horncolour, and the pit does not project so far inwards: *hinge-line* obtusely angular: *hinge-plate* thick: *teeth*, a strong recurved and rather short triangular lateral in the right valve, and only an obscure and blunt laminar lateral in the other valve: *inside* glossy, porcellanous, and nacreous, indistinctly striated lengthwise; it is furnished on the posterior side in each valve with a thick rib, extending from below the beak half-way across to the indentation that

---

* Pointed.

defines the snout-like process: *pallial scar* well marked, with a semicircular sinus: *muscular scars* rather deep; anterior irregular, posterior triangularly oval. L. 0·55. B. 0·8.

Var. 1. *curta*. Rostral or snout-like process shorter.

Var. 2. *cinerea*. Shell ashcolour, and thinner.

HABITAT: Land's End (M'Andrew); Northumberland and Durham (Brown, Thomas, Alder, and Mennell); Aberdeen (Macgillivray); Firth of Forth (Gerard and Thomas); throughout the west of Scotland (Smith and others); Shetland (M'Andrew and others); off Cape Clear (M'Andrew); Arran Isle, Galway (Barlee); in muddy sand, at depths varying from 12 to 82 f. Var. 1 and 2. Hebrides (Barlee). Searles Wood has recorded this species as fossil in the Coralline Crag, Risso from Nice, and Philippi from Sicily; upper miocene bed near Antibes (Macé). Its foreign distribution in a recent state comprises Spitzbergen and South Greenland (Torell); Scandinavia, 22-180 f. (Lovén and others); Carthagena and Gibraltar, 45 f. (M'Andrew); Provence, in a gurnard's stomach (Martin); Italian coasts of the Mediterranean (Maravigna and others); Adriatic (Olivi and Chiereghini); Malta, 40 f. (M'Andrew); Ægean, 12-185 f. (Forbes); Algeria (Deshayes and others); Madeira, in 18-24 f., and Teneriffe, in 20-35 f. (M'Andrew). Mr. Hinds, after giving some European localities, remarks, " Nor can I perceive any difference in the valve of a shell obtained from 84 f. in the China Sea; the temperature below being 66°, and at the surface 83°."

It is much more globular and obliquely twisted than *N. rostrata*, and it is more finely striated; the snout in all specimens is considerably shorter; the front or ventral margin is more curved; and the posterior dorsal side is abruptly truncated, and not so rounded and prominent as in that species. The young of Loch Fyne specimens are proportionally more slender than the adult, and more elongated in the line of the major axis; but they essentially differ from *N. rostrata* of the same age or size. A valve which I dredged in deep water off the east coast of Shetland is nearly an inch broad, and coarsely wrinkled: it agrees with specimens which I examined in the Museum at Christiania, described by Sars as *N. arctica*, as well as with some dredged by Torell in the Arctic Sea.

Brown called the present species *Anatina brevirostris* and *Thracia brevirostris,* and Nardo *Cuspidaria typica.*

Genus III. COR'BULA*, Bruguière. Pl. II. f. 5.

BODY oval, rather thick: *tubes* seldom protruded; orifices fringed: *gills* 2 on each side, unequal-sized: *palps* corresponding with the gills in number and position, but equal in size: *foot* tongue-shaped and thick.

SHELL oval, nearly equilateral, rather solid; posterior side wedge-shaped: *teeth*, a short and strong cardinal in each valve, and a ridge-like lateral on both sides of the right valve.

The structure of the shell is like that of the *Anatinidæ*: according to Carpenter "the outer layer is composed of large fusiform cells, whilst the inner is nearly homogeneous." Searles Wood informs us that fossil species have been found as early as in the lower Oolite.

Mühlfeldt called this genus *Aloides*; and modern systematists have invented for it other equally ill-compounded names, such as Spenser, in his 'Teares of the Muses," designates

"Heapes of huge words uphoorded hideously,
With horrid sound though having little sence."

---

* A little basket.

CORBULA GIBBA[*], Olivi.

Tellina gibba, Olivi, Zool. Adr. p. 101. *C. nucleus*, F. & H. i. p. 180, pl. ix. f. 7-12, and (animal) pl. G. f. 3.

BODY whitish, with often a tinge of yellow : *mantle* thick ; its edges minutely ciliated : *tubes* contiguous, very short, and scarcely protruded beyond the valves, edged with narrow lines of pink or orange a little below the extremities ; orifices fringed with conical and rather slender cirri or tentacles (from 8 to 12 round each), having truncated points ; these cirri are transparent, and spotted with a few flake-white marks, and each is encircled at its base by a line of red dots ; hyaline apparatus of the upper tube bell-shaped, retractile, and in frequent action : *gills* very unequal, hanging obliquely, the upper one narrow, and the lower one larger and more triangular ; they are brown, smooth outside and finely striated within : *palps* long, narrow, pointed, pendulous, and brown, pectinated strongly on both surfaces : *foot* large and thick, very fleshy, bent near its junction with the rest of the body, sometimes forming an elongated cone and byssiferous : *liver* dark green.

SHELL triangularly oval ; right valve much larger and more gibbous than the left, which it overlaps to a considerable extent ; left valve compressed towards the front and sides ; the substance is thick and opaque, and the surface of the right or deeper valve is more glossy than that of the other, and occasionally iridescent : *sculpture*, numerous concentric striæ, which in the smaller valve are slight and irregular, and are often crossed by a few raised lines radiating from the beaks, but in the larger valve these striæ usually become cord-like and close-set ribs : *colour* white, with more or less of a yellowish or reddish-brown tinge, sometimes varied by longitudinal rays or streaks of the latter hue on the larger valve : *epidermis* brown, thick, and somewhat fibrous, mostly abraded and wanting on the larger valve : *margins* rounded on the anterior side and in front, truncated on the posterior side (which is depressed and diagonally separated in the smaller valve, and twisted in the other valve), with a slight groove or fold proceeding from below the beak ; dorsal margins straight : *beaks* calyciform, obliquely incurved to the anterior side ; umbones prominent and contiguous ; dorsal excavation generally deep, but not distinctly defined : *cartilage* small, narrow, and triangular, composed of several leaflets, which represent the successive accretions of

* Gibbous.

growth; it is contained in a cavity or depression of the cardinal tooth in the left valve: *hinge-line* obtusely angular: *hinge-plate* rather broad and strong: *teeth*, in the right valve a thick, pyramidal, and recurved cardinal, besides a long ridgelike lateral on each side; in the left valve a thick cardinal, which resembles in shape the bowl of a spoon, and may be considered the cartilage-pit, although it is not horizontal and it slopes upwards from the beak; close to it on the anterior side of the same valve is a cavity for the reception of the opposite tooth: *inside* porcellanous and glossy, microscopically and closely wrinkled, more or less stained with coffeecolour; edges somewhat bevelled: *pallial scar* slight, with an extremely shallow sinus: *muscular scars* distinct; anterior oval, posterior nearly circular. L. 0·5. B. 0·6.

Var. *rosea*. Shell rather more oval and glossy, with a purplish streak on either side of the beak in each valve, and the rays on the larger valve of a more vivid hue. *C. rosea*, Brown, Ill. Conch. p. 105, pl. xlii. f. 6; F. & H. i. p. 185, pl. ix. f. 13, 14.

HABITAT: Gregarious in sand, mud, and gravel on every part of our coasts. I once found live specimens burrowing in the sand at Oxwich Bay, Glamorganshire, on the recess of an unusually high spring tide; and it occurs as deep as 72 f. in Shetland. It usually frequents the laminarian zone. The variety is equally diffused in the British seas, and ranges from Norway to the Mediterranean; Weinkauff has taken it at Algiers in brackish water. *C. gibba* is not uncommon in post-pliocene and pliocene deposits, *e. g.* at Belfast (Grainger); raised beach at Moel Tryfaen (Darbishire); Scotch and Irish glacial beds (Smith); Norwich Crag at Bramerton (Woodward); Red and Coralline Crag (Wood); "glacial" formation near Christiania (Sars); Nice (Risso); Belgian tertiaries (Nyst); Sicily (Philippi); and I noticed it in M. Macé's collection of upper miocene fossils from Antibes. In a recent state it is universally distributed throughout the North Atlantic, from the Loffoden

Isles to the Ægean and Canaries, at depths of from 4 to 80 f.

Our northern shores seem to produce the largest specimens, those from the Channel Isles being more brightly coloured. The fry have a squarish outline, and are highly polished. This species varies both in shape and sculpture, from oval to round, and from ribbed to smooth. The shell is subject to the attacks of predatory mollusks, which do not always succeed in perforating it; in such cases the white outside layer only is removed, exposing the succeeding layers, which are of a firmer texture and coffeecoloured. Aucapitaine states that he found specimens of a smaller size and paler colour than usual, living abundantly in brackish water at Rochelle, often floating on grasses half covered with water, and sometimes buried in mud to the depth of their siphons.

It is the *Cardium striatum*, &c., of Walker, *Mya inæquivalvis* of Montagu, *Corbula nucleus* of Lamarck, and *C. olympica* of Costa; several other specific names have been given to it by palæontologists.

Among the shells collected by Mr. J. D. Humphreys at Cork were a few specimens of *C. mediterranea*, Costa, mixed with *C. gibba*. Philippi referred this species to the *Tellina parthenopæa* of an unpublished work by Delle Chiaje; and it appears to be also the *C. physoides* of Deshayes's 'Mollusques d'Algérie.' The Irish specimens may have been imported (as well as *Petricola lithophaga*) in ballast, and I therefore merely indicate the possibility of its being indigenous; but this species is interesting in connexion with another shell, which I have now to mention. In the 'Malacologia Monensis' of Forbes will be found a short description, but characteristic figure, of a species named by him *C. ovata*. It was established on a single specimen "taken from the root of a fucus cast ashore at Ballaugh.'' Dr. Mörch gave me the same species, which he had procured from Greenland. It is undistinguishable from *C. mediterranea*, except in its much larger size and the absence of coloured streaks; in shape, sculpture, and peculiar dentition it corresponds exactly with the Irish specimens, and with some from the Gulf of Lyons, for which I am indebted to the kindness of M. Martin. I cannot help conjecturing that the Manx shell might have been brought to this country with others from the Arctic seas, and have afterwards become accidentally mixed in Forbes's collection; especially when I remember that he sent me about the time of his publishing the 'Malacologia,' and when he was almost a tyro in British conchology, another shell for my opinion. This was *Venus fluctuosa*, a native of the North-American seas. The memorandum accompanying the last-mentioned shell stated that it had been received by Forbes, as picked up on the shore at Leith, but not by himself. The difference of size between Greenland and Mediterranean specimens of the same species further exemplifies my remarks in the first volume on this subject.

The late Dr. Lukis sent me specimens of *C. labiata*, a handsome South-American species, with which the tide-mark in a small bay in Guernsey had been strewn in November 1859, immediately after the wreck of a ship in ballast from Buenos Ayres. Along with this *Corbula* were found a small *Melania* and other tropical shells. This shows the importance of carefully studying the geographical distribution of the Mollusca, in order to avoid errors likely to result from accidents of the above kind. Otherwise all these shells might be described or enumerated as British.

Family XX. MYIDÆ, (MYADÆ) Fleming.

BODY oval: *mantle* rather thin, except at the edges: *tubes* united, and wholly enclosed in a tough, leathery, brown sheath; orifices fringed: *gills* of moderate length, unequal on each side, and striated: *palps* triangular, striated like the gills: *foot* tongue-shaped, furnished with a byssal groove.

SHELL oval or oblong, somewhat inequivalve, usually gaping at both ends, but more widely on the posterior side: *epidermis* membranous: *beaks* more or less contiguous, not prominent, turned towards the anterior side: *cartilage* internal, contained between a perpendicular spoon-shaped and fixed receptacle, lying under the beak in the right valve, and a cavity of the cardinal tooth or process in the left valve: *hinge* strong, furnished with a small cardinal in the right valve, and with an erect triangular tooth in the left valve, which latter tooth is strengthened by an inside flange on the posterior side: this tooth is not inserted into the hinge of the right valve, but is merely attached by the cartilage to the sunken receptacle above mentioned: *pallial scar* broad and deeply sinuated: *muscular scars* large and strongly impressed; anterior elongated, posterior triangular.

The typical genus *Mya* is the only one that I consider British. There seems to be no valid reason for separating *Sphenia* (Turton) from it, either in respect of the animal or of the shell. The so-called *Panopea Norvagica* has a very different kind of hinge, besides an external ligament: it belongs to *Saxicava*. So far as is at present known, the *Mya* or "gaper" family is restricted to the northern hemisphere. They inhabit sand and mud, usually in the lowest part of the littoral zone.

Genus MYA*, Linné. Pl. III. f. 1.

The characters have been already given in the description of the family.

* So named from a supposition that it was the μῦς of ancient writers.

It is impossible to say what were the μύες of Aristotle, except that they were not our shells; nor is it probable that the latter could have come within the scope of his observation, inasmuch as they are not natives of the Archipelago. The μύες were included by him with the κτένες (or Pectens) among the bivalves, but they were said to produce spawn-capsules, like the πορφύρα or *Murex trunculus*. Æschylus, Athenæus and other Greek writers also mention μύες, but only in such a way as to show that they were an eatable kind of shell-fish. The Myes of Pliny, that indefatigable naturalist with so little originality, were described by him as "rufi ac parvi." They may have been *Mytilus edulis*. The hinge in the present genus resembles that of *Thracia* in structure, but not in position. In the last-named genus the process in each valve is horizontal, and projects inwards; but in *Mya* it is perpendicular or erect in one valve, and depressed in the other. In each case the office is the same, namely to contain the cartilage. Messrs. Alder and Hancock have carefully investigated the nature of the "branchial currents" in *Mya* as well as *Pholas*, produced by the action of cilia, and admitted and discharged by different apertures; and the following extract from their excellent paper on the subject, which appeared in the 'Annals and Magazine of Natural History' for November 1851, will explain to those who have not studied the economy of the Bivalve Mollusca how this operation is performed. "We lately had an opportunity of observing *Mya arenaria* in its native haunts, and watched the play of its siphonal currents under very favourable circumstances. This species, at the mouth of the Tyne, buries itself to a depth of 6 or 8 inches in a stiffish clay, mixed with shingle; and in shallow pools left by the tide, the siphonal tubes may be seen just level with the surface of the muddy bottom in full action. The mud lies closely packed against the walls of the tubes, so that nothing is to be seen but the internal surface of the expanded lips of the siphonal orifices fringed with numerous tentacles. When it happens that the surface of the water is only a little above these orifices, a strong current can be distinctly seen to boil up from the anal siphon, and another, with a constant, steady flow, to set into the branchial one. These currents were quite visible to the naked eye without the aid of a glass, so long as the mollusk remained undisturbed. We watched one individual for nearly a quarter of an hour, and no interruption of them took place, and it was not until the siphon was touched, that the tubes were withdrawn and the current ceased to play. But the siphon soon made its appearance again at the surface, and the orifices once more expanding, the currents commenced to play as strongly as ever. . . . . . . On removing these animals from their concealed abodes, and placing them in a vessel of fresh sea-water, the two siphonal currents were generally found in action when the individuals were undisturbed. And further, on placing the shell with its back downwards and the pedal gape raised above the surface of the water, these currents still continued to play; the excurrent and incurrent being as distinctly observed as before." The authors of this paper also ascertained that the currents communicate through minute openings in the laminæ of the gill-plates, which are sieve-like, filtering and collecting all the nutritious particles imbibed through the inhalant tube, in order that they may be carried to the mouth by the labial palps. Mr. Clark opposed the above view of the case, and endeavoured to prove that the water was mainly, if not altogether, introduced through the pedal opening; but although this mode of introduction may take place to a certain extent when the *Mya* or *Pholas* is removed from its hole, and placed in a vessel of water (after having ejected the greater part of its fluid contents, so as to create a vacuum), it is difficult to conceive how the requisite supply of food and water can be thus procured while the *Mya* is imbedded several inches in impervious clay or the *Pholas* is enclosed in its stony cell, or what in either of the above cases would be the use of the larger tube. I have repeatedly witnessed in many kinds of Bivalve Mollusca a current charged with animalcula or molecules being absorbed by this tube in a continuous stream, and a limpid current discharged at the same time by the smaller tube, occasionally together with pellets of fæcal matter or other rejectamenta. The structure of the shell has been investigated by Dr. Carpenter, and found to consist of variously formed cells: in the tooth or hinge-process is seen a group of large cells, the calcareous contents of which are arranged on a very regularly radiating plan, like that of the mineral called Arragonite or Wavellite. Neither in the shell nor in the tooth is there animal matter enough to give anything more than a delicate membranous residuum, in which no vestige of cell-walls can be detected.

This genus is modern in a geological sense, and does not occur in any formation older than the upper tertiaries. Only three species live in the European seas, the larger two of which are edible.

1. MYA ARENARIA*, Linné.

M. arenaria, Linn. S. N. p. 1112; F. & H. i. p. 168, pl. x. f. 4-6.

BODY fleshy, yellowish-white; tubular sheath covered by an extension of the epidermis of the shell; orifices of the tubes tinged with red, and fringed with tentacles of different sizes.

SHELL oblong (the right valve a trifle larger than the left, the inequality being more observable in young specimens), equilateral, gaping considerably at both ends, compressed, rather solid, opaque, usually lustreless: *sculpture*, coarse and irregular concentric striæ, diversified by stronger marks of growth: *colour* ashy-grey, with often a ferruginous tinge, or variegated by radiating lines of a brownish hue, which are caused by slight longitudinal folds of the *epidermis*: the latter is thin, yellowish brown, fibrous at the sides and in front, and imparting an oblique striation to the surface of the shell: *margins* rounded on the anterior side, slightly curved in front, and wedge-like on the posterior side; dorsal margins sloping more on the posterior than anterior side; posterior side obscurely keeled: *beaks* small, inflected, placed close together, that of the left valve being worn away or broken by continual pressure: *cartilage* triangular, strong, horncolour: *hinge-line* almost straight: *hinge-plate* broad and thick: *teeth*, in the right valve a slight and oblique cardinal on the anterior side of the cartilage-pit; the left valve has the complicated process described as one of the characters of the family, which in this species is very large, and irregularly shaped, convex within and concave without; the spur-like flange on the posterior side is placed obliquely, and there is a deep groove next to the hinge-plate for the reception of a blunt tooth-like fold on the same side in the opposite valve: *inside* chalky-white: *scars* distinct and deep. L. 2·5. B. 4.

Var. *lata*. Shell dwarfed, more oval and solid. *M. lata*, J. Sowerby, Min. Conch. t. 81.

Monstr. Furnished inside with foliaceous plates, showing a laminated structure.

HABITAT: Common on many parts of the coast, at low-water mark; chiefly in estuaries, where there is an admixture of fresh water with the sea. The variety is

* Inhabiting sand.

from the Firth of Forth and Oban, and the abnormal form from Exmouth. Fossil in all our newer tertiaries up to the Red Crag inclusive ; Nieuwerdiep, Friesland, in excavating the Royal naval dock (J. G. J.) ; newer beds of the "glacial formation" at Christiania, 50-200 feet above the level of the sea (Sars) ; Belgium (Nyst). In a living state *M. arenaria* is universally spread over the shores of the western hemisphere as far south as New York (de Kay), and the eastern hemisphere as far south as Rochelle (D'Orbigny, père), and between the 30th and 40th degrees of latitude in China (Debeaux). Dr. Walker records it from South Greenland at depths of from 10 to 120 f. ; and on the coast of Norway it is enumerated by Danielssen as taken in 2–15 f., and by M'Andrew and Barrett in 20–40 f. It is, however, in the main a sublittoral species.

M. arenaria received its name from Baster, and its habits are well described in his 'Opuscula subseciva.' He says that the foot, with which it penetrates the sand or mud, is wonderfully flexible, and assumes various shapes—now a trepan or pointed graving-tool, then a sharp wedge, a bent hook, or else a spade or dibble—each shape being adapted to some mode or other of boring, displacing, or removing the material in which this mollusk makes its abode. It is eaten and relished by man and fish in Europe, Asia, and America. At Southampton the fishermen used to call it "old maid" according to Montagu ; and at Belfast it has the equally strange name of "Cockle-brillion." It forms one of the numerous articles of Chinese diet, being brought to market after having been boiled for a long time, and cooked with a seasoning of which onion is the base ; the people call it "Tse ga." The occurrence of this circumpolar shell-fish so near the tropic of Cancer probably indicates the most southern limit in space of the glacial epoch. In the United States it goes by the general name of "clam"; and Gould informs us that it is more important, in an economical point of view, than the oyster. About 5000 bushels are annually brought to Boston market alone as food for man; and much more than ten times that quantity is salted and used as bait for fish. Its capability of living in brackish and even fresh water is well known. Lindström has given the following list of Mollusca associated with it in the Baltic: *Neritina fluviatilis, Bythinia tentaculata, Physa fontinalis, Limnæa stagnalis, L. auricularia, L. peregra, Tergipes lacinulatus, Limapontia nigra, Mytilus edulis, Cardium edule,* and *Tellina balthica.* To these may be added several kinds of Crustacea and Hydrozoa. Multitudes of young *M. arenaria* may be seen in the Loch of Steunis, about 5 miles from Stromness in the Orkneys, attached by byssal threads to the under side of loose stones: *Neritina fluviatilis* lives with them and deposits its spawn on the same stones. Full-grown individuals of the *Mya* are found (with *Littorina obtusata*) in the lower part of the loch, which is open to the sea. The fry are squarish-oval, decidedly inequivalve, and not unlike *Corbulæ.* My finest specimen is 3 inches by 5. Lapland seems to produce much larger.

Gould considers the *M. mercenaria* and *M. acuta* of Say synonyms of the present species.

2. M. TRUNCA'TA*, Linné.

M. truncata, Linn. S. N. p. 1112; F. & H. i. p. 163, pl. x. f. 1-3, and (animal) pl. H. f. 1.

BODY somewhat elongated and compressed, pale brown: *tubes* very long; tentacular filaments alternately large and

* Lopped.

small, marked with a brown spot at the base of each; valve of excurrent tube conspicuous: *gills* pale brown, their points entering the lower tube: *palps* large, excessively thin, and rather sharp-pointed: *foot* narrow and straight, yellowish-white.

SHELL oval, less inequivalve than *M. arenaria*, nearly equilateral, gaping widely at the posterior end but very little at the anterior end, rather convex (especially towards the beaks), solid, opaque, and lustreless: *sculpture* as in the last species: *colour* greyish-white, with often a yellow or ochreous tinge: *epidermis* rather thick, irregularly wrinkled or puckered, and minutely striated in a transverse direction: *margins* rounded on the anterior side, nearly straight in front, and truncated on the posterior side; dorsal margins sloping equally on both sides: *beaks* small, sharp-pointed and inflected, more or less contiguous, and sometimes abraded by mutual pressure: *cartilage*, *hinge-line*, and *hinge-plate* as in *M. arenaria*; but the hinge-plate is narrower: *teeth*, in the right valve an oblique spur-like cardinal, which is more conspicuous in young and immature specimens; in the left valve a nearly upright triangular plate, with a central cavity for the cartilage and a ridge-like process or wall on the posterior side; this plate is not so large as in *M. arenaria*, compared with the size of the shell: *inside* chalky-white, but occasionally nacreous and exhibiting a few minute pearls within the pallial line: *scars* strongly marked. L. 2. B. 2·65.

Var. *abbreviata*. Shell not so broad, abruptly truncated at the posterior end.

HABITAT: Littoral in muddy gravel and sand; but frequenting more the open sea than *M. arenaria*. It is sometimes found at considerable depths: I dredged a young live specimen of the variety on the Antrim coast in 80 f. about 10 miles from land. This variety has also been taken by Professor King on the Dogger bank, and by Mr. Barlee in Shetland. *M. truncata* occurs in every upper pliocene bed, including Moel Tryfaen (Darbishire); boulder-clay at Wick, Whitby, and Scarborough (Peach, J. G. J., and Leckenby); Sussex raised beach (Godwin-Austen); Norwich,

Red and Coralline Crag (Wood). It is dug up in such quantities on a farm near the Crinan Canal, as to be carted and used for manure. "At Lochgilphead the syphon is preserved in the clay filling the interior of the shell" (Geikie). I have also seen specimens *in situ* at Tufvoe near Gottenburg, about 200 feet above the present level of the sea. In clay near Palermo (Philippi); glacial deposits throughout Scandinavia; "aldre leer" at Christiania, 90–470 feet (Sars); Hudson's Bay (Drexler); Canada (Bell). Its foreign range in a living state extends from Spitzbergen (Phipps) and Kamtschatka (Steller), to the Black Sea (Siemascho), but probably subfossil in the last locality, as Middendorff believed; Misquer in lower Brittany (Cailliaud); Quibéron (Hémon); Bay of Biscay (Aucapitaine), in the old world: from Greenland (Scoresby and others) to Massachusetts (Gould), and Vancouver's Island (P. Carpenter) in the new world. It is possible that *M. truncata* may serve as a link in the chain of evidence to support the hypothesis of Professor Unger, that Europe was once connected with North America through the space now represented by the Atlantic Isles. Olivi enumerated it as an Adriatic species, and even gave a short description which leaves no doubt of its being our shell; but he may not have had recent specimens. The same remark applies also to Brocchi's statement, repeated by Risso, that it is found on the shores of Tuscany. The *M. truncata* of Chiereghini from the Adriatic has been identified by Nardo with *Thracia papyracea*. On the Scandinavian coast its bathymetrical limits lie between low-water mark and 100 f.

Its vernacular name is "smyrsling" in Iceland, "smirslingur" in the Faroe Isles, and "smirslin" in Shetland and the west of Scotland, all these being evidently derived from the Danish word "smor," or butter, which is expressive of the rich flavour of the animal. It is eaten and much esteemed not only by the natives of all northern countries, but by the walrus, arctic fox, and the grey-headed duck or King Eider in Greenland; and there, according to Fabricius, the shell is sometimes used as a spoon. Torell informs me that when he was last at Spitzbergen he took from the stomach of a walrus, that had been recently killed, a great number of the feet of *M. truncata*, the other parts having been apparently digested or got rid of. He is of opinion that the walrus rakes up the *Mya* from the mud by means of its long tusks, and that, after crushing the shell between its molar teeth, it spits out the fragments, as well as the leathery tube. The cod on the North-American fishing-banks seem to be equally fond of this mollusk; but it is not so easy to say how they procure it. *M. truncata* is often buried from 8 to 10 inches below the sea-bottom; and it does not seem to be capable of changing its habitation. The young occasionally occupy the deserted holes of *Saxicavæ*. They are more active than their parents, and exhibit a remarkable precocity of instinct. In Mr. Osler's well-known paper "On Burrowing and Boring Marine Animals" (Phil. Trans. 1826) he says, "On examining a *Mya truncata*, dug up on the preceding day, and which, when grown, will not attempt to burrow, I found two young ones, entangled in the cuticle at the extremity of the syphon, scarcely more than a line in length, and apparently but just excluded. Being placed on sand in a glass of sea-water they buried themselves immediately." In this and a later stage of growth the shell has a distinct keel on the posterior angle; the beaks are calyciform and resemble a *Kellia*, so that the fry must be of that shape. The half-grown shell is wedge-like on the longer side, with the terminal edges reflected outwards: until it arrives at maturity the truncation is incomplete. This alteration of shape is not caused by absorption, but by the formation of additional layers in front, which make the shell proportionally longer or deeper than it previously was. The Arctic variety, to which Forbes gave the name of *Uddevallensis*, is the usual form in glacial deposits; it is more depressed in the middle, obliquely truncated inwards, and excavated at the posterior end, frequently to so great an extent, and in such a fashion, as if there were cut

"A huge half-moon, a monstrous cantle out."

The internal structure of the shell is distinctly seen in fossil specimens of this variety which have been perforated by the *Cliona*. A section thus exposed shows at least 18 layers, and is unequally eroded, so as to resemble in miniature a perpendicular rock of oolite with caverns at its base. A specimen of an intermediate form, which I lately dredged in Dourie voe, Shetland, measures 3½ inches in breadth, and is of proportionate length.

Petiver called this shell "The broad Pholade-muscle"; when half-grown it is the *M. ovalis* of Turton, and *M. pullus* of S. Wood; the young is the *Sphenia Swainsoni* of Turton, and *M. Swainsonii* of Lovén.

3. M. Bingha'mi[*], Turton.

Sphenia Binghami. Turt. Dith. p. 36, t. 3. f. 4, 5, and t. 19. f. 3. *Sphænia Binghami*, F. & H. i. p. 190, pl. ix. f. 1–3, and (animal) pl. T. f. 3.

Body elongated and compressed, pale yellowish-white: *tubes* short, especially the incurrent one; mouth of each encircled by 5–10 rough white cirri; valve of excretal tube large and

---

[*] Named after the late Lieut.-General Bingham, an assiduous collector of British shells.

very long, subhyaline, and delicately frosted: *gills* pale brown; lower one of each pair much larger than the other, lying horizontally, and obliquely overlapped by the upper one: *palps* somewhat triangular and pointed: *foot* small, narrow, subcylindrical, of a bluish transparent hue; it produces a byssus of a few coarse filaments.

SHELL wedge-shaped, decidedly inequivalve and inequilateral, gaping at the posterior end, but not to the same extent as the young of *M. truncata*, compressed, rather solid, opaque, and somewhat glossy: *sculpture*, numerous fine but irregular concentric striæ, and occasional stronger marks of growth: *colour* milk-white under the *epidermis*, which has a brownish-yellow cast, and is often strongly wrinkled on the posterior side, and extends over part of the pallial sheath: *margins* obliquely truncated on the anterior side, usually straight in front, and narrowing to an abrupt and straight edge on the posterior side; this latter part has in each valve a blunt angle or keel running diagonally from the beak to the lower point of the posterior extremity; dorsal margins extremely short on the anterior side, long and mostly straight (although sloping) on the opposite side: *beaks* small, incurved, not contiguous: *cartilage* yellowish-brown: *hinge-line* slightly curved: *hinge-plate* narrow: *teeth*, in the right valve a small and blunt but distinct cardinal, besides the cartilage-pit, which is placed as usual in this genus; in the left valve the erect triangular tooth is flatter and less elevated than in the preceding two species, and considerably elongated on the posterior side: *inside* porcellanous: *muscular scars* extremely large, and placed near the edges of the shell. L. 0·25. B. 0·5.

Var. *elongata*. Shell considerably broader in proportion to its length, which is nearly equal throughout and gives a cylindrical appearance; posterior dorsal margin sometimes concave and turned up at the extremity.

HABITAT: In the cavities of limestone rocks and old oyster-shells perforated by *Saxicava rugosa* and *Cliona celata*, as well as among the roots or bases of *Laminaria saccharina*, and in other places of shelter and concealment; Channel Isles northward to Scarborough (Bean) and Skye (Barlee), and all the coast of Ireland, in 5–25 f. The variety is found in the deserted cases of

Serpula triquetra at Guernsey, Lulworth, and other places. Coralline Crag (coll. S. Wood). Its extra-British localities are the Boulonnais (Bouchard-Chantereaux); Croisic in lower Brittany (Cailliaud); coast of Spain (M'Andrew); Gulf of Lyons (Martin); Cannes (Macé); Spezzia (J. G. J.); and Tunis, in 25 f. (M'Andrew).

M. Binghami does not appear to have the power of excavating stones or shells, because specimens thus enclosed are frequently distorted or constricted, so as to fit the holes which they occupy. Its habits in this respect are the same as those of *Tapes pullastra* var. *perforans, Thracia distorta,* and some other bivalves. My largest specimen is scarcely three-quarters of an inch in breadth; but the late Dr. Lukis obtained much larger ones from the cavities left by *Saxicavæ* at Guernsey. It differs from *M. truncata* of the same size in being more inequivalve, inequilateral, and compressed; in the anterior side being invariably and abruptly truncated, instead of rounded; in the posterior extremity being more straight, and having a smaller gape; in that side being distinctly angulated, especially in the left valve; and the tooth in the left valve is less raised. It, however, belongs unquestionably to the same genus.

Family XXI. SAXICA'VIDÆ, Swainson.

BODY oval or oblong: *mantle* thick: *tubes* more or less united; orifices fringed with cirri: *gills* unequal on each side: *palps* triangular: *foot* finger-shaped, occasionally byssiferous.

SHELL rhomboidal, more or less inequilateral, and in the genus *Saxicava* sometimes inequivalve, always gaping at the posterior end (where it is obliquely truncated), and sometimes also towards the other end: *epidermis* membranous: *beaks* usually separate, not projecting, turned towards the anterior side: *ligament* external: *hinge* strong, furnished with cardinal teeth, which are in some cases small, indistinct, or obliterated, and an upright ledge to support the ligament: *pallial scar* placed far inside, and having a broad sinus: *muscular scars* large and conspicuous.

A family having close affinities with the last, but different in possessing an external ligament instead of an internal cartilage, and in the consequent structure and apparatus of the hinge. Some burrow in sand or mud like *Myæ*; others perforate certain rocks and hard substances, to a depth equal to the breadth of the shell and length of the tubes when fully extended. The mode by which these various objects are effected appears to be the same in every case, viz. by the propulsion or attrition of the muscular foot, which is always placed near the posterior end of the shell, assuming when in action the form of a cone or disk, and occupying the space to be excavated. Having already discussed at some length the latter part of the subject in the introduction to the first volume, I will not here say more than that occasional notices of this remarkable operation will be found in subsequent pages, while treating of particular genera and species comprised in the Lamarckian group of "Lithophages," as well as of the *Teredines*. Most of the *Saxicavidæ* pass their lives in a hermit-like seclusion, each immured in its own cell, content with the food brought by the waves or minute currents produced by the siphonal cilia, as well as with a certain degree of immunity from outward foes. Having no means of mutual intercourse, the nature of their sexual organization may be easily inferred; and the analogy in this respect between them and many flowering plants, which are rooted to the soil, cannot be very remote. Individuals of the same species of *Saxicava* which excavates holes in calcareous rocks or sandstone will, failing such materials, or for other reasons which are at present unknown to us, spin a byssus and thus fix themselves in the chinks and crannies of harder rocks, or now and then inside old bivalve shells.

Genus I. PANOPE'A*, Ménard de la Groye.
Pl. III. f. 2.

Body oval, fleshy: *tubes* very long, united nearly throughout, and enclosed in a tough leathery sheath: *gills* long: *foot* short, stout and muscular.

Shell equivalve, wrinkled transversely, gaping widely at both ends but much more so at the posterior end: *epidermis* thin: *ligament* short, prominent, attached to a process of the hinge-plate, which extends as the shell increases in size, and is sometimes triangular or represents the arc of a circle: *tooth*, a small conical cardinal in the right valve fitting into a cavity in the left valve: *pallial scar* entire, not deeply sinuated.

Most British conchologists are better acquainted with the large and scarce shell usually known as "*Panopæa Norvegica*" (but which, as I have before remarked, is a species of *Saxicava*), than with the small shell which I consider a true *Panopea*. Although the animal of this latter species is as yet unknown, the peculiar form of the shell, the structure of the hinge, and the pallial scar present the same characters which belong to *P. glycimeris* (or *Aldrovandi*) and its numerous congeners. The animal of *P. australis* was described by Valenciennes in the 'Archives du Muséum d'Histoire naturelle' for 1839, and that of *P. glycimeris* by Woodward in the 'Proceedings of the Zoological Society' for 1855. The former likened it to that of *Mya arenaria*, and was of opinion that the labial palps are olfactory organs.

* A Sea-Nymph.

But neither of these zoologists appears to have seen it alive. A great many species are known, both recent and fossil, some of the latter being Oolitic, and others (according to D'Orbigny) Permian.

The name of this genus has been spelt in various ways. Besides the original and correct one which I have given, Goldfuss and others called it *Panopæa*, Swainson *Panopia*, and Nyst *Panopœa*. *P. glycymeris* is the type of Klein's genus *Glicimeris*, which name has precedence of *Panopea* by more than half a century; but *Glycimeris* is now used for another well known genus.

PANOPEA PLICA'TA*, Montagu.

Mytilus plicatus, Mont. Test. Brit. Suppl. p. 70. *Saxicava rugosa*, young?, F. & H. i. p. 149, pl. vi. f. 1–3, and app. iv. p. 248.

SHELL rhomboidal, considerably dilated towards the posterior end (where the gape is very long, although not much wider than that of the anterior side), compressed, especially in front, thin, of a nacreous texture, semitransparent, and somewhat glossy: *sculpture*, numerous fine but irregular concentric striæ or plaits, and the surface in perfect specimens is minutely and partially granulated: *colour* milk-white: *epidermis* extremely thin, pale yellowish-white: *margins* rounded on the anterior side, nearly straight in front, expanding and arched (although obliquely truncated) on the posterior side, and forming a high shoulder at the back, with a distinct but blunt keel or ridge from the beak to the lower angle; anterior dorsal margin very short: *beaks* small, slightly inflected and calyciform as in *Mya Binghami*: *ligament* yellowish-brown: *hinge-line* nearly straight: *hinge-plate* rather narrow but reflected, and forming in the left valve a slight groove on the outside; it is furnished with a triangular process for the ligament, which slants a little inwards obliquely, like the tooth or cartilage-pit in *Mya*; this process varies in position, as well as in shape and size: *tooth* very minute, and not always present: *inside* porcellanous, and somewhat iridescent: *pallial scar* very distinct, with a shallow sinus: *muscular scars* irregularly triangular. L. 0·25. B. 0·4.

HABITAT: Skye (Laskey); among trawl refuse from

* Plaited.

Plymouth, and dredged in muddy sand off Skye, and in the voes of Deal, Dourie, and Basta, Shetland, at depths ranging from 5 to 40 f. (J. G. J.); small living specimens were also dredged by Mr. Barlee in Loch Fyne, and single valves by Mr. Hanley near the pier at Ryde in the Isle of Wight; Moray firth (Dawson); Stonehaven (Macgillivray); Walton-on-the-Naze (S. Wood). It is a common shell in the Coralline Crag at Sutton; and Nyst found it in the corresponding formation near Antwerp. M'Andrew dredged it in 40 f. off Gibraltar and in Vigo Bay, Lilljeborg in 70 f. at Bergen; and it has also been found at Hellebæk in Zealand.

I hope the animal will at some future time be made known. The shell may be distinguished from *Mya Binghami* by its nacreous texture, the extreme dilatation of the posterior side, and having a ligament instead of a cartilage, with a different hinge. Some specimens are partially incrusted by a mineral or fæcal deposit, showing the sedentary or inactive habits of the animal. The largest in my cabinet is nearly half an inch broad. Fossil specimens are rather more oblong, and the posterior dorsal margin is straighter and less arched than in recent specimens.

If the present species, or my description of it, is compared with Montagu's account, and with the figure given by the original discoverer, Laskey, in the 'Memoirs of the Wernerian Society' (vol. i. pl. viii. f. 2), their identity will, I think, be found undeniable. It is the *Sphenia cylindrica* of S. Wood, and *Saxicava fragilis*? of Nyst. The *Mytilus carinatus* of Brocchi may possibly be a variety. Philippi proposed for this last and another species the generic name *Arcinella*, which had been previously used by Oken and Schumacher for two other kinds of bivalve shells.

The evidence that *P. glycimeris* has been found in our seas is not satisfactory; this species inhabits the Lusitanian and Mediterranean coasts.

Genus II. SAXI'CAVA*, Fleurian de Bellevue.
Pl. III. f. 3.

BODY muscular: *tubes* extensile, diverging at their extremities, and covered by a leathery or membranous sheath: *gills* prolonged into the cavity of the branchial tube: *foot* furnished with a byssal groove.

SHELL often inequivalve, wrinkled transversely, gaping at the posterior end, and sometimes also in front (or what may be termed the antico-ventral part): *epidermis* thick: *ligament* short, prominent, attached to an elongated process of the hinge-plate: *teeth*, a small conical cardinal in the right valve, locking between two similar ones in the left, but frequently obsolete or wanting: *pallial scar* interrupted or broken up into separate spots, not deeply sinuated.

The doubtful position which this genus formerly occupied among bivalve shells appears from the circumstance that Linné called the typical species (*S. rugosa*) and its variety *arctica* respectively *Mytilus* and *Solen*, Fabricius *Mya*, Ström *Chama*, Poli *Donax*, Solander *Venus*, Bruguière *Cardita*, and Turton *Anatina*; and that the variety constituted the genera *Hiatella* of Daudin, *Clotho* of Faujas St. Fond, *Byssomya* of Cuvier, *Byssonia*, *Rhombus*, and *Rhomboides* of De Blainville, *Didonta* of Schumacher, *Biapholius*, *Coramya*, and *Pholeobia* of Leach, and *Agina* of Turton. Gray makes *Hiatella* and *Saxicava* distinct genera. The former name was published in 1799, and the latter in 1802; but Daudin did not sufficiently characterize his genus, and *Saxicava* may be considered as now established by general usage. According to Chenu the geological age of the present genus dates from

* Rock-perforator.

the Jurassic epoch; Searles Wood, however, believes that it was not born before the tertiary formation.

1. SAXICAVA NORVE'GICA*, Spengler.

Mya norvegica, Spengl. Skrivt. Nat. Selsk. iii. (1). p. 46, t. ii. f. 18.
Panopæa Norvegica, F. & H. i. p. 174, and app. iv. p. 249, pl. xi. and (animal) pl. W. f. 1.

BODY oblong, pale pinkish drab: *mantle* covered with a black skin: *tubes* protected by a dark-brown leathery sheath, somewhat unequal in length, the upper or excretal tube being the shorter and smaller of the two; orifice of the larger tube encircled by 30–40 short tentacular cirri, of a brick-red colour, alternately large and small, and sometimes folded back on the edges of that tube; the smaller tube is also fringed, but much less distinctly: *gills* irregularly pectinated: *palps* long, delicate, slender and pointed, united around the mouth: *foot* very small when contracted: *liver* green.

SHELL oval with a somewhat oblique and irregular outline, the right valve a trifle larger than the left, moderately convex; it has a broad furrow in the middle gradually enlarging towards the front, a considerable gape between that part and the anterior side, and a remarkably large and wide opening at the posterior end; it is thick, opaque, and lustreless: *sculpture*, coarse, distant, and irregular concentric wrinkles: *colour* whitish, occasionally stained with a ferruginous tinge: *epidermis* pale yellowish-white, puckered in every direction, not continued over the tubular sheath, which is of a fibrous nature: *margins* rounded or obtusely angular on the anterior side, nearly straight or slightly incurved in front, obliquely truncated on the posterior side, and a little reflected outwards in adult examples; dorsal edges sloping gradually on each side, the posterior one being usually more than twice the length of the other: *beaks* blunt and much inflected: *ligament* large, horncolour: *hinge-line* almost straight: *hinge-plate* broad and thick, excavated for the reception of the teeth, and furnished with a short but solid process for the ligament, which is reflected outwards and callous in younger shells: *teeth*, in the right valve a comparatively minute cardinal, and in the left valve two others of even a smaller size, which are placed so near together as scarcely to allow space for receiving between them the opposite tooth:

* Norwegian.

inside whitish, with a faint iridescent hue in certain parts: *pallial scar* exhibiting about a dozen spots of different sizes; *muscular scars* deep; anterior triangular or semioval, posterior elongated. L. 2. B. 3.

HABITAT: The Dogger bank, off the coasts of Yorkshire, Northumberland, and Durham, deeply imbedded in muddy ground at about 30 f.; and Mr. M'Andrew dredged a valve in 82 f. east of Shetland. It is very difficult to procure, and is consequently scarce. Fossil in most of our newer tertiaries up to the Red Crag: at Chillesford it is found in pairs (Woodward), and at Wick in a fragmentary state in the boulder-clay (Peach); raised beach at Moel Tryfaen (W. Drury Lowe); tolerably common in the Clyde district; Palermo, in clay (Philippi); near Christiania in the older part of the glacial formation at 460 feet, and in the younger or post-glacial group at 60–100 feet above the sea-level (Sars); Greenland (Rink). Recent in Iceland (Steenstrup); Finmark, 68° 45' (Blix); Drontheim (Spengler); Cattegat (Lovén); Bohuslän (Malm); Hellebæk in Zealand (Mus. Copenhagen); White Sea (Lamarck); Coasts of Russian Lapland (Baer and Middendorff); Sea of Ochotsk (Middendorff); Newfoundland fishing-banks (Gould); Labrador (Mighels); New England (Stimpson).

This is probably the strange shell which Donovan in 1802 noticed as having been "fished up between the Dogger bank and the eastern coast of England"; but his knowledge of it appears to have been derived from hearsay, and he mistook *Panopea glycimeris* for the present species. The first reliable announcement of its being British was made by Doctor Turton in the 'Zoological Journal' for 1826 on Mr. Bean's authority, with the addition that a single valve had also been found on Aberlady sands in the south of Scotland. Mr. Bean's relation of the circumstances connected with his discovery is amusing. He says, "To some of the fishermen of our coast it is well known by the name of the 'baccabox,' from a fancied resemblance to one of their most useful household gods. All the specimens [which he obtained] were rescued from destruction in a singular manner. The first was destined for a tobacco-box; the second had the honour of holding the grease belonging to the boat establishment; and the third—'Tell it not in Gath, publish it not in the streets of Askalon'—was inspected alive for several days by the officers and members of a modern philosophical Society (all of them unconscious of its value), and after amusing them by squirting water to the ceiling, was at last seen by a learned friend, purchased for a trifle, and generously placed in our cabinet." The long-line fishermen, every now and then, capture living specimens, by means of their hooks becoming fixed in the tough leathery sheaths of these enormous *Saxicavæ*; and they thus increase not a little their precarious earnings. The shell is much sought after by collectors, and is never likely to be so plentiful in their repositories as it evidently is in that of Nature—unless some adventurous zoologist, like Milne-Edwards or the unfortunate Barrett, should be tempted and able to explore, with the aid of a diving helmet or dress, the comparatively deep sea-bottom inhabited by these mollusks. Dr. Mighels says that the specimens which he obtained were taken from the stomachs of cod fishes. *S. Norvegica* is gregarious, and lives in company with *Mytilus modiolus*, whose byssal fibres may be occasionally seen adhering to the shell of the present species. Sessile Foraminifera (*Truncatulina lobatula*) may also be detected on the outside of the tubular sheath, even at its base, showing that this part is habitually left exposed, and not merely protruded at rare intervals. No portion of this appendage can be withdrawn into the shell; and the same is often the case also with *S. rugosa*. The structure of the shell must be cellular, because in fossil specimens the surface when abraded or worn appears under the microscope to be studded with circular pits. As Spengler well remarked, the shell is not unlike *Mya truncata*, especially in the large opening at the posterior end. Clark pointed out its close relation to *Saxicava*, and Woodward has satisfactorily ascertained and shown its generic place.

It is the *Glycimeris arctica* of Lamarck, *Panopæa Spengleri* of Valenciennes, *P. Bivonæ* of Philippi, and *P. Middendorffii* of A. Adams.

2. S. RUGO'SA*, Linné.

Mytilus rugosus, Linn. S. N. p. 1156. *S. rugosa*, F. & H. i. p. 146, pl. vi. f. 7, 8, and (animal) pl. F. f. 6.

BODY varying in shape from oval to cylindrical, greyish-white more or less tinged with yellow, sometimes brownish-yellow or orange: *mantle* very thick, coarsely and deeply wrinkled: *tubes* very extensile, enclosed in a brown membranous sheath to within a short distance from their extremities, where they separate and slightly diverge; orifices often pinkish, fringed with a double row of short whitish cirri with truncated points; each tube has from 16 to 20; those in the outer row are much smaller than the inside ones; excretal valve bell-shaped, widely open: *gills* very narrow: *palps* small: *foot* finger-shaped, rather long, extremely flexible and muscular.

SHELL oblong, usually somewhat inequivalve but especially in its free and younger state, slightly compressed except towards the beaks, frequently gaping in the front or on the antico-ventral side, as well as at the posterior end, thick, opaque and lustreless: *sculpture*, coarse, distant, and irregular concentric wrinkles; the posterior side is marked in young and free specimens by a double ridge, which is usually spinous or imbricated, and diverges from the beak in each valve

* Wrinkled.

towards the siphonal extremity: *colour* whitish: *epidermis* light brownish-yellow, more or less puckered: *margins* rounded on the anterior side, nearly straight in front, either curved or bluntly truncated on the posterior side; dorsal edges gently sloping on each side, the posterior one being three or four times as long as the other: *beaks* small and blunt, inflected, and inclining considerably to the anterior side: *ligament* yellowish-brown, proportionally longer than in the last species: *hinge-line* slightly curved: *hinge-plate* broad and thick, excavated externally to receive the ligament, so as to form in some specimens an elongated ledge or process, which is reflected outwards and callous in younger shells; it is occasionally also excavated (but slightly) internally: *teeth* often wanting; but when they occur, the right valve has a very small erect cardinal, closely interlocking between two others in the left valve: *inside* porcelain-white and glossy: *pallial scar* exhibiting in dead and fossil specimens a few spots of different sizes, which are indistinct in fresh specimens: *muscular scars* more conspicuous, triangular. L. 0·6. B. 1·4.

Var. 1. *arctica*. Shell more angular, and having distinct ridges; *beaks* less worn; *teeth* usually more developed: this variety never burrows in stone, but is attached by a byssus. *Mya arctica*, Linn. S. N. p. 1113. *S. arctica*, F. & H. i. p. 141, pl. vi. f. 4–6.

Var. 2. *minuta*. Shell smaller, and having prickly ridges: this is the younger or immature state of the first variety. *Solen minutus*, Linn. S. N. p. 1115.

Var. 3. *præcisa*. Shell smaller, abruptly truncated close to the beaks at the anterior end. *Mytilus præcisus*, Mont. Test. Brit. p. 165, t. 4. f. 2.

Var. 4. *pholadis*. Shell gaping widely in front, and wedge-shaped. *Mytilus pholadis*, Linn. Mant. Plant. p. 548.

HABITAT: On every part of our coast, from the Shetland to the Channel Isles, where there is limestone, chalk, or new-red sandstone, all of which this species excavates. It usually inhabits the lowest verge of spring-tides, and the Laminarian zone; but Mr. Peach procured live specimens from a rock perforated by them in

30 fathoms, 4 or 5 miles off the Deadman in Cornwall; and a piece of primitive limestone similarly excavated was brought to me by a fisherman, having been hooked up from more than twice that depth about 30 miles eastward of the Whalsey Skerries in Shetland. Vars. 1 and 2. Universally diffused from low-water mark to 145 f. (Beechey). Var. 3. Confined in narrow crevices of rocks, and beneath the hinges of old bivalves. Var. 4. In siliceous limestone. This very common species is found everywhere in upper tertiary strata, as far back in time as the Coralline Crag, and it frequently denotes arctic conditions. Glacial formation at Christiania, 50–470 feet (Sars); Subapennine and Sicilian beds (Brocchi and Philippi); Antwerp (Nyst); newer miocene near Antibes (Macé). The extent of its geographical range is almost unparalleled in the history of the Mollusca. It appears to have spread over the greater part of the globe, from one pole to the other. I cannot distinguish Australian from Greenland specimens by any character except that of size, those from the north being much larger.

The animal was well described by Fabricius. He said that it was cooked and eaten by the Greenlanders, and that on being touched or alarmed it squirts out water and contracts itself like an *Ascidia*. He found the variety *pholadis* with other shell-fish from deepish water in the crop of the King Eider-duck. The fact of its being byssiferous of course did not escape his notice, and it has been since mentioned by Mr. Osler and Mr. West. It is equally notorious that trias or new-red sandstone (which is not calcareous) as well as limestone, is perforated by the typical form. I can fully corroborate Mr. Clark's observations on this point. Lister noticed nearly two centuries ago that the holes are considerably larger than these shell-fish require in order that they may freely open their valves. This gives room for the foot to expand and work. The side of the shell in all such cases is often more or less rubbed or worn, in the same way as the spinous fringes of *Pholas dactylus*, in which the last-formed rows are uninjured; and the epidermis is seldom preserved on that part. In specimens of the typical form of *S. rugosa*, excavating limestone in Shetland, the ventral and exposed border of the mantle has sometimes delicate sessile Foraminifera (*Truncatulina lobatula* and *Discorbina globularis*) living on it, which proves that the mantle is not the organ of attrition. If an acid were employed by the *Saxicava* in dissolving calcareous rocks, it would assuredly destroy that portion of the shell from which the epidermis had been removed, as well as the shells of the Foraminifera. The edges of the excavation are sharply defined, and present an appearance very unlike that which would be produced by a solvent action. Therefore, either the shell or the foot must be the operative agent. Were it the former, the epidermis in front would be entirely abraded; and such is never the case. The *Saxicavæ* do not work, if they can meet with ready-made holes. The late Dr. Lukis, in one of his letters to me, said, " Successive generations will occupy the same hole. The last inhabits the space between the valves of its predecessor. In this way four or five pairs of shells may be frequently seen nested one within the other, and not unusually a *Sphenia Binghami* in the centre of all." Cailliaud observed a *Saxicava* within a specimen of *Venerupis Irus*, which it had perforated. Malm found a cylindrical variety in the burrows of *Limnoria lignorum*. The form of the shell is so variable and dependent on habitat, that (as the late M. Bouchard-Chantereaux remarked) it is possible to discover almost as many species as individuals. I am sorry to differ from Turton and the authors of the 'British Mollusca'; but I do not believe *S. arctica* to be a distinct species. The characters given, in the same terms, by these writers, are equally applicable to both forms. The "lunule-like excavation in front of the beaks" arises from the anterior side being more contracted than the other. Specimens enclosed in stone are generally symmetrical, and less angular than those which are free or attached by a byssus. The present species differs from *S. Norvegica* in being oblong instead of oval, not having a wide furrow in front, gaping much less at the posterior end, and in being furnished with a double ridge, which is often serrated in young individuals. It is, besides, comparatively a dwarf.

It would be tedious and unnecessary to particularize all the synonyms. I have collated seventeen, the specific names of which are different, in addition to those quoted above in describing the principal varieties.

Genus III. VENERU'PIS*, Lamarck.   Pl. III. f. 4.

Body oblong, thick: *mantle* bilobed: *tubes* united for about two-thirds of their length, naked; longer cirri pinnate: *gills* and *palps* small: *foot* compressed, byssiferous.

Shell equivalve, cancellated: *ligament* elongated, and sunk within the dorsal margins: *teeth*, 3 in one valve, and 2 or 3 in the other: *pallial scar* rather deeply sinuated.

Although the shell described by Lamarck as the type of *Venerupis* is *Tapes pullastra* var. *perforans*, the characters by which he defined the genus are sufficiently comprehensive to apply also to *V. Irus*, which he included in it. There is undoubtedly a great similarity

* Rock-Venus; per syncopen for *Venerirupis*.

of shape between this genus and *Tapes*; but the shell of *Venerupis* is regularly cancellated, while that of *Tapes* is nearly smooth or marked only by concentric flattened ribs and obscure or microscopical longitudinal striæ. Perhaps *Venerupis* is here scarcely in its place. It is impossible to make a linear or graduating arrangement. An oak tree in the course of its growth will have many interlacing boughs, and will spread out: so with the system of natural history in passing through successive stages of development. The *Venerupes* occupy holes made by *Saxicavæ*, or attach themselves by byssal threads to rocks and other substances. The genus does not claim a greater antiquity than the miocene period.

Venerupis Irus*, Linné.

Donax Irus, Linn. S. N. p. 1128. *V. irus*, F. & H. i. p. 156, pl. vii. f. 1-3, and (animal) pl. G. f. 2.

Body white with a pinkish tinge: *tubes* slender, unequal in length, pellucid, speckled with flake-white, diverging near the extremities, which are of a pink colour; longer cirri of the orifice erect and feathered, shorter ones reflected and plain; retractile valve of excretal tube conspicuous.

Shell oblong, compressed, slightly gaping at the posterior end but in no other part, solid, opaque, and usually lustreless: *sculpture*, 15-20 thin laminar concentric ridges, which become broader and foliaceous towards each end of the shell; these ridges and their interstices are crossed by numerous fine longitudinal striæ, radiating from the beaks; the surface is also covered with minute close-set transverse striæ: *colour* pale yellowish-white, with sometimes (especially in the young) a stain of reddish-brown or purple on the umbonal area and posterior side: *epidermis* thin and seldom visible: *margins* rounded or obtusely angular on the anterior side, usually straight in front, truncated or wedge-shaped on the posterior side, which is at least twice as large as the opposite side, more or less straight behind, with an abrupt slope from the beaks

* The name of a beggar in the Odyssey.

to the anterior end, so that the posterior dorsal margin occupies nearly one side of a parallelogram: *beaks* small, inflected, and inclining considerably to the anterior side: *ligament* yellowish-brown or horncolour, contained in a groove or excavation with shelving sides, which extends for some distance from the beaks: *hinge-line* obtuse-angled: *hinge-plate* thick and broad: *teeth* erect, placed on the anterior dorsal line, the outermost and innermost diverging; in the right valve 3, of which the outside one is much smaller than the other two, and these are cloven; in the left valve are also 3 similarly arranged, the innermost being the smallest: *inside* porcellanous, often partially stained with reddish-brown, particularly the hinge-plate and posterior side: *pallial scar* uneven, with a tongue-shaped sinus: *muscular scars* deep; anterior roundish, posterior oval and situate close to the margin at the lower angle. L. 0·5. B. 1.

HABITAT: Laminarian zone on the coasts of Dorset, Devon, Cornwall, Glamorgan, Pembroke, and Ireland (west, south, and east). Red Crag (S. Wood); and Philippi has recorded it from the South-Italian tertiaries. It has not been noticed in Scotland or further north; its southern distribution extends from Brittany to the Adriatic, Black Sea, and Ægean, both sides of the Mediterranean, and the Canary Isles, at depths ranging from the shore to 20 f.

It is attached by a byssus to gneissic rocks in Connemara (Farran), and to the roots of *Laminaria bulbosa* in the west of Ireland (Harvey); it occupies holes made by *Saxicava rugosa* in new-red sandstone at Exmouth (Clark), in limestone at Tenby (J. G. J.), and Cannes (Macé); and it inhabits crevices of rocks (but never perforates them) on the coast of Sicily (Philippi). The shell being frequently distorted shows that the *Venerupis* is not a borer, but accommodates itself to any place of shelter: when thus enclosed it is occasionally incrusted with nullipore and *Spirorbis granulatus*. The very young are square, and the fry nearly circular. In perfect specimens the laminar ridges are fringed and resemble those of the lovely *Venus lamellata*. We learn from the interesting experiments made by M. Beudant, as to the capability of marine mollusks living in fresh water, that out of 20 individuals of *V. Irus* 16 remained alive at the end of 22 days after the sea-water in which they were placed had been gradually mixed with fresh, so as to render the proportions equal, and that all the survivors were instantly killed on being immersed in quite fresh water. Of *Mytilus edulis* 30 individuals lived for 5 months in the admixture, and for 15 days in fresh water.

Its specific name may have been derived from the ragged look of the shell, compared with that of its original congeners, the *Donaces*. Irus was a common beggar and messenger at Ithaca, who insulted Ulysses in his own palace upon his return home incognito, and was knocked down by him with a blow of his fist. Among the Roman poets the name symbolized wretched poverty and that reverse of fortune which Ovid expressed in the following line,

"Irus et est subito, qui modo Crœsus erat."

This characteristic shell is the *Tellina Cornubiensis* of Pennant, *Cuneus foliatus* of Da Costa, *Venus cancellata* of Olivi (but not of Linné), and *Venus Bottarii* of Renier.

I have a specimen of *Venerupis Lajonkairii*, Payraudeau (*V. decussata*, Philippi), which came from the collection of Mr. George Humphreys, with the undermentioned name and note of its locality, "*Venus striata*, H. Brighthelmstone W. H. 1768." It is a native of the Mediterranean, and resembles the young of *Tapes decussatus*; but, besides the difference of hinge-structure, it is beaded lengthwise, and has a more square outline with rounded angles. Philippi calls it "rotundato-quadrangula," a definition at which mathematicians might be inclined to smile.

Family XXII. GASTROCHÆ'NIDÆ, Gray.

BODY oblong: *mantle* large and thick: *tubes* long, united throughout; orifices cirrous: *gills* unequal, prolonged into the branchial tube: *palps* small, somewhat triangular and thick: *foot* finger-shaped, sometimes byssiferous.

SHELL enclosed in a testaceous flask-like case with a narrow bipartite or divided neck: it is oblong, equivalve, very inequilateral, wedge-shaped and widely gaping in front; anterior end pointed: *epidermis* membranous: *beaks* nearly terminal: *ligament* long, external: *hinge* straight, toothless but furnished with a horizontal plate or shelf-like process: *pallial scar* broad, and deeply sinuated: *muscular scars* strong.

This family connects the *Saxicavidæ* and *Pholadidæ*. Its smooth shell is not unlike that of *S. rugosa* var. *pholadis*, which has also a ventral gape, and is most frequently toothless; although its reflected hinge-plate and pointed anterior extremity remind us of *Pholas crispata*. The foot is more characteristic of the first-named family, and the pallial tubes of the latter. But its distinctive and peculiar feature is the outer case or covering, in which all the *Gastrochænidæ* enwrap themselves on arriving at maturity. The body of this case appears to be formed, like the shell, by a secretion of the mantle; and the enormous opening in front, between the valves, must greatly facilitate the work. The animal probably uses its flexible foot, turning it round on every side, to mould the walls of the case. There can be no doubt that the neck is elaborated by the siphons, in the same way that the *Teredo* makes the outer points of its testaceous sheath. In *Clavagella* and *Aspergillum* the valves are united with the case, being apparently soldered to it. Mr. Tryon has lately published a careful monograph on the Order '*Pholadacea*,' comprising the present family, *Pholadidæ*, and *Teredinidæ*.

Genus GASTROCHÆ'NA*, Spengler. Pl. III. f. 5.

See the description of the family for that of the solitary genus which inhabits the British seas.

Placed by Lamarck in his family *Pholadaires*. It was included in the genus *Uperotus* of Guettard, *Chæna* of Retz, *Fistulana* of Bruguière, and *Trapezium* of von Mühlfeldt. Swainson erroneously spelt the name *Gastrochina*. Mörch, H. and A. Adams, and Tryon consider the *G. mumia* of Spengler (a tropical shell) the type of the present genus, and refer the European species to Fleurian de Bellevue's genus *Rocellaria*. The only species which we possess excavates and encases itself in limestone, new-red sandstone, and old shells; sometimes the case is found free, and incrusted with fragments of shells and grains of sand. There can be no question as to its perforating powers. The case is occasionally lodged in the valve of a large *Pecten maximus* or *Lutraria elliptica*, half of it on one side of the valve and half on the other. If an acid or chemical menstruum were necessary for this operation, it would either dissolve the calcareous matter of the envelope, or not act on the uncalcareous sandstone. The shell is invested by a delicate epidermis, which is more perfect on the posterior or outer end, but is sometimes worn

* Gaping in the ventral part.

away in the line of excavation. Several exotic kinds are known. One of these excavates coral, and when full-grown encases itself; but the coral soon outgrows, smothers, and completely envelopes the *Gastrochæna*. Instinct seems in this case to be at fault. The Jurassic, cretaceous, and tertiary formations contain many species.

GASTROCHÆNA DU'BIA*, Pennant.

Mya dubia, Penn. Br. Zool. iv. p. 82, t. 44. f. 19. *G. modiolina*, F. & H. i. p. 132, pl. ii. f. 5–8, and (animal) pl. F. f. 5.

BODY club-shaped, pale brown with a reddish tinge: *mantle* corrugated: *tubes* slender and transparent, capable of being extended to three times the length of the shell; terminal cirri short, dark purplish-brown: *gills* narrow, and of a thin texture, finely pectinated on both surfaces: *palps* smooth without and striated within: *foot* very expansible, conical when at rest.

SHELL elongated and obliquely twisted from the beaks to the posterior end, so that its proportions and course of growth are those of a *Mytilus*, convex, rather thin, opaque, and lustreless: *sculpture*, distinct but irregular concentric striæ, which are slightly imbricated in front: *colour* whitish: *epidermis* yellowish-brown: *margins* narrow and acutely angular at the anterior end, largely excavated in front (exhibiting an oval gape), with a bold sweep towards the posterior side, which is broad and obtusely angular; dorsal margin on the latter side long and high-shouldered or raised; anterior dorsal margin extremely short and truncated: *beaks* blunt and inflected: *ligament* semicylindrical, somewhat prominent, yellowish-brown; posterior end attenuated: *hinge-line* nearly straight: *hinge-plate* rather broad, thin, and reflected outwards; it is thickened within, where it projects downwards, occasionally resembling a large and blunt triangular tooth; the anterior edge is also callous: *inside* porcelain-white, glossy, and faintly iridescent: *pallial scar* usually indistinct: *muscular scars* irregular on the anterior side; the posterior one is large and triangular. L. 0·8. B. 0·35.

Var. *ovalis*. Shell shorter, broader, and thinner.

* Doubtful.

HABITAT: Southern coasts of England, the Channel Isles, South Wales, Barmouth, and the south and west of Ireland, in 8–20 f. The variety was found by Mr. Clark at Exmouth. Red and Coralline Crag (S. Wood); Italian tertiaries (Brocchi and Philippi). It does not appear to be a northern shell; but it is tolerably common on the coasts of France, Spain, Italy, Greece, Algeria, Madeira, and the Canary Isles, at depths between 2 and 60 f.

This remarkable shell, as well as its animal and case, were described by Montagu with his wonted accuracy. He states that he had specimens not only in limestone, but in granite; and he modestly observes, "How the siliceous part of this last is destroyed, we do not pretend to determine." Cailliaud has ascertained that all the lithophagous bivalves secrete a corrosive liquid, at least in the months of May and June; and, not content with the usual test of litmus-paper, he tasted some of the animals—thus exemplifying the saying of Seneca ('De vita beata'), "curiosum nobis natura ingenium dedit." We are told by the French philosopher that the *Gastrochæna* affects the throat with an insufferably acrid flavour, like that of a bitter cucumber. But the oyster, which is not lithophagous, and *Pholas* are equally provided with the same acid. This fact seems to militate against the chemical theory. I have a cluster of a dozen *G. dubia* in a single oyster-shell. The case or crypt is thick and composed of a great many layers. The exposed part of it is formed of tubercular concretions of different shapes and sizes; its neck resembles a double cylinder joined together but open on the inner side, and it is frequently curved. The anterior part of the shell is evidently subject to much friction during the process of excavation, and is invariably divested of the epidermis: it does not show the slightest indication of any corrosive action; and the inside is highly polished, although in close contact with the mantle.

It is the *Chama parva* of Da Costa, *Pholas pusilla* of Poli and Olivi (but not of Linné), *P. faba* of Pulteney, *Mya Pholadia* of Montagu, *P. hians* of Brocchi (but not of Chemnitz), *G. modiolina* of Lamarck, *Mytilus ambiguus* of Dillwyn, *G. pelagica* of Risso, *G. cuneiformis* of Philippi (but not of Lamarck), as well as his *G. Polii* and *G. Poliana, G. fulva* of Leach, and *G. tarentina* of Costa.

Family XXIII. PHOLA'DIDÆ, Gray.

BODY conico-cylindrical: *mantle* thickened at its outer edges, and reflected behind, where it covers the hinge of the shell: *tubes* large, extensile; orifices of both or of one of the tubes more or less cirrous: *gills*, a pair on each side, narrow, for the most part adherent on one of their sides, and prolonged into the branchial tube: *palps* also two on each side, coarsely pectinated as well as the gills: *foot* short and sucker-like, never byssiferous.

SHELL wedge-shaped, convex, equivalve, inequilateral, widely gaping in front (except in the adult *Pholadidea* and allied genera, which have the gape closed by a shelly layer), and at the posterior end in all the genera but *Xylophaga*: *epidermis* membranous, thin: *beaks* not prominent: *hinge* connected by the anterior adductor muscle, which supplies the place of a ligament; it is covered by a thickened fold of the mantle, which is to a greater or less extent protected externally by one or more testaceous shields or plates: the hinge is inarticulated, but sometimes furnished with laminar or tubercular processes: *apophyses*, as in *Terebratula*, falciform, springing forwards from beneath the hinge, one in each valve: *pallial* and *muscular scars* indistinct.

These burrow in stone, clay, mud, sand, wood, peat, and other mineral and vegetable substances. In the holes thus excavated they dwell at ease, never of their own accord removing from one place to another, in this respect unlike the *Solen* and other bivalves which make only a temporary sojourn in sand or mud. The extensor muscle, aided by the prickly surface of the shell, serves to keep the *Pholas* fixed in its case when it rises to the surface in search of food. The depth of the hole excavated by *P. candida* is between 5 and 6 inches; its shell and tubes, the latter being fully extended, measure only 4 inches. The ascent must be effected by stretching out the foot; and by contracting it the *Pholas* can descend to the bottom and retreat for shelter. I have observed the latter fact; and I believe the above to be the correct explanation of it. The *Pholas* gets rid of the excavated material by closing the valves of its shell, and forcibly expelling the detritus by a spasmodic action, through the larger or incurrent tube, together with the water contained in the body. The detritus is not removed to any distance; and some of the finer particles are occasionally washed by the waves into the hole, and line its sides. Mr. Osler and M. Cailliaud account in other and different ways for this phenomenon. According to the latter naturalist, the maternal care of the *Pholas* for its young is very peculiar. He says that, like the *Gastrochæna*, it makes with its acidulated siphons small oval holes in the surface of the calcareous rock which it inhabits, and inserts in these holes a portion of its brood! This remarkable instinct has its parallel in the case of *Teredo*, if we place equal confidence in everything that Sellius wrote on that subject. Most of the *Pholadidæ* can entirely withdraw into their shells: *Pholas crispata* is an exception among the British kinds. If a layer of peat, mud, or shale inhabited by *Pholades* is too thin to contain them, they will either perish or their growth will become stunted.

They usually burrow in a slanting direction. When several individuals occupy the same layer, one of them seldom interferes with another by breaking into the hole of its neighbour; but it pursues a parallel course. Cases, however, now and then occur in which no such forbearance is shown. The avoidance of each other's burrow is probably owing to the extreme sensitiveness of the foot or perforating organ, which is always pushed out in advance to feel its way. The structure of this organ is similar to that of the foot in *Teredo* and *Patella*, being thinner and of finer texture in the middle than at the circumference; it is nearly circular and truncated. Dr. Fischer believes that the foot in *Pholadidea*, when it has ceased to perforate, becomes atrophied. It is then hindered from further action by the shelly wrapper which closes the front gape of the shell, and is therefore useless. As far as has been hitherto observed, all the members of this family possess the hyaline " style " or cuspidated process, which is found in many of the other Conchifera. The use of this curious internal apparatus is unknown. Some physiologists consider it a digestive appendage of the stomach; but Cailliaud is of opinion that it is connected with the fecundation of the eggs, in consequence of these mollusks being hermaphrodite. Lacaze-Duthiers, however, regards them as of distinct sexes. The splendid work of M. Emile-Blanchard, now in course of publication, and entitled " L'organisation du Règne Animal," ought to be consulted with respect to the internal structure of the *Pholadidæ*. He has shown that each lobe of the mantle on the anterior side is extended, and reflected behind, where they are united and form a long and muscular expansion apart from the rest of the body. Siebold thought that at the base of the siphonal tentacles in *Pholas* there were eyes analogous to those with which the scallop is furnished; but Blanchard could not detect any such organ, although he had traced all the nerves in this part of the body to their extremities. Born's view, that the hinge is connected by a ligament, was adopted by Clark. It is incorrect. *Pholas* has no ligament, conchologically speaking; and its proper function, that of opening and closing the valves, is performed by the anterior adductor muscle. The shells are white or colourless, owing to their confined position. Their composition is exceedingly firm, and partakes of the nature of arragonite. This is sometimes necessary, in order to sustain the almost constant pressure of the shell against hard rocks. Adanson seems to have mistaken the nature of the dorsal shields (or "accessory valves," as they have been also called) when he used the same word, "palettes," to designate these appendages and the opercular bars of *Teredo*. The falchion-shaped shelly processes which issue from the hinge were first observed by Lister, and called "apophyses." Klein afterwards applied the same term to the dorsal shields. The true apophyses were regarded by Deshayes as cardinal teeth. But it seems to me that they have nothing to do with the hinge, and that they are formed by a different part of the mantle. They are probably of service in keeping the viscera in their proper place, and protecting them from the strain caused by the muscular exertions of the animal in the act of boring.

The notion that the shell is the instrument of perforation originated with Bonanni, in 1684. It was adopted in the last century by Adanson, Born, and others; and in the present century most zoologists of note and experience have favourably entertained it. No one, however, can be compared to M. Cailliaud of Nantes in respect of the zeal, ability, and conscientious labour with which he has investigated the subject; and he may be justly termed the apostle of this theory. His researches have been carried on, with scarcely any intermission, for more than twenty years; and although I have ventured to disagree with him in the present instance, I still entertain a profound respect for his opinion. Were I to become converted, it would be solely by his arguments. The careful and precise experiments made by him leave no doubt that the shell can be used by *man* as an instrument of perforation; it by no means follows that it is so used by the *mollusk*. It is easy to scrape with the edge of a limpet-shell a cavity in chalk or shale, such as *Patella* occupies; but can it be imagined that in this case the shell, instead of the foot, is naturally employed for that purpose? I believe that all the phenomena which have been attributed by Cailliaud to the mechanical action of the shell may be accounted for by the theory of Sellius, or rather of his predecessor Reaumur. For instance, the fine and regular striæ, which are observable on the *sides* of the cell of a *Pholas*, are unquestionably caused by the friction of the spinous ridges that ornament the shell. These striæ are wanting at the *bottom* of the cell, and are replaced there by a far more delicate elaboration, which I am of opinion is produced by the suckerlike motion of the foot. Assuming the latter to be the instrument of perforation, the shell would partake of its motion, and would rasp the walls of the cell while the foot was doing the work of excavation. The prickly surface of the shell, stretched to its full extent by the adductor muscle, is pressed against the sides or walls of the cell, and acts as a fulcrum. Born's idea, repeated by Cailliaud, viz. that the foot acts as the fulcrum or point of leverage, is exactly the reverse of mine. Besides, let us not lose sight of the fact that the shells of many mollusca which are not borers have also prickles or spines like those which cover the shell of a *Pholas*. *Anomia, Pecten, Lima, Arca, Cardium, Venus,* and *Psammobia* offer examples among the British shells of this sort of ornamentation. Such asperities appear to result from a superfluous secretion of shelly matter, which it is convenient to dispose of in this way; they strengthen the fabric of the shell, but are of no further service to its constructor. To my desire of doing justice to the investigation of M. Cailliaud must be added an apology for having, in the introduction to the first volume of this work, misinterpreted his views as to the mode in which the *Pholades* excavate rocks. The mistake was participated by Dr. Fischer, and arose from the following passage in M. Cailliaud's 'Mémoire sur les Mollusques perforants' (1856)—"les siphons des pholades, coupés en pièces et morceaux, attestent (dans les temps voulus) la présence de cette liqueur acidulée. Elle est donc faible puisqu'elle ne paraît sous aucun rapport nuire à l'organisation de ces animaux, et cependant elle dissout des test de coquilles très-durs, les calcaires les plus compacts." (p. 27.) I certainly understood by this, that M. Cailliaud was of opinion that the *Pholades* employ an acid or corrosive solvent in excavating calcareous rocks; but he has since distinctly asserted that they make use of their shells only. It had been generally supposed that *Pholas* does not secrete an acid. M. Thorrent, however, in the 'Journal de Conchyliologie' for 1850, proved that an acid exists in *P. crispata*; and this important discovery has since been confirmed by M. Cailliaud as to other species of *Pholas*. But how do the fry perforate? Are we to suppose, with M. Cailliaud, that the parent makes with its acidulated siphons minute holes in the rock for the reception of its progeny? It does not appear that he has ever witnessed so extraordinary a proceeding; and even if it is offered as a probable explanation of the method by which the fry effects its entrance into the stone, this would only apply to the case of chalk or limestone, which being calcareous can be dissolved by an acid, and not to that of gneiss, sandstone, peat, or wood, which are not liable to be thus acted upon, and in which the *Pholades* more frequently take up their abode. If the foot is to be considered merely as a *point d'appui*, the motive power would be altogether wanting while the young *Pholas*, encased in its tender shell, remains outside the much harder material which it has to penetrate. The same remark holds good with regard to other mollusca which excavate stone and wood. I am more than ever convinced that the *modus operandi* is similar in all these cases, and that the laws of Nature are more simple and uniform than those which direct human actions; nor do I infer from the case now before us,

"That many things, having full reference
To one consent, may work contrariously."

When the *Pholas* has to make its habitation in clay or sand, instead of in stone, no great amount of force seems requisite. Reaumur in 1712 stated, from his own observation, that *P. candida* uses its lozenge-shaped and comparatively large foot for this purpose. He took several of them out of their holes, and placed them on a clay as soft as mud; each soon put out its foot, and in a few hours made a fresh hole deep enough to contain the *Pholas*, which met with so little resistance and was evidently anxious to conceal and shelter itself without delay. Is it likely that the *Pholas* uses its foot or shell according to the nature of the material which it seeks to inhabit? Even where it has to erode the solid rock, the quantity of water it takes in, and with which all its tissues are saturated, cannot fail to render the process more easy. On the importance of this latter agent Valenciennes lays considerable stress in advocating the theory that the rock is worn away by the continual friction of the foot. The work of perforating gneiss, in which M. Cailliaud discovered living *Pholades*, must be extremely slow and gradual. It takes probably a year and a half before a *Pholas* arrives at maturity; by that time it has made a hole 5 or 6 inches deep. One hundredth part of an inch may therefore be reckoned its daily task. Time is of course a necessary element in all operations; and it serves no less to advance the labours of the persevering and patient shell-fish, than to scoop out valleys by the agency of running waters and yielding glaciers,

"And waste huge stones with little water-drops."

The *Pholadidæ* are distributed over the greater part of the globe; but the species, although prolific, are not numerous. According to Searles Wood "*Pholades* have been found fossil as early as the Lias"; and Chenu says that they occur in the Jurassic, cretaceous, and tertiary formations.

They comprise with the *Teredinidæ* the multivalves of Adanson and other writers of a later date; but neither the dorsal shields possessed by some species of *Pholas*, as well as by *Pholadidea* and *Xylophaga*, nor the sheath and pallets of *Teredo* are "valves" in a conchological sense, any more than the opercula of many univalves. Nor are any of these appendages homologous, or formed by similar organs. Linné also considered *Pholas* a multivalve, and placed it with *Chiton* and *Lepas*. Pulteney was not so far from the mark when he conjectured that the Tunicata were shell-less *Pholades*. Schumacher ranged *Pholas* with the pedunculated Cirripeds, and *Teredo* in another division of his medley collection of Monothalami. In Blanchard's system the present family is regarded as closely allied to the *Myidæ*. The number and position of the dorsal shields are useful characters to distinguish sections of genera, but they do not appear to be of any greater value. Ever since the groups called families were instituted in classifying the animal kingdom, conchologists have been busy in framing synonyms for the one of which we now treat. These synonyms, with the exception of two (*Adesmacea*, De Blainville, and *Cladopoda*, Gray), were compounded out of the generic name *Pholas*; and the ingenuity of the systematists may well excite our admiration, or some other feeling of perhaps not a laudatory kind, when we find no less than fourteen of such compositions.

Genus I. PHOLAS*, Lister.   Pl. IV. f. 1.

BODY oblong or oval, usually incapable of being altogether contained within the shell: *tubes* united except at their extremities, and enveloped in a membranous retractile sheath, as in *Mya*; both orifices cirrous: *gills* nearly equal: *palps* large and broad: *foot* truncated, but expansible to a certain extent.

SHELL shaped like the body, nearly opaque and lustreless, more or less covered with rows of prickles: *beaks* concealed by a fold of the hinge-plate in each valve: *apophyses* long and partly concealed within the hinge: *pallial scar* narrow and deeply sinuated: *muscular scars* widely separated; anterior elongated, posterior short: *dorsal shields* usually present, and varying in number, size, and position; when absent, their place is supplied by a tough integument of the mantle.

The present genus is not so ancient as has been gene-

* Lurking in dens.

rally supposed. The φωλὶς of Aristotle was a kind of fish, which he classes with the mullet; and the φωλὰς of other Greek writers appears to have been *Lithophaga dactylus*, which is certainly the *Pholas* of Rondeletius and Aldrovandus. Our species of *Pholas* are the "Piddocks" of old English naturalists, and the "Pitaux" or "Dails" of the French. Mr. W. Wood remarks that on the coast of Normandy they are eaten in abundance, well seasoned and cooked with bread-crumbs and fine herbs. They are also reckoned a delicacy when pickled in vinegar. In the neighbourhood of Dieppe a great many women and children, each provided with an iron pick, are employed in collecting them, either to sell in the market, or for fishermen's bait. They are almost entirely littoral,

"Entomb'd upon the very hem o' the sea."

The property which they possess of shining in the dark is very remarkable. It was mentioned by Reaumur in the Mém. de l'Acad. Roy. for 1723; and his communication, "Des Merveilles des Dails, ou de la lumière qu'ils répandent," shows his power of accurate observation. He says that this property is not confined to the skin or outer membrane of the *Pholas*, but that every part of the body is imbued with it, and when the *Pholas* is cut into pieces, each portion is luminous. Much of the water that drops from them sparkles brilliantly. The phenomenon is visible only when the *Pholas* is in a moist state. He dried several specimens, and after four or five days moistened some with common or fresh water, and others with water in which sea-salt had been dissolved. In every case the phosphoric light reappeared, but with less intensity than at first. When the *Pholas* was put into brandy, the luminosity almost instantly disappeared. No light is emitted by them in a dead or putrid state. He attributed the phenomenon (which he considered a "vrai phosphorus naturel") to a fermentation, resulting from the breeding-season; and he supposed that it was analogous to the cases of the male glowworm and centipede. These experiments were made in autumn, and at other times of the year when the weather was not very warm. Dr. J. M. Davis examined *P. dactylus* at Tenby in the autumn of 1840; but although he kept it alive and in a vigorous state for many weeks, it never was luminous or phosphorescent. Out of fifteen living individuals of this species obtained by M. Cailliaud at the end of April and in December 1854, ten or twelve shone in the dark. In none of these did the foot exhibit any light; only the mantle and siphons, which when rubbed with the finger were extremely phosphorescent, and shone even through the shells. The siphons were furnished with it in such quantity, that he was able to trace with them bright marks on a table. He endeavoured, but in vain, to find the same property in other perforating mollusks. I am disposed to believe that this luminosity is caused not by the *Pholas*, but by extraneous microscopic organisms. The subject ought to be further investigated. M. Necker has shown that the shell of *Pholas*, as well as of several other mollusca, is formed of arragonite; and inasmuch as that mineral slightly exceeds calc spar in specific gravity (the proportion being 2·9 to 2·7 or 2·8), he came to the conclusion that *Pholas* excavates calcareous rocks by means of the prickles with which the shell is furnished, aided by an acid. But he placed *Helix nemoralis* and *Mytilus edulis* in the same mineralogical category with *Pholas*, and ascribed a still greater density to the common oyster. It is also important to notice that the impurity of most calcareous rocks increases their hardness, and that the admixture of organic matter with the mineral ingredient in the shell diminishes the specific gravity of the latter.

The animal is partly the *Hypogæa* of Poli. Three or four genera have been proposed by Leach and Gray for the shells of certain species. *Pholas,* being derived from the Greek, is feminine.

A. Shell oblong: hinge-plate furnished behind with a layer of cells: dorsal shields 4, viz. 2 anterior, placed side by side; 1 cardinal, and complicated; 1 posterior, and elongated. *Dactylina,* Gray.

1. PHOLAS DAC'TYLUS*, Linné.

P. dactylus, Linn. S. N. p. 1110; F. & H. i. p. 108, pl. iii.

BODY oblong, whitish, sometimes tinged with blue or yellow: *tubes* more or less covered with short papillæ; orifice of longer tube margined with about a dozen fringed tentacles, besides as many intermediate smaller ones which are ciliated on the sides; the excurrent tube has its orifice either plain or margined with a few short cirri; the points of the siphonal tentacles or cirri are brownish; outer sheath brown or of a pepper-and-salt colour: *foot* rather obliquely fixed to the rest of the body by a long, cylindrical, thick, fleshy, white stalk.

SHELL elongated, somewhat obliquely twisted on the anterior side, moderately solid: *sculpture,* 40-50 longitudinal rows of small prickles or vaulted scales, which are formed by the intersection of slight longitudinal ribs and wavy transverse striæ; these prickles extend over the greater part of the shell, but they are much stronger and more crowded on the anterior side, and less so in front, and, especially, towards the posterior side, where they are altogether wanting; this latter part is often coarsely and irregularly granular, as if from an imperfect consolidation of the shell; the whole surface also is closely puckered: *colour* whitish: *epidermis* pale yellowish-brown, more persistent at the edges: *margins* narrow, angular, and more or less attenuated or beaked at the anterior end, widely

---

* Shaped like a finger; formerly, but erroneously, supposed to be the δάκτυλος or dactylus of the ancients.

open and exhibiting an oval gape towards the front, whence there is a regular slope both above and below to form the posterior end, which is rounded, and has a sharp edge, with a decided gape; dorsal margin on the anterior side short and obliquely convex: *beaks* very near the anterior end: *hinge-line* flexuous: *hinge-plate* extremely broad; it forms a double fold, one of which has a free cutting edge and projects outside in the middle of the hinge-plate, and the other adheres for the most part to the anterior side, its outer edge being likewise free; the interspace between these folds is fitted with about a dozen transverse plates, besides occasionally a few short intermediate processes in the opposite direction; the hinge-plate is sometimes crossed in its thickest part by two or three oblique tooth-like ridges: *apophysis* strong, broad, and curved, concave and expanding outwards: *dorsal shields*, two on the anterior side, large, irregularly lance-shaped, broader in the line of the beaks, and often cracked in a direction radiating from outside; another in the middle is morticed into the two anterior shields, and is of an irregularly triangular shape, twisted, and very solid, lying perpendicularly across the valves; the fourth or posterior shield is long, narrow, and slightly bent, so as to fit the slope of the shell on that side. L. 1·75. B. 5.

Var. 1. *gracilis*. Shell smaller, more slender, and of a finer and thinner texture.

Var. 2. *decurtata*. Shell stunted or truncated at the posterior end, and of a coarser and more solid texture; sculpture closer and usually effaced.

HABITAT: Slate rocks, coal-shale, new-red sandstone, chalk, marl, peat, and submarine wood in Guernsey, the south of England, and Bristol Channel; Seacombe, Lancashire (Dr. Walker); north, east, and south of Ireland. Var. 1. At extremely low tides below the Warren, Exmouth, in pure sand (Clark). Var. 2. Occasionally met with in hard rocks. Fossil at Belfast (Grainger); Sussex (Godwin-Austen); in the Scotch glacial beds at Ayr and Stevenston (J. Smith and Landsborough); Tarento (Philippi): and the variety *gracilis* was found by M. Cailliaud on the faluns of Touraine. Its exotic range in a recent state extends from Norway to Sicily and Algeria. M'Andrew describes his Spanish specimens as being of small size. Cailliaud has noticed it as perforating micaceous schist at Croisic in Lower Brittany.

The " Pierce-Stone " of Petiver. In Da Costa's time it was reckoned " a very excellent and dainty food." Philippi says that it is esteemed in Sicily by all classes; and at Rochelle it is sold in the market and served at the best tables. I am not aware, however, that it is now eaten in Great Britain; although it is often dragged out of its hole by our fishermen to entice and capture their finny prey. It buries itself eight, ten, or even twelve inches; and its tubes, when fully extended, are three times the breadth of the shell. Like all its congeners this species is very prolific. In a spot three feet square at Saundersfoot near Tenby, Mr. Jordan dug up 100 living specimens. He calculated, that owing to the removal by the waves of a foot in depth of mud during the autumn equinox of 1863, no less than 15,000 individuals perished; their empty shells remained below the surface. Some of them might also have been choked and destroyed by a silting up, as well as by the mud being disturbed in the course of its removal. The late Dr. Lukis took a *P. dactylus* out of peat, and kept it alive in clear sea water for four or five days. At the end of that time it died. The shell had become so thin from excessive absorption of its calcareous substance, that he was unable to lift it with the animal out of the water in a perfect state. Another intelligent and indefatigable naturalist, Mr. Peach, endeavoured to discover the way in which this *Pholas* makes its cell. He carefully and patiently watched 15 or 16 of them in a slab of clay-slate, and placed marks in order to see if they had any rotatory motion; but he found that they all invariably retained the same lateral position, and that the movement was vertical only. When the shell has been abraded or worn by rubbing against the sides of its stone cell, the new layers formed in front have of course their prickles, when they exist, quite perfect and sharp. Specimens now and then occur which measure about 6 inches in breadth.

The synonyms are antiquated; and two only are post-Linnean, viz. *P. muricatus* of Da Costa, and *P. hians* of Pulteney. The animal is the *Hypogæa verrucosa* of Poli.

B. Shell oblong: dorsal shield single, posterior, and elongated.
Barnea, Leach.

2. P. CAN'DIDA *, Linné.

P. candidus, Linn. S. N. p. 1111. *P. candida*, F. & H. i. p. 117, pl. iv. f. 1, 2.

BODY oblong, dirty white with a faint tinge of brown: *tubes* more narrow, slender, and elongated than in *P. dactylus*; larger tube funnel-shaped, grooved inside lengthwise like the barrel of a rifle, and appearing as if marked with white or light-brown stripes; its orifice is surrounded by about a dozen papillæ which terminate the grooves; smaller tube cylindrical, and contracted or bell-shaped at the top, with its orifice either plain or surrounded by a few papillæ; sheath minutely tuberculated: *foot* small, oval, attached by a compressed stalk.

SHELL elongated, tumid, and thin: *sculpture*, 25–30 longitudinal rows of sharp thorn-like prickles, which cover all the surface except at each end, and radiate from the hinge outwards; on the anterior side the prickles are stronger but not crowded: *colour* chalky-white: *epidermis* light-brown, somewhat fibrous on the posterior side, and forming delicate thread-like lines to connect the rows of prickles: *margins* rounded

* White.

or slightly angular at the anterior end, exhibiting a long and rather narrow gape towards the front, whence there is a regular slope (less above than below) to the posterior end, which is rounded and has a sharp edge, with a moderate gape; dorsal margin on the anterior side short, concave, and smooth: *beaks* very near the anterior end: *hinge-line* flexuous: *hinge-plate* extremely broad, and forming a single fold on the umbonal area, to which it adheres, the outer edge being free; the centre is marked across by a few indistinct furrows, resembling the walls of the cells in *P. dactylus*, as sometimes seen in that part of the shell; and it is furnished with a sharp ridge, that winds obliquely from above the apophysis to the posterior side, and ends in a projecting spur-like process; this is more prominent in the right than left valve: *apophyses* strong, narrow, curved, and concave at the point: *dorsal shield* slightly bent, and shaped like a lance-head with the point outwards; it has a small boss near the broader end, from which a shallow groove runs in the middle to the other end, with a slope on each side; the lines of growth are distant, diagonally arranged, and numerous. L. 1. B. 2·75.

Var. *subovata*. Shell smaller, and somewhat oval, in consequence of the posterior end being shortened or less developed.

HABITAT: Coal-shale, Great Oolite, and Oxford clay, chalk, marl, peat, submarine wood, and sand, from Guernsey to Oban and the Moray Frith, as well as throughout Ireland. Fossil at Belfast (Grainger, who has recorded a specimen from that deposit measuring 3 inches by 1¼); Bracklesham (Dixon); Christiania district in the newer beds, 100–120 feet above the present level of the sea, and at Drontheim, 30–40 feet (Sars). Abroad, it ranges from Iceland (Olafsen and Povelsen, *fide* Müller) and Norway (Lovén) to the Black Sea (Nordmann, *fide*, Middendorff); Sicily (Philippi); and Algeria (Deshayes and others).

Mr. Clark found it living in sand at Exmouth, and M. Cailliaud in gneiss at Croisic. It occurs in company with *P. dactylus* and *P. parva* at Guernsey. This species differs from *P. dactylus* in its more convex shape and thinner texture; the front gape being much narrower; not having any dorsal cells, nor more than a single shield; and in possessing a strong and remarkable fold on the hinge-plate.

The specific name was given by Lister. Spengler described and figured the present species as the *P. papyraceus* of Solander; but his description, quoted by Spengler, is more like that of a young *P. crispata*. It is also the *P. dactyloides* of Delle Chiaje, and *Barnea spinosa* of Risso. The *P. cylindrica* of J. Sowerby, from the Red and Coralline Crag, appears to be intermediate between the present species and the next.

3. P. PARVA *, Pennant.

P. parvus, Penn. Brit. Zool. iv. p. 77, pl. xl. f. 13? *P. parva*, F. & H. i. p. 111, pl. iv. f. 1, 2, pl. ii. f. 2, and (animal) pl. F. f. 3 & 3 A.

BODY oval, milk-white: *mantle* invested at its edges by a thin membrane: *tubes* marked inside lengthwise with alternate brown and white stripes; orifices scalloped, but neither are cirrous; sheath thick, reddish-brown, covered with numerous granular papillæ; these become larger towards the extremity of the sheath, which is encircled by a fine pile or fringe: *foot* oval when at rest, rounded in front and pointed behind when protruded, and attached by a long cylindrical fleshy stalk.

SHELL oblong, somewhat compressed, rather solid: *sculpture*, very numerous transverse rows of imbricated and flexuous ridges, which are puckered or flounce-like on the crests formed by the intersection of slight and less numerous longitudinal ribs; these markings are more crowded on the anterior side, and in the adult gradually disappear towards the posterior side, which is smooth or only exhibits some irregular lines of growth; there are seldom prominent and sharp prickles as in the preceding two species: *colour* white, sometimes slightly stained with brick-red from the matrix in which the shell is imbedded: *epidermis* light-yellowish and irregularly fibrous,

* Small.

more persistent on the posterior side: *margins* acutely angular or beaked at the anterior end, with a wide oval gape towards the front, whence there is a regular slope above and below to the posterior end, which is broad and rounded, with sharp edges and a moderate gape; dorsal margin longer than in *P. candida*, concave and sculptured like the rest of that side: *beaks* placed at a distance of about ⅔ths from the anterior end: *hinge-line* flexuous: *hinge-plate* extremely broad, folded over the umbonal area but not adhering to any part of it; the centre is marked as in the last species, and furnished with a thick knob or tubercle, which apparently serves by its intervention to prevent the valves from being squeezed too closely together; the crown of this tubercle is consequently more or less worn by continual pressure, and it is connected with the dorsal posterior margin by a sharp ridge, so as to give it additional strength: *apophyses* of moderate breadth, not much curved, and nearly flat: *dorsal shield* somewhat curved, and lanceolate with the point outwards; it has a small boss close to the broader end, which is bent inward; there is a slight depression down the middle, and the lines of growth are distinct, diagonally arranged, and numerous: *inside* polished and occasionally iridescent, usually showing the external sculpture, and having the edges notched on the anterior side. L. 0·8. B. 1·85.

Var. *quadrangula*. Shell smaller and more contracted at each end, with closer and finer sculpture.

Monstr. *tuberculata*. Shell divided into two nearly equal parts by a longitudinal irregular furrow. *P. tuberculata*, Turt. Conch. Dith. p. 5, t. 1. f. 7, 8.

HABITAT: New-red sandstone, marl, clay, and submarine peat, at Guernsey and on the southern coasts of England; Oxwich Bay near Swansea (J. G. J.); Abergelly, Denbighshire (Pennant); Dublin Bay (Warren); near Belfast (Hyndman); St. Cyrus, Kincardineshire (Brown). The North Welsh and Scotch localities are doubtful; because Pennant's shell was probably the young of *P. crispata*, and the single specimen said to have been found at St. Cyrus may have been from ballast. The variety is from indurated clay, and the monstrosity from the same material as well as from sandstone. The furrow or groove in the latter case is quite accidental, and does not even extend to the beaks. It was probably caused by an injury or obstruction of the mantle in front. I have already noticed similar cases in other conchiferous mollusks; and the univalves are also subject to this kind of partial deformity. *P. parva* has been observed by De Gerville and many other conchologists in the north of France, by M'Andrew at Malaga (of small size), and by Weinkauff at Algiers.

On a fine living specimen, which I took out of its burrow in sandstone at Exmouth, was a *Truncatulina*, full of sarcode. It still adheres to the crest of one of the ridges on the most exposed part of the anterior side of the *Pholas*. Is it possible that this part of the shell could have been employed in grinding the stone, and that the delicate Foraminifer remained uncrushed? In the instance just mentioned the posterior side of the *Pholas* was more worn than the other. Sometimes the entire sculpture of the shell is quite perfect, and appears not to have suffered the slightest attrition. The oval shape, smaller size, close and delicate sculpture, wide gape in front, large tubercle on the hinge-plate, and more central position of the hinge will readily serve to distinguish *P. parva* from *P. candida*. My largest specimen is $2\frac{1}{4}$ inches in breadth.

Da Costa, Boys, and Donovan mistook the young of *P. crispata* for the present species; and it is not unlikely that they were misled by Pennant, judging from his ambiguous description and figure. The last-named author confounded his species with *Martesia striata*. Our shell may have been known to Lister, who says, with reference to *P. crispata*, that sometimes it has a third small shell at the hinge. Solander called it *P.*

crenulatus. Perhaps it is the *P. callosa* of Lamarck from the neighbourhood of Bayonne. His diagnosis, and especially the words "valvarum callo cardinali prominulo globoso," are more applicable to *P. parva* than to *P. dactylus*. It certainly is his *P. dactyloides*. Although the very specimens which he thus described were received by him from Dr. Leach as the *P. parva* of Montagu, he capriciously rejected that name, and substituted an inappropriate one of his own. It is the *P. ligamentina* of one of the earlier works of Deshayes, and *Anchomasa Pennantiana* of Leach.

C. Shell oval; valves furrowed lengthwise: dorsal shield single, placed centrally, extremely small and triangular. *Zirphæa*, Leach.

4. P. CRISPA′TA*, Linné.

P. crispata, Linn. S. N. p. 1111; F. & H. i. p. 114, pl. iv. f. 3–5.

BODY very thick, reddish-brown: *tubes* long, encircled with branched papillæ: *sheath* velvety: *palps* much smaller than in the other species: *foot* oval.

SHELL convex with a slight depression in the middle, solid, and of a coarse and rugged aspect; it is divided into two nearly equal parts by a rather broad furrow, which runs obliquely from the beak in each valve to the front margin: *sculpture*, about 20 longitudinal rows of imbricated prickles, formed by the intersection of the ribs with numerous transverse scalloped ridges; these markings are on the anterior side only, and do not extend to the separating furrow: the rest of the surface is nearly smooth, or exhibits the usual irregular lines of growth: *colour* dull white with a slight tint of yellow: *epidermis* whitish, becoming brown towards the edges, wrinkled obliquely, and leaving its impress on the surface of the shell: *margins* acutely angular or beaked at the anterior end, with a very wide heart-shaped gape towards

* Curled.

the front, where there is an upward curve to the posterior end, which is broad and rounded, with sharp edges and a large gape; dorsal margins sloping almost equally on each side, the posterior being the larger of the two: *beaks* placed at a distance of about ⅔ths from the anterior end: *hinge-line* flexuous: *hinge-plate* extremely broad, folded over the umbonal area, and adhering to the greater part of it; it has no protuberance or other process, and is consequently more or less worn away in the centre by continual contact: *apophyses* curved, somewhat dilated, and concave at the points: *dorsal shield* triangular, with the apex downwards and the sides turned in; it is almost rudimentary, and covers only the angle formed by the meeting of the hinge-plate in each valve on the posterior side; the lines of growth are strong: *inside* marked with a ridge, which corresponds to the outside furrow, and terminates in a blunt tubercle: *pallial scar* narrow, very deeply sinuated, and extending far within the shell: *muscular scars* conspicuous; posterior pear-shaped, lying near the edge of the dorsal slope. L. 1·6. B. 2·8.

HABITAT: Mica-schist, coal-shale, Great Oolite, Oxford clay, gypsum, and peat, on various parts of the coast from Unst in Shetland (Edmondston and Dawson) to Weymouth (Metcalfe), and throughout the greater part of Ireland. Da Costa gives Cornwall also as a locality. It is found in all our upper tertiaries from the Belfast bed to the Coralline Crag, and especially in boulder-clay and other deposits of the glacial period. Uddevalla (Malm); Christiania, in newer deposits, 100 feet above the sea-level (Sars); Montcleone in Calabria, as *P. vibonensis* (Philippi). Its extra-British range in a recent state is chiefly northern. Iceland (Mohr and Spengler); Scandinavia (Müller and others); Heligoland (Frey and Leuckart); coasts of Holland (Waardenburgh); north of France (De Gerville and others); Charente-Inférieure (Aucapitaine); Marseilles (Matheron, *fide* Philbert); it is also extensively distributed in the New World, *e. g.* Canada and the United

States (Bell, Gould, and others); N.W. America, Vancouver's Island, and California (P. Carpenter).

Captain Bedford informs me that it is eaten by the poor at Oban. Inside the mantle of several specimens Sars found a large parasite, about an inch long, which he believed to be the *Malacobdella grossa* of Müller. The shells imbedded in stone are often stunted and much rubbed; but some which Bouchard-Chantereaux took from the trunk of a tree entangled in a fisherman's net at sea, and others noticed by Mr. Wright from turf at low-water mark, were in a remarkably fine state of preservation, as well as more convex. They seldom exceed on our coasts 3 inches in breadth. Mighels, however, mentions a specimen brought up on the fluke of an anchor in Portland Harbour, U. S., that was $4\frac{1}{2}$ inches; and Grainger found valves in the Belfast deposit of the same size.

Lister suspected that it might have been the *Peloris* of the ancients. Was not that the *Lithophaga dactylus* of modern naturalists? Petiver gave our shell the name of "Furrow-rib'd Pholade-Muscle," and Da Costa that of *Pholas bifrons*; Gmelin called it *Solen crispus*. In the tenth edition of the 'Systema Naturæ' it was placed in the genus *Mya*.

The hulls of ships returning from South America, off which the copper has been accidentally stripped, and pieces of mahogany drifted to these shores by the Gulf Stream are occasionally drilled by *Martesia striata*. This is more nearly allied to *Pholadidea* than to *Pholas*, and rejoices in the following synonyms: *Pholas lignorum*, Rumphius, *P. conoides*, Parsons, *P. nanus*, Solander (*fide* Pulteney), and *P. clavata*, Lamarck, besides *P. pusilla*, Linné, which is the young state.

The *P. sulcata* of Brown, from Dunbar, appears to

be an exotic species of *Parapholas*, perhaps the *ovoideus* of Gould. That genus is distinguished by having *two* furrows.

Genus II. PHOLADI'DEA*, Goodall. Pl. IV. f. 2.

BODY oblong, rather thin, capable of being contained within the shell: *tubes* united throughout and terminating in a disk, enveloped in a fine membranous retractile sheath; the orifice of the larger tube is cirrous, that of the smaller one plain : *gills* very unequal: *palps* long and narrow : *foot*, in the young and half-grown state very large, truncated, and springing from a long stalk in the centre of the body; in the full-grown state it becomes atrophied, and is reduced to a mere point.

SHELL oval, semitransparent but lustreless; anterior part covered with prickly ridges; in the adult the front gape is closed by a shelly dome or convex plate, and the posterior end is furnished with a cup-shaped appendage, which has a texture between shell and membrane : *beaks* much inflected, and concealed (but not covered) by a fold of the hinge-plate: *teeth* conspicuous, triangular : *apophyses* long, and partly concealed within the hinge: *dorsal shields* two, formed in the adult only; they are very small and triangular, placed close to the hinge on the anterior side, and in a line with the fold of the hinge-plate.

The distinctive characters of this genus are rather physiological and conchological than malacological; they are not developed until the *Pholadidea* has attained its full growth. In the young and immature state it does not differ from *Pholas*. The same peculiarity is found in *Martesia, Jouannetia*, and other allied genera. Mr. Berkeley has suggested to me that the cup-shaped appendage may be the homologue of the pallets in *Teredo*. It certainly occupies the same place in the animal; and both serve to protect the entrance of the hole, although less efficaciously in *Pholadidea* than in

* Having the shape of a *Pholas*.

Teredo. This hypothesis seems preferable, in a biological point of view, to that of Deshayes, who likened the appendage in question to the sheath of *Teredo.* Very few species of *Pholadidea* are known; and only the typical species (*P. papyracea*) is fossil.

Pholadidea papyra'cea*, Turton.

Pholas papyracea, Turt. Conch. Dith. p. 2, t. i. f. 1–4. *Pholadidea papyracea,* F. & H. i. p. 123, pl. v. f. 3–6, pl. ii. f. 1, and (animal) pl. F. f. 4.

Body somewhat conical, bluish-white, mottled in the centre with white roundish spots: *tubes,* when fully extended, often twice as long as the shell is broad, at other times more or less strongly wrinkled across; orifice of larger tube encircled by about 20 white cirri of different lengths; sheath of a pale reddish-brown hue, terminated by a fringe of short white cirri: *foot* clear white or almost transparent: *liver* green.

Shell convex, thin, and of a delicate texture, depressed in the middle, and divided into two nearly equal parts by a rather narrow groove or constriction, which runs obliquely from the beak in each valve to the front margin: *sculpture,* numerous transverse scalloped ridges on the upper half of the anterior side of the groove, the lower half being nearly smooth, much thinner, and forming an oval-shaped dome; the crests of the ridges are sometimes prickly but not much raised; the posterior half is marked only by irregular lines of growth: *colour* dirty white: *epidermis* very thin, partly fibrous at the posterior end, light yellowish-brown: *margins* rounded (in the young obtusely angular) on the anterior side, straight (in the young widely gaping) in front, squarish (in the young rounded) at the posterior end; anterior dorsal margin upturned, doubled, and folded back; posterior one pinched up and nearly straight (in the young sloping, so as to give a wedge-like appearance to that part of the shell): *beaks* placed at a distance of about ⅗ths from the anterior end: *hinge-line* flexuous: *hinge-plate* extremely broad, folded over the anterior side, and forming a free angular projection above that part of the hinge; from the posterior part of the hinge issues an oblique triangu-

---

* Paper-like.

lar plate in each valve (somewhat longer in the right), which
interlock and seem analogous to cardinal teeth in other bivalves:
apophyses curved, frequently twisted, narrow, and rather short:
dorsal shields often united, so as to form a single plate only,
which in that state is not unlike the shield in *Pholas crispata*;
it is also deeply scored by the lines of growth: *inside* porcellanous and glossy, showing on the anterior side the impressions of the outside sculpture, and marked with a strong ridge,
which corresponds to the outside groove and terminates in a
blunt tubercle: *scars* as in *Pholas crispata*: the *calyciform
appendage* is capacious, expanding considerably outwards, with
the edges slightly reflected: it is divisible into two parts, one
belonging to each valve. L. 0·75. B. 1·5.

Var. *aborta*. Shell stunted and sometimes distorted, vary-
ing in size from $\frac{1}{4}$th to $\frac{3}{4}$ths of an inch, exclusive of the
terminal process.

HABITAT: New red sandstone or trias, at low-water
mark on the South Devon coast (Turton and others);
Hayle (Miss Hockin); peat, at Ballycotton, co. Cork
(Wright); submarine forest, Clonea near Dungarvan
(Farran); Dublin Bay? (Thompson); sandstone at low
water, Castle Chichester near Belfast (Hyndman). The
variety has been taken from lumps of hard clay dredged
in deep water off Exmouth (Clark); in a piece of reddish
sandstone from deep water on the Cornish coast, drawn
up by a fisherman's line (Couch); in soft sandstone
dredged in 80 f. off the coast of Antrim (J. G. J.); in
indurated clay from 25 f. near Lismore in the west of
Scotland, with *Nucula sulcata* (Bedford). Mr. Searles
Wood detected some shelly fragments which he referred
to *P. papyracea* in the Coralline Crag at Sutton; otherwise it appears to be unknown as a fossil. No foreign
locality has been recorded.

The burrows are occasionally flexuous. One of these
in sandstone has near its opening a piece of silex much
larger than the rest, which the animal appears to have

been unable to remove, and the passage is partially obstructed by it. The immature shell (which Turton described and figured as *Pholas lamellata*) is not unlike the young of *Pholas crispata*; but it is more expanded breadthwise, and the sculpture is much finer. This form can always be traced in the earlier lines of growth of every adult specimen.

The *Pholas papyraceus* of Solander is only known to us by Spengler's quotation; it probably was the young of *P. crispata*. Turton, in his 'Conchological Dictionary,' first indicated the present species, and stated that Dr. Goodall had given it the name of *Pholadidea Loscombiana*; but in his 'Conchylia Dithyra' he retained it in *Pholas*, and altered the specific name to *papyracea*, on the authority of the sale catalogue of the Portland Museum. In this catalogue occurs "*Pholas pypyraceus* S." without any further particulars. I think the name proposed by Dr. Goodall ought therefore to stand; but I hesitate to restore it, because the other name, *papyracea*, is generally recognized. Blainville called the present species *Pholadidea Goodallii*; and in Griffith and Pidgeon's edition of Cuvier's 'Règne Animal' it bears the fearful name of *Pholadidoides Anglicanus*, which, however, is matched by one in Leach's 'Mollusca of Great Britain,' viz. *Cadmusia Solanderia*.

Genus III. XYLO'PHAGA*, Turton. Pl. IV. f. 3.

Body globular, all but the tubes, which, according to Dr. Landsborough, are not included within the shell: *mantle* puckered around the sides of the foot: *tubes* slender, covered by a single sheath, very extensile, marked lengthwise with

---

* Wood-eating.

crested ridges, which are pectinated at the edges, and separate at the extremities: *foot* large, pillar-shaped, capable of being protruded to some length.

SHELL globular, semitransparent, and somewhat glossy, divided lengthwise by a double ridge and furrow, which latter is terminated inside by a small knob or tubercle in the middle of the front edge; anterior part triangular and sculptured by numerous fine transverse striæ; middle area or strip narrow and covered with oblique, finer and more crowded striæ; posterior part on the other side of the ridge nearly smooth, and having the end closed: *beaks* as in the last genus: *apophyses* short and prominent: *dorsal shields* two, similar to those in *Pholadidea*, but proportionally much larger and more conspicuous as well as more complicated in structure.

Although *Xylophaga* resembles *Teredo* in the shape and sculpture of its valves, and forms a connecting link between the *Pholadidæ* and *Teredinidæ*, it is more nearly related to the former than to the latter family. Its habits are those of *Pholas*, in never perforating wood or vegetable matter (its only habitat) to a much greater depth than is necessary for the reception of its shell. It has no testaceous sheath or pallets like *Teredo*; but, instead of these processes, its shell is provided with dorsal shields or plates, similar to those possessed by other members of its own family. In fact it is a short *Pholas*, and not a long *Teredo*. More information as to the animal is desirable: I believe it can be entirely contained in the shell. The epidermis is conspicuous, and closely invests the anterior side of the shell; this affords an additional proof that the valves in the present case cannot be the instrument of excavation, otherwise the epidermis would be the first thing to be removed, from the continual friction to which that part must be subjected. Only two species have been described, one inhabiting the North Atlantic, and the other South America; both are recent.

XYLOPHAGA DORSA′LIS*, Turton.

Teredo dorsalis, Turt. Conch. Dict. p. 185. *X. dorsalis*, F. & H. i. p. 90. pl. ii. f. 3, 4.

BODY white, with the exception of the foot, which is tinged with buff at its extremity.

SHELL helmet-shaped, convex, thin, parted in the middle (but not equally, owing to the wide anterior gape) by a broad longitudinal groove, which is margined on each side by a sharp narrow ridge: *sculpture* as described in the generic characters; the striæ which cover the anterior and middle areas, as well as their interspaces, are exquisitely crenulated or crossed obliquely by still more numerous and microscopical striæ (giving the edges of the main striæ an exquisitely beaded appearance); these main striæ become more crowded or close-set as the growth of the shell increases, being at first comparatively few and remote; there is a distinct line of demarcation between the two sets of main striæ; the marks of growth on the posterior area are concentric and tolerably regular: *colour* white: *epidermis* yellowish-brown, more persistent on the anterior side of the separating groove: *margins* obtusely angular on the upper part of the anterior side, with a large triangular excision on the lower part, so that when the valves are united the opening is broadly heart-shaped; they are curved in front with a notch for the groove, and rounded at the posterior end: dorsal margins sloping abruptly and equally on each side: *beaks* much incurved, somewhat nearer to the anterior end: *hinge-line* projecting and pointed in the middle, by reason of the abrupt inflexion of the beaks, with a deep curve on either side: *hinge-plate* very broad on the anterior side, over which it is folded, adhering to the umbonal area but free towards the extremity, where the edges are turned up; it is narrow in the middle and on the posterior side: *apophyses* curved and projecting outwards; that of the right valve is larger than the other; in aged individuals they are thick and tusk-like: *dorsal shields* not unlike the opercula of *Neritina fluviatilis*, but having a less decided spire and doubled underneath at the wider end; they lie close to the beaks, on the outside of the dorsal anterior margin: *inside* glossy, marked with a broad and strong rib, which corresponds to the external groove, and sometimes also with a slight and indistinct ridge, which is

---

* From its being furnished with plates on the back.

impressed by the line of demarcation between the striæ on the anterior side: *pallial scar* narrow, withdrawn and deeply sinuated on the posterior side: *muscular scars* well marked: posterior oval and large; anterior covering the fold of the hinge-plate on that side. L. 0·375. B. 0·4.

HABITAT: Oak, pine, and birch wood, submerged between tide-marks or floating in the sea, on different parts of the coast from Unst to Torbay. Although its distribution is extensive, it has not been noticed in many localities. I will therefore enumerate them. Torbay (Turton); Exmouth (Clark); Gravesend (Crouch); Scarborough (Bean and J. G. J.); Northumberland and Durham coast (Backhouse and Abbes, *fide* Alder); Marsden Bay on the Northumberland coast (Howse); Bantry Bay and Waterford (Humphreys); Skerrie Islands in the south of Ireland (Walpole); Dublin Bay (Harvey and Warren); Loch Fyne (M'Andrew); in dock gates at Ardrossan, Ayrshire (Martin); Moray Firth (Macdonald); in a wooden shipping-stage at the Whalsey Skerries, Shetland, and a single valve dredged in 80 f., 30 miles north of Balta Sound (J. G. J.). It has also been taken at Drontheim in 30–40 f. by M'Andrew and Barrett; at Dröbak in 10–15 f. by Asbjörnsen; at Bergen and Christiansund by Lilljeborg; in other parts of Norway by Lovén; on the coast of Bohuslän in 22 f. by Malm; in the Cattegat by Mörch; at Brest by Dr. Daniel; in the Gulf of Lyons by H. Martin; and Professor Huxley gave me young specimens which had penetrated the outer coating (tarred hemp) of the Mediterranean electric telegraph cable on the coast of Spain at a depth of from 60 to 70 f.; some of these last were about to attack the gutta-percha tube, that formed the inner case or covering of the wire, when the cable was taken up.

This curious little mollusk attacks and injures submarine timber, but not to anything like the extent that *Teredo* does. Its burrow only extends 1½ inch in depth. The course of its perforation is diagonal or slanting, and therefore is partly against the grain of the wood. Its cell is flask-shaped with occasional concavities, the edges of which are sometimes sharp to receive the sides of the shell during the progress of the animal.

It is the *Pholas xylophaga* of Deshayes.

Family XXIV. TEREDI'NIDÆ, Fleming.

BODY worm-shaped and almost gelatinous, more or less enclosed in a testaceous sheath, which is usually flexuous: *mantle* very thin and cylindrical, enveloping the whole body, open only for the passage of the foot at the anterior end, and for the orifices of the tubes or siphons at the posterior end; it is folded back over the hinge of the shelly valves at the anterior end, as in the *Pholadidæ*; and it adheres to the sides of the sheath at the base of the pallial tubes, by means of a muscular ring: these *tubes* are short in proportion to the length of the body, but extensile; they are united near their origin, and forked towards their extremities; orifices fringed with short cirri: *gills*, a pair on each side, long, ribbon-like, and distinctly laminated; they are separate in front, adherent for the greater part in the middle, and prolonged behind to the base of the larger tube: *palps* consisting of two pairs, short and pectinated: *foot* large, truncated, muscular and expansile, not byssiferous; it is attached to the rest of the body by a thick and powerful cylindrical stalk.

SHELL or principal valves placed at the anterior extremity of the animal, helmet-shaped, equivalve, the valves touching each other only at the hinge and in front, but elsewhere widely gaping: each is divided and sculptured as in *Xylophaga*: *epidermis* membranous and thin: *beaks* not prominent, when viewed in front, owing to their being inflected: *hinge* connected by the anterior adductor muscle, which supplies the place of a cardinal ligament; it is covered by a thickened fold of the mantle, but there are no shelly plates or shields, such as the *Pholadidæ* have: the hinge is inarticulated or jointless, although sometimes furnished with tubercular processes: *apophyses* falciform, springing outwards from beneath the hinge, one in each valve: *scars* seldom distinct; the posterior is large and fixed to an ear-shaped expansion of the valve at that end: *pallets* or bars (set in the muscular ring at the base of the pallial tubes) paddle-shaped, with a narrow stalk; the blades are covered with an epidermis, and are either simple or compound: *sheath* tubular, often flexuous, usually open at both ends, and always at the posterior or outer end, which is conical and has the throat lined with a series of slight concentric plates.

Nearly all these burrow in hard vegetable substances; none in stone. A species allied to *Teredo* (*Kuphus arenarius*), which inhabits tropical seas, lives in sand; its sheath is closed at the anterior or broader end when the animal has attained its full growth. Deshayes, Quatrefages, and Emile Blanchard (all eminent physiologists) consider the *Teredinidæ* a distinct family, on account of their peculiar organization; according to Gray and the authors of the 'British Mollusca' they ought to be comprised in the *Pholadidæ*. The extremely elongated shape of the body, and its being enveloped in a testaceous sheath or cylinder, as well as possessing a pair of paddle-shaped bars to protect the tubes of the mantle, seem to be characters not less important than those which distinguish any other two allied families of the Conchifera.

Genus TERE'DO*, Sellius.   Pl. IV. f. 4.

Characters included in those of the family. All our native species have *simple* pallets.

1. *General remarks.*—The " shipworm " of British

* A borer, from τερέω.

sailors, "taret" of Adanson and the French, "zeeworm" or "paalworm" of the Dutch, "see-wurm" of the Germans, "trœmark" of the Norwegian fishermen, and formerly the "bysa" or "bruma" of the Italians, and "broma" of Peter Martyr and the Spaniards. I do not know any conchological study more interesting and important, and at the same time more difficult, than that of the *Teredo*. Although I have investigated its natural history for many years, have carefully examined a multitude of specimens, alive and dead, in order to learn something of their habits and forms, and have consulted perhaps every book or treatise published on the subject, I feel as if I still knew but little of this wonderful creature. Its biographers have been by no means wanting for the last century and a half; so, like the complete traveller in one of Bacon's essays, I "shall suck the experience of many." The information I have thus acquired, and the result of my own investigations will be embodied in the following remarks; and I hope that other observers will take up the thread of my discourse, and make it more complete. The *Teredo* is an anomaly. It consists of a long and nearly gelatinous worm-like body, without rings or segments, terminating at one end in a pair of hemispherical valves, that somewhat resemble the two halves of a split nutshell which has had a large slice cut off at each side, and at the other end in a pair of symmetrical shelly paddles with handles of different lengths, which close this extremity at the will of the animal. The open part of the bivalve shell is placed at the further end, and receives a circular disk, of a fleshy or rather muscular nature, which may be termed the foot; this is the broadest or widest part. Inside each valve is seen a curved process, like a bill-hook, that projects from the hinge at a right angle. The shell covers and protects the mouth, palps, liver, and other delicate organs. The body tapers gradually to the outer or nearer end, where it becomes quite small and attenuated; it contains the gullet, intestine, and gills, and is enveloped in a thin membrane or mantle, which forms at the outward point two cylindrical tubes, mostly of unequal length. The larger tube takes in infusoria or similar animalcules, which constitute the food of the *Teredo*, as well as imbibes water charged with air for the purpose of respiration and keeping the whole fabric moist; while the smaller tube is employed in the ejection of the water which has been exhausted or deprived of its aëriferous qualities, and also serves to get rid of the woody pulp that is excavated by the *Teredo*. Both tubes form a kind of hydraulic machine. At the base of each lies one of the paddles, often termed " pallets," and which may be translated into scientific language as " claustra." When the *Teredo* is alarmed, or not feeding, it withdraws its tubes into the neck of its sheath or shelly cylinder; and the pallets, which had been previously kept pressed against the sides, then spring forward and close the opening, so as to form an efficacious barrier against all foes, whether crustacea or annelids. This complicated animal mechanism is entirely enclosed in the sheath or cylinder above mentioned, which is secreted by the mantle and varies considerably in thickness and extent. The inside of the sheath is at its outer or narrower end divided into short strips or ledges, arranged in an imbricated fashion; the last-formed of these ledges serves as a *point d'appui* for the blades of the paddles, and it greatly assists the *Teredo* in closely shutting its doors. The whole of what I have above endeavoured to describe is found only within some hard vegetable substance, either the hull of a vessel or boat, a harbour-pile, a shipping-stage, a floating tree or the roots of one growing on the banks of an estuarine river, a piece of balk timber, a fisherman's cork, a cocoa-nut, a bamboo rod, a walking-stick, a beacon or buoy, a mast, rudder, oar, plank, cask, hencoop, or other ligneous waif or stray of the ocean. These the *Teredo* perforates, like a rabbit or mole in the earth, for the purpose of making its burrow and protecting its soft and sluggish frame. It is never free, nor found living anywhere except in its wooden gallery; and it may be cited as a teleological example. Without entering much into the doctrine of final causes, I consider that the *Teredo* shows an exact adaptation of means to the end or object, viz. its existence. If it were not endued with this or a similar power of self-preservation, it would fall an easy and dainty prey to fish, crabs, and sea-worms; and the race would be soon exterminated. Such is the general aspect of the *Teredo*.

2. *History*.—The ancient history of this mollusk is involved in much obscurity. Homer did not mention in any of his works the word τερηδών. It occurs for the first time in the Knights of Aristophanes, where the chorus reports a conversation that is said to have taken place among some triremes, in which the eldest of them declared to her companions that, sooner than be engaged in a rumoured expedition, she would remain where she was, grow old, and be consumed by Teredines. Now as it was the custom of the Greeks, as well as of the Romans, to lay up their vessels high and dry on the beach, until they were wanted for service, the word τερηδών, used by the great comedian, may have signified the wood-boring grub of a beetle or moth, and not a shipworm. Nor does it appear that Aristotle was acquainted with it. The word is only to be met with once

in his history of animals, when he describes the τερηδὼν as a grub, which is bred in bee-hives. Possibly he meant a young honey-bee. His τενθρηδὼν (which Casaubon incorrectly rendered *teredo*) is another kind of bee. However, his friend Theophrastus, who succeeded him in the Lyceum at Athens, mentioned the τερηδὼν in such precise terms as to leave no doubt of its being the mollusk in question. In the history of plants, written by this great naturalist and philosopher about 350 B.C., he restricted the name to a marine destroyer of wood, distinguishing the terrestrial kinds as σκώληκες and θρῖπες, which may be designated worms and grubs. His observations were made in his native island of Lesbos; and he says that the τερηδὼν lives in the sea only, and is of small size but has a large head and teeth. This description was probably taken from *Teredo minima*. He remarked that wood attacked by grubs might be easily restored and made useful, by dipping it into the sea; but there was no remedy for wood infested by the *Teredo*. In the same restricted sense the word "teredo" was mentioned by Ovid; and in his first epistle from Pontus occur the well-known lines which were quoted by Sellius, and were considered by Forbes and Hanley applicable to his own sad case. The kind alluded to by Ovid was in all probability the *T. navalis* of Linné, because after the Crimean war I received specimens of this species, which had been extracted from one of the Russian vessels sunk at the entrance of Sebastopol. Pliny gave no information of his own on the subject; and even the meagre account which he gleaned from Theophrastus and others was very confused. Natural history was at a considerable discount during the "dark ages;" and the *Teredo* does not appear to have attracted the attention of our remote ancestors. They were perhaps too much engaged in waging open war with their neighbours, to notice the secret and insidious attacks which the shipworm made on the few vessels which then traversed the ocean. Literature of every kind was confined to the cloisters of the monks, who had few opportunities, if any, of studying marine animals. A curious piece of information, however, has accidentally fallen in my way on reading one of the poems in the "Black book of Carmarthen," which, according to Mr. Skene, a learned antiquary, was compiled or written in the twelfth century, and is of unquestionable authenticity. It seems to show that the *Teredo* was at that time indigenous to our seas. Yscolan, a monk and scholar, gives an account in poetical and of course hyperbolical terms, of a penance which he endured for some ecclesiastical offence; and the following is a literal translation of the lines :—

> A full year I was placed
> At Bangor, on the pole of a weir.
> Consider thou my sufferings from sea-worms.

One kind of *Teredo* (*T. Norvegica*) is still found in the stakes of fishing weirs on the Welsh coasts. After the revival of letters Hooft, a Dutch historian, appears to have been the first to notice the *Teredo*. He says the dykes in Zealand had been destroyed by these vermin before the close of the 16th century. We learn from Johnston's 'Thaumatographia (Historiæ naturalis de Insectis,' 1653), that Drake's flag-ship was found on his return from circumnavigating the globe to be completely riddled by it. In the 'Ephemerides' for 1666, Nitzschius recorded its appearance at Amsterdam in ships which had been in the Indies, where it was supposed to have originated. He describes the method adopted by the Portuguese to get rid of it. This was to scorch their vessels, so as to form a crust of charcoal an inch thick; but he observes that the process was not "sine periculo," for it not unfrequently happened that the fire would spread and the whole of the vessel be burnt down. In the same century Bonanni and Dampier briefly alluded to it; but it seems to have escaped the attention of Aldrovandus and Lister. In 1715 Vallisnieri, and in 1720 Deslandes published some observations on the subject; those of the first-named writer were made at Venice, of the other at Brest. In each case more fancy than philosophy is exhibited. The "ver de mer" of Deslandes was a fabulous production, compounded of the *Teredo* and a well-known annelid which accompanies and preys on it. He believed that some of these "vers de mer" lived in wood, and others in the sea, and that the latter copulated in the water and afterwards entered into the wood, where the reproductive power ceased. One remark of Deslandes is more correct, and at all events is quaint. He says that it is difficult to imagine how an insect, which has such a phlegmatic air, can be so wonderfully active in its malice. In consequence of the excessive devastations which Holland suffered from this cause in the last century, and especially in 1730, 1731, and 1732, the history of the "Zee-worm" was then assiduously investigated by a crowd of native writers, who would seem to have been actuated by their patriotic feelings; and innumerable remedies were invented to stop the plague. In 1733 eight different treatises, of more or less merit, appeared. Preeminent among these was a monograph by Godfrey Sellius, a celebrated lawyer of Utrecht, and a fellow of our Royal Society. His 'Historia Naturalis Teredinis seu Xylophagi marini,' in quarto, contains 366 pages, besides two well executed plates. It is written in Latin. The work is a masterpiece of learned research, and replete with classical allusions; and it evinces far greater knowledge of the organization of the mollusca than that shown by any of his predecessors with the exception of Reaumur. He describes the external shape of the *Teredo,* then its internal structure, its peculiar habitat, the method of its perforating wood, the arrangement and uses of its different parts, its sexual nature and propagation, its teleological relations, its history, name, and definition, together with an explanation of its sudden appearance on the coasts of Holland; and lastly he details all the recipes which were known in his time to prevent its destructive operations, and he suggested others in addition. Nor did he share the erroneous notions entertained by most of his contemporaries as to its place in the animal kingdom. He proved that it was a true mollusk, and closely related to *Pholas;* and he insisted on the advantage, if not the necessity, of studying the animal as well as the shell—thus anticipating, by nearly a quarter of a century, the much lauded views of Adanson in both these respects. He distinguished no less than three European species, viz. his *T. marina* (which was perhaps the *T. navalis* of Linné), *T. navium* of Vallisnieri (*T. Norvagica,* Spengler), and *T. oceani* of the same author or *T. megotara,* Hanley. The subject appears to have fascinated him, much in the same way as a capricious mistress does her lover, who now deprecates the cruelty of his fair tormentor, and then extols to the skies her beauty and gentleness. He calls the *Teredo* a wicked beast, the worst plague that angry Nature could inflict on man; but he defends it against the calumnies of certain anonymous writers who had preceded him, and he expresses in enthusiastic terms his admiration of its symmetry, economy, ingenuity, social harmony (especially in avoiding controversy and litigation!), and its wonderful perfection in every particular. His account would almost persuade us that its dwelling is a model for the architect, and its mode of life a rule for the Christian. The observations of Sellius with respect to *T. navalis* are so interesting, and on the whole so correct, that I trust I may be here permitted to republish some of them, although they are antiquated, with such comments and explanations as I may deem necessary. If the perusal should occasionally provoke a smile, may it be one of charity; and let the disadvantages under which the Dutch naturalist laboured at the time of his writing be fully taken into account. He says that the *Teredo* varies greatly in dimensions, from the minutest point to a foot or more in length, and that specimens had been recorded which were even a foot and a half and two feet long. The pallets (which he styles "pinnæ") are likewise of unequal size in different individuals, the larger ones being more soft, and of a chalky consistency and dull aspect, not unlike morsels of old yellow cheese; they are frequently mutilated or distorted. The *Teredo*, when taken out of the wood, soon dies, although it be immediately placed in clear sea-water. This observation does not agree with those made by Professor Laurent in 1845 and 1847 with respect to *T. Norvegica*; and M. Eydoux ascertained that the last-named species, after having been taken out of the wood and kept in sea-water, actually secreted and formed a new calcareous sheath, although very thin and more or less incomplete, into which the animal retreated, closing the larger end with an hemispherical epiphragm (like those made by individuals in wood), and constructing at the smaller end two distinct apertures, for the passage of the siphons. Quatrefages, too, extracted specimens of *T. pedicellata* from their cases, and kept them alive in sea-water for more than fifteen days. Experiments tried by Sellius in putting *Teredines* into rain-water, beer, milk, and similar fluids resulted (as might have been expected) in their becoming feeble, and ultimately in their death. The fecundity of the *Teredo* next attracted the attention of its biographer. He computed that the eggs contained in a portion of one ovary were 1,874,000 (a number exceeding the then population of the eight chief cities of Christendom, namely London, Paris, Amsterdam, Venice, Rome, Dublin, Bristol, and Rouen), and that the entire ovary contained nearly seven times as many, and considerably exceeded the population of the seven United Provinces and all Great Britain to boot. He minutely described the ova and fry, which latter he found in different parts of the body. But Quatrefages has recently investigated this branch of the subject with very great care, aided by the light of modern science; and the result of these investigations will be given in the proper place. The knowledge of comparative anatomy possessed by Sellius was of course somewhat imperfect. Perhaps the phrase which he used in describing the ovary, "materia formatrix ovulorum," is not recognized by physiologists of the present day; at any rate it is intelligible. Deshayes has pointed out two or three more errors of this kind; but certain modern naturalists, whose opportunities were far greater than those which Sellius enjoyed, have committed mistakes of a not less grave character. I need only allude to the published accounts of the organization of *Dentalium*, as an instance of such inaccuracies. Sellius goes on to say that the sheath is testaceous, and annulated or divided into ring-like segments; it is highly polished inside. The larger or inner extremity is concave; the other extremity is conical. Adanson considered this appendage to be a part of the shell. The *Teredo* is gregarious, although not of a sociable habit; and, in the prosecution of its burrowing operations, it is actuated by a conscientious anxiety not to infringe on its neighbour. When a collision is imminent, it secretes a cup-shaped dome or plug in front, of a thinner texture than the rest of the sheath; and it shuts itself up. Sometimes it makes several of these outer walls, one after another. Young and old equally do this. It then, being unable to eat its way through the wood and thus procure a supply of food, dies of starvation, preferring suicide to the alternative of invading and injuring its companions! This sacred duty, he assures us, is performed with almost a reverential care. He evidently considered his "hero" (as he called the *Teredo*) the pink of chivalry and honour. The wood is often so completely honeycombed, that the party-walls which separate the burrows of the *Teredines* consist of mere films. Rousset compared the wood in this state to an extremely light and porous kind of rusk or biscuit. Sellius stated truly the object and mode of the curious dome-like fabrication which I have above described; but there was no foundation for the consequences pictured by him, except in his fertile imagination. The progress and further growth of the *Teredo* would necessarily be arrested by the barrier which it had interposed in front. But that was all. The food of the *Teredo* consists entirely of minute organisms, that are introduced with the water into the incurrent or branchial tube; and it does not consume the wood as any part of its nutriment. Nor do I believe that the eroded material undergoes any chemical change, either in the stomach of the *Teredo* or in the passage outwards through its intestine, although in the latter receptacle it is closely compressed. When it is voided or expelled by the excurrent tube, and separated in the water, it becomes a flocculent mass or pulp, like that of paper, composed of extremely minute and fine particles of an irregular size and shape, but still retaining its fibrous structure. It does not exhibit any appearance of having been digested. The notion that the *Teredo* feeds on the wood which it excavates originated in the lignivorous habit of the grubs of certain insects. It was lately revived by Laurent to a qualified extent. He tells us that the water, imbibed by the larger siphon, holds constantly in suspension particles arising from the decomposition of organic matter, as well as living animal and vegetable bodies, and that these particles, coming from outside, are united with the ligneous molecules which are produced by the wood being rasped and continually softened or macerated by the water, in order to form the usual food of the *Teredines*. But, independently of what I have above stated with reference to this question, the cases of *Saxicava* and the *Pholades* must be considered. It can hardly be imagined that these are stone-eaters. Sellius found that the *Teredo* did not attack a pile below fourteen feet. Further information is desirable as to the depth at which it is capable of living. He observed that it commonly follows the grain of the wood; and that consequently its tunnellings in fir and alder are straighter and longer than in oak, which is tougher and more knotty. It usually works round knots in a curved direction; but occasionally it drives right through them. The odour emitted by the *Teredo* is different from that of the oyster and other shell-fish, and is derived from the kind of wood in which it lives. I can answer for its being very disgusting and almost insupportable. The valves of the shell found in fir- and alder-wood are white, almost pearly, and marked with pale ash-coloured striæ and dots; whereas those taken out of oak are almost entirely yellow, sometimes of the darkest shade of black with striæ and dots of the latter hue. This remark applies to the external surface only, and not to the inside, which is uniformly pure white and pearly. The pallets or " pinnæ " have a yellowish tint, and their stalks are invariably of the same colour and lustre as the inside of the valves. The colour of the sheath varies in like manner according to the kind of wood. The outside tints appear to be extraneous, and not inherent in the *Teredo* or secreted by it. Rousset having succeeded in keeping *Teredines* alive in his own house, Sellius thought that oysters, mussels, and other kinds of eatable testacea might be profitably cultivated in tanks or reservoirs. A small crustacean, called " Springertje " or " Snel " (*Limnoria lignorum*, Rathke), is generally seen in company with the *Teredo*, and with its horny mandibles gnaws away the surface of the wood. With regard to the mode of perforation by *Teredo*, I have already stated the views of Sellius in the ' Introduction ' to the first volume of the present work. I would, however, add that I am now inclined to differ from him in the supposition that the *adult* shell is not strong enough or adapted to rasp the wood. Cailliaud has shown practically that this can be done ; and I have lately repeated, with success, the same experiment. But the improbability of the *young* or newly born shell being able to effect a lodgment in this way seems to me as great as ever. By examining the *Teredo in situ*, it will be manifest that the foot is closely applied to the larger end of the tunnel, and that it occupies the whole of the front or hemispherical cavity. That part of each valve which may be supposed to have a rasping power is placed at the side, and not at the bottom. I believe that the valves, instead of the foot, serve as a fulcrum, and that they are pressed equally against both sides, while the tissue of the foot is employed in absorbing and detaching, slowly but gradually, minute particles of the moistened wood. If the shell were the instrument of perforation, it would be applied to the bottom, and not to the sides of the tunnel; and no muscle has yet been detected which could effect such a change in the relative positions of the valves and foot. Mr. Osler strongly advocated the theory that the wood is rasped away by the shell; yet he admitted that, owing to the shortness of the lateral muscles in *Teredo*, it was not probable that this mollusk could bore, like the *Pholas*, by the action of these muscles alone. Quatrefages agrees with Deshayes in considering the muscular apparatus by no means adapted for putting the valves in action as perforating-instruments, by either a rotatory or a twisting movement. He attributes this agency to the anterior fold of the mantle, especially that part which lines the back or beaks of the shell (called by him the "capuchon céphalique") aided by continual soaking of the water, and perhaps also by some secretion of the animal, as well as possibly by the siliceous particles observed by Hancock in the mantle of certain other perforating mollusks, and by Deshayes in the integuments of the *Teredo*. But no part of the mantle is placed in contact with the excavated end of the tunnel or canal, which is entirely occupied by the foot. In a memorandum which I received from the late Dr. Lukis on this subject, he says (after summarily dismissing the chemical theory), "Mechanical force seems also scarcely probable or even possible; for it is not very evident how this can be employed whenever a lateral opening is to be made in the side of the tunnel. This opening is usually at some distance from the inner or further end, and its edges are often very sharply defined. If force were required to be exerted, these sharp edges would be a serious inconvenience to the *Teredo*, whose body is bent at this point into often considerably less than a right angle; such angles occur more than once in the same specimen." The marks at the extremity of the tunnel, when examined under a microscope, resemble in miniature those which are left in mowing a grass lawn with a scythe; but they are arranged in a circular manner, and are continuous. These marks are very numerous and narrow; they do not correspond with the anterior and striated part of the valves, which (although rounded) are never bent at such an angle as would produce the sharp lines exhibited on the eroded cavity of the wood. The notorious fact that the valves are covered with an epidermis is evidently a stumbling-block in the way of M. Cailliaud; because it would be difficult to understand why this slight film is not rubbed off, if the valves are used in scraping the wood. He endeavours, with considerable ingenuity, to dispose of the difficulty by assuming that the epidermis is only formed temporarily and provisionally, to protect the valves from the effect of the acid which the *Teredo* employs in dissolving its sheath or outer case, in order to make a new one. I am not aware that any part of this assumption has been verified by observation. M. Cailliaud was even unable to detect the presence of any acid in *Teredo*, although he has given us a long list of other mollusks which secrete it, including not only

Saxicava, Gastrochœna, and *Pholas,* but also the common oyster. I now take leave of this curious subject, believing that it has been sufficiently discussed or ventilated ("soaked" is the term which an English statesman lately invented); and all naturalists, who take an interest in it, may adopt whichever theory they prefer, be it chemical, conchological, or malacological—in other words, that the excavation is caused by the solvent power of an acid, the rasping action of the shell, or the sucker-like application of the foot. This is a very long commentary, and I am afraid it will terribly "bore" many of my readers; so I will resume the analysis of Sellius's monograph. The quantity of water taken in and retained by the *Teredo* is prodigious: Sellius not inaptly compared the animal to an hydraulic machine. I feel the same admiration that he avowed of the wonderful sagacity shown by the *Teredo* in making its way through a piece of wood, so as to avoid the tubes of other individuals. Every one pursues its own course with unerring instinct; and it must be gifted with some organ of sense or apprehension, more delicate than we can conceive, in order that it may be aware when it approaches another *Teredo.* The sheaths are never contiguous, but in every instance separated by an intervening layer of wood. The *Teredo* uses its pallets as a means of defence against its enemies, by closing the opening of the canal, thus

". . . . omnem aditum custode coronans."

He rightly described them as inserted in a sphincter-like ligament at the base of the siphons. The function of these processes is identical with that of the operculum in many univalves—although they are not homologous, or produced by similar organs. He next considered the sexuality of the *Teredo.* His assertion that it is hermaphrodite (in which he followed Fontenel and Massuet) has been within the last few years maintained by Laurent, in opposition to the opinion of Quatrefages that it is bisexual. The last-named author, indeed, stated not only that the sexes are separate, but that the proportion of males was 5 or 6 out of 100 individuals of *T. pedicellata* which he examined, the rest being females. Baster had fancied, more than a century before this, that coition took place between the *Teredines* by means of their siphons! Laurent informs us that he found in an hermaphrodite gland of *T. Norvegica* eggs and spermatic capsules at the same time, and that the internal organization of the animal did not offer any character to distinguish one sex from the other. I will not pretend to decide such a controversy, which in all probability concerns the whole of the Conchiferous mollusca; but I have already (vol. i. introd. p. xxv) given my reasons for concurring with Milne-Edwards in the belief that all the members of this class are monœcious. The period of propagation, according to Sellius, extends over the greater part of the year, even as late as December, although the summer would seem to be the most favourable season. In the month of February he found the ovaries flaccid and empty. Sellius states that the eggs are never produced inside the wood, but excluded by one of the siphons. He suggested that the latter might have a peculiar (we may say strange) function, namely that of moistening the outside of the wood, and agglutinating the eggs to its surface, or even excavating minute holes in it for the purpose of assisting the fry in effecting an entrance. He was also mistaken in supposing that the fry were hatched only when the eggs adhered to the wood. It has since been ascertained that this process takes place inside the mantle of the *Teredo*, and that the fry are ejected into the water in a larval or metamorphic state. He was not aware that the fry have eyes and can therefore select their own habitat; or he would not have attributed their position in the wood to the maternal care of their parents, under the idea that they are at the mercy of the winds and waves. Massuet, moreover, had previously put forth a notion that the fry crept about the surface of the wood, and sought out convenient spots where they could burrow. Our author observed that the *Teredo*, in its earliest stage, underwent a kind of metamorphosis by the method called "epigenesis," which is now recognized by most physiologists. This remark is followed by an inquiry into the mystery of Creation, in which he discusses the common opinion of certain neoteric writers of his time that all living beings had descended from original forms or types. The solitary nature of the *Teredo* was not overlooked by him. Although surrounded on every side by companions, it has no means of communication with them. Each lives alone in a crowd. Nevertheless Sellius gives his favourite credit for a generous and unselfish disposition, which its fellow creature, man, might well endeavour to emulate. Nor did the Dutch philosopher exhibit less industry in his examination of the nomenclature of *Teredo*. He ransacked the works of many a classical author and naturalist, from Plato and Aristotle to Oppian and Reaumur, with a view to elucidate its history; but he appears to have got rather bewildered by the gossip of Pliny, who confounded the *Teredo* with the grub of an insect. Sellius did not share the credulity of some of his contemporaries in supposing that *T. navalis* was introduced into Holland by vessels (or in any other way) from foreign parts; for he unquestionably knew that the European species are different from those which inhabit tropical seas. Although the Dutch shipworm also infests the coasts of the Crimea, there is just as much reason for believing that it had been imported from the German Ocean into the Black Sea, as that it had been exported in the opposite direction. Linné's assertion, made seventeen years after the publication of the work now under consideration, that the *Teredo* was " ex Indiis propagata," had no other foundation than common rumour. He ought to have known better. Sellius, however, was inclined to suspect the recent origin of *Teredo*, as a native of the German Ocean, and to agree with his pious countrymen that it was a scourge in the hand of an offended Deity, and inflicted on them for their sins. It is mentioned by Smollett, in his chronological medley of home and foreign news, called a 'History of England,' that in 1732 " the Dutch were greatly alarmed by an apprehension of being overwhelmed by an inundation occasioned by worms, which were said to have consumed the piles of timber work that supported their dykes. They prayed and fasted with uncommon zeal in terror of this calamity, which they did not know how to avert in any other manner. At length they were delivered from their fears by a hard frost, which effectually destroyed these dangerous animals." Among the enemies of the *Teredo*, which serve to check its increase, Sellius enumerates the smaller fishes, which prey upon the fry in their free state, and many insects (annelids and crustacea) which attack and devour the adult. Foremost among the latter class of natural foes he ranks the *Nereilepas* (or *Lycoris*) *fucata*, which he calls a marine *Scolopendra*. This is frequently found in the empty canal of the *Teredo*, of which it has taken possession, after insinuating itself and clearing out the original occupant. His account of the voracity of *N. fucata* is confirmed by a most valuable and instructive report presented to the Academy of Sciences at Amsterdam, in 1860, by Professor Vrolik, the Secretary of a Commission which was appointed to inquire into the natural history of the *Teredo* and the best mode of preventing its ravages on the coasts of Holland. It was there stated that the larvæ of the *Nereilepas* and *Teredo* live together; and it is probable that, instead of the annelid entering in an adult state the canal of the shipworm, as Sellius conjectured, it deposits its eggs in the open siphons of the latter, whence they afterwards find their way into the body and are developed. The larvæ of some dipterous insects have been also observed by Dr. Verloren, as well as Sellius, to prey on the Dutch shipworm. *Cochleoctonus vorax* disposes in nearly a similar way of certain snails. I have seen shells of *Helix strigella* and *H. incarnata*, each of which was occupied by a grub of that beetle, coiled round in a spiral shape like the snail which it had supplanted. The name of the artificial remedies which were known at the time when Sellius wrote was legion. He reckoned about 600 kinds of ointment, or preparations of an oily nature; and he proposed one, which we now call creosote, to penetrate the pores of wood by some hydrostatic power, and which would have the effect of hardening and preserving the timber. He had no faith in the efficacy of any poison, being fully impressed with the idea that the *Teredo* feeds on wood only; nor did he believe that, even if this were not the case, the wood could be saturated or imbued with poison by the most expensive process that it was possible to discover. A thick and durable coat of varnish, applied to the surface of the wood, was in his judgment the best preventive, because it would keep out the fry. He especially noticed a balsam of wonderful virtue, and kept a secret, which was patronized by Peter the Great. Possibly this was the resin now extracted by the Cochin Chinese from a gigantic tree called "cay-dan," and lately noticed by M. Mariot, a lieutenant in the French navy. Native canoes, hollowed out from the trunks of this kind of tree, are said never to be worm-eaten. Among other means of protection that had been long in use and were still in vogue in his day, were the following :—for ships, an inner layer of calf-skins, cow-hair, pounded glass, ashes, glue, chalk, moss, or charcoal; for piles, large iron nails driven in close together; and for both, hard and close-grained woods. By the first of these methods, however (which is still partially made use of by the Turks and Arabs in the Mediterranean), the ship's course was apt to be retarded; and the latter remedy was expensive and not always efficacious. He said that the application of pitch or coal tar to the surface of the wood had been recommended by a Londoner of some repute. We find in the 'Philosophical Transactions' for 1666 an announcement by an anonymous writer that "a very worthy person in London suggests the pitch, drawn out of sea-coals, for a good remedy to scare away these noisome insects." The late Lord Dundonald little suspected that the boasted discovery of his father had been so long forestalled. Nor did Sellius overlook the patent, granted by Act of Parliament in the reign of Charles II. (1671) to Sir Philip Howard and Major Watson for preserving the hulls of ships from worms by a sheathing of lead mixed with some other metal, a composition now superseded by copper. The conclusion arrived at by Sellius was that the surest remedy consisted in trying to propitiate the wrath of the Almighty by constant prayer and praise. Many succeeded Sellius in investigating the natural history of the *Teredo* ; but Adanson, Home, Montagu, Deshayes, Quatrefages, Laurent, Clark, Fischer, and Harting are perhaps all whose observations are worthy of being noticed. If I have omitted the name of any other writer, I offer by anticipation the most ample apology for my neglect.

3. *Habits and organization.*—The opportunities which I have had of examining this villanous animal (as Massuet calls it), and of observing its habits, were not so many as I wished ; but I will relate faithfully what I have witnessed. On my return in 1860 from the Continent, through Holland, I had the pleasure of meeting Dr. Verloren at Utrecht, and of carefully inspecting at his house living specimens of *T. navalis*, enclosed in pieces of the dyke-piles, which he had kept in long glass jars for about ten months. They appeared to have become habituated to the loudest noise ; and even when the jar was moved, or the light suddenly obstructed, they did not withdraw their terminal tubes or siphons. The longer (or alimentary and incurrent) tube was in frequent motion, and bent in various directions, as if in search of food, while a current of water, full of animalcula, continually passed into it. The shorter (or fæcal and excurrent) tube performed its functions at intervals, expelling the woody pulp by a spasmodic action, and occasionally withdrawing, in order the better to effect its purpose when any stoppage occurred. Each tube was transparent, and fringed with cilia at its orifice. The *Teredines* seemed to prefer the sunny side of the jar ; they are said to be very sensitive to cold. But the most interesting peculiarity which I observed, and to which my attention was directed by Dr. Verloren, was that each of the tubes was protected or enveloped exter-

nally by a very thin, pellucid, and film-like membrane or sheath. These tube-sheaths are irregularly annular, like the testaceous sheath or case which lines the excavation in the wood; and they bear a considerable resemblance in shape to the stem of *Tubularia indivisa*, though differing from it in texture and colour. The sheath of the alimentary tube is about an inch long, and the other is half that length. Their annular structure evidently arises from successive accretions of growth. The use of these membranous sheaths may be either to prevent the delicate tubes, which they cover for about half their length, being choked or obstructed by the accumulation of the flocculent pulp lying outside, or else to protect them from the attacks of minute predacious animals. They are renewed from time to time; and in one of the specimens two separate pairs of these membranous sheaths were attached to the outer opening of the testaceous sheath in the wood, one pair having been apparently disused and a new set formed. The *Teredines* grow and multiply with astonishing rapidity. Quatrefages has given us an instance. A ferry-boat plying between two villages on the opposite sides of the mouth of Guibuscoa harbour in the Bay of Passages, on the north coast of Spain, was accidentally sunk in the beginning of spring. Four months afterwards some fishermen raised the boat, hoping to make use of the materials. But in this short space of time the *Teredo* (*T. pedicellata*) had made such ravages, that the planks and beams were quite worm-eaten and destroyed. Sailors have given me some interesting accounts of hair-breadth escapes which they had, while engaged in boat duty for a few weeks at a time on foreign stations, in consequence of the paint having been rubbed off the sides of the boat below the water-line:

wherever this was the case the ship-worm got in, and speedily reduced the thickness or strength of the plank to little more than that of an egg-shell. I have not unfrequently noticed crowds of very young individuals in a small and thin strip of deal, which could not accommodate any one of them if it grew larger: in fact each had gone to the very end of its tether; and another step would have laid bare its foot, and thus have exposed the most vulnerable part of the body to its rapacious enemies. Not having room to grow, or the power of removing to a larger piece of wood, all these individuals must necessarily perish without arriving at maturity. This fact apparently illustrates a law of nature, which might be termed blind; but it may also be regarded as one of the numerous methods by which various races of animals and plants are kept under, so as to prevent an excessive multiplication of any of them to the exclusion or detriment of the rest. If no such checks were imposed, all the wood on the face of the earth, if placed in the sea, would probably not suffice to contain the *Teredines* produced in a single year. The natural span of life allotted to the *Teredo* is unknown to us: perhaps it may be ascertained by means of the aquarium. It is supposed that they attain their full growth in the course of a few months. Extreme cold is fatal to them. According to the observations of Quatrefages on the north coast of Spain, nearly all appear to perish in the winter; a few only survive to continue the breed. Vrolik believed that they hybernate on the Dutch coast. Warm and dry seasons are favourable to them. In Holland, where their proceedings have been watched with so much anxiety, it was noticed that the greatest ravages are made in July and August, and that the most destructive years during the last and present centuries were

1731, 1770, 1827, 1858, and 1859. Very little rain fell in those years. Laurent showed that they are suffocated and destroyed by oil being poured on the water in a vessel containing *Teredines* in a piece of wood. He also proved that they could not live in the "Salines" at Hières, too much salt being as injurious to them as fresh water. But it appears that certain species live in brackish or even fresh water. The *T. Senegalensis* of De Blainville was discovered by Adanson in the roots of the mangrove and another kind of tree lining the banks of the Niger, Gambier, and other rivers on the west coast of Africa, which were only subject to an influx of sea-water for a few months in each year. According to Adanson the water of these rivers is quite fresh or sweet during the remaining months; and *T. Senegalensis* not only exists, but retains its full vigour throughout the whole year. This statement, however, must be received with some qualification. I am told by Dr. Welwitsch, the great botanical traveller, that in the tidal rivers of South Africa the water in the middle of the stream is fresh, while that on the sides is brackish, and that no kind of mangrove has been known to live in fresh water. Another sort of shipworm (*Nausitora Dunlopei* of Perceval Wright) has been lately found in India, inhabiting the river Comer, one of the branches of the Ganges, and a perfectly freshwater stream, that returns to the main river at a distance of about 70 miles from the sea. Dr. Kirk, the friend and companion of Livingstone, informs me that he picked up a piece of ebony (*Dalbergia melanoxylon*) on a sandbank in the Zambesi river, the water of which was there always fresh and drinkable, 100 miles from the sea—very far beyond the influence of the tide, which never comes above 10 miles up the creeks of the delta. This piece of ebony was pierced in all directions by a species of *Teredo* having a calcareous sheath. The kind of wood mentioned by Dr. Kirk resembles the ebony of commerce, but is utterly worthless, except as fire-wood; and therefore it is not at all likely that the piece in question could have been accidentally brought inland, after being perforated in the sea by the *Teredo*. It sinks in water, is rather brittle, much harder and far more compact than either mahogany or teak, and is full of some mineral matter that quickly deadens the edge of any tool. It does not grow on the coast, nor within 50 miles of it on the Zambesi. Dr. Kirk adds that in the bottom planks of the pinnace belonging to the expedition the shipworm was also found, with its soft parts attached to the finely sculptured valves. The boat was so riddled that the quartermaster pushed a paint-brush through her double planks. This was at Tete, 250 miles from the sea, after the pinnace had remained there six months at anchor. I regret not having space to give *in extenso* Dr. Kirk's interesting account of all the circumstances connected with this discovery. Unfortunately the specimens were lost on the way home; but not the slightest doubt can be entertained that the *Teredo* observed by him inhabits water which is at all times perfectly fresh and sweet. The habits of the *Teredo* are littoral. When they are met with far from land, the piece of wood which contains them has been accidentally detached and carried out to sea by some marine current. Dr. Lukis noticed that, at Sark, *T. Norvegica* and *T. pedicellata* pass more than half their time out of water, during the recess of each tide, when the shipping-stages in which they live are left high and dry. Sir Everard Home confirmed the observation of Sellius, by saying that " the worm appears commonly to bore in the direction of the grain of the wood, but occasionally it bores across the grain, to avoid the track of any of the others." Although this is the direction which it usually takes, it is by no means uncommon to see perforations inclined at various angles, and sometimes even made right through a tough knot in a piece of oak. Montagu also remarked, with his usual acuteness, that " the *Teredo* bores across the grain of the wood as seldom as possible ; for after it has penetrated a little way, it turns and continues with the grain, tolerably straight, until it meets with another shell, or perhaps a knot which produces a flexure; its course then depends on the nature of the obstruction ; if considerable, it makes a short turn back in form of a syphon, rather than continue any distance across the grain." The same kind of siphonal bend takes place when the piece of wood, being shorter than the average length of the *Teredo*, is nevertheless broad enough to admit of its abruptly turning and doubling like a coursed hare. If the space is not sufficient for its complete development, the *Teredo* shuts itself up and closes the front with a cap-shaped epiphragm; it never penetrates that end of the wood, so as to make the canal pervious. The *Teredo* possesses the same cartilaginous styliform process which I noticed in the account of *Pholas*. The imbricated plates, or septa, that line the neck of the sheath in probably every species, serve as ledges to support or strengthen the pallets, which are withdrawn further into the sheath as the *Teredo* increases in length and bulk ; the last formed plate is consequently innermost. Fischer counted twenty-five of these plates in a sheath of *T. Norvegica*. I do not agree with him in believing that the *Teredo* goes on perforating the wood beyond what is required for its habitation, nor that it abandons by slow degrees the narrower end of the canal. The pallets of course increase in size relatively to the growth of the body; and as the sheath enlarges inwards, new plates are formed in that direction to accommodate the increased size of the pallets. Although the body is contractile to a certain extent (as we see in dead specimens), it is fixed to the sheath by the muscular ring which contains the pallets, and therefore cannot be withdrawn into the canal beyond that line; the other extremity is employed in excavation, until the canal has been completed. When a *Teredo* has ceased to excavate before attaining its full growth, and has interposed a barrier in front, its valves become stunted and somewhat altered in shape, although their sculpture is similar to that of ordinary specimens. The same fact is observable in many other bivalves that inhabit cavities or confined spaces, whether they are of a boring or of a free nature. The cap-shaped plug, often formed in front of the valves by individuals of every age, serves as a partition wall between adjoining canals, as well as indicates that the animal has ceased working; it is formed like the sheath, but its substance is thinner. Sometimes two or more of these plugs may be seen, one after another, at various distances apart, as if the animal had withdrawn and thus strengthened its inner line of fortifications. Fischer was disposed to regard this secretion as analogous to the epiphragm of land shells. That, however, is only constructed for a temporary or occasional purpose, and can be dissolved by the snail at pleasure. It does not appear that the *Teredo* can do this and resume its work of perforation. Laurent believed that the plugs or caps of the *Teredo* are made for hybernation, an idea that is open to the same objection as that of Fischer. The tubes or siphons, when in action and extended, diverge considerably; so that the excretal tube discharges the exhausted water, fæces, and woody pulp backwards, or in such a direction as not to interfere with the current which passes into the branchial and alimentary tube. Clark insists that the anterior adductor muscle in *Teredo*, as well as in *Pholas*, is a "genuine cartilage, which is a secretion from glands." This notion is opposed to that of other physiologists; and I merely mention it to show how difficult it is for one not conversant with such matters to decide the question, or even to understand how a cartilage or ligament can be secreted in the manner suggested by my late friend. He also stated that the pallets act as a sort of force-pump, to facilitate the flow of water through the long canal. M. Cailliaud supposes, on the other hand, that the use of those appendages is to macerate such food as is too bulky to enter the tube. I cannot accept either view. The one is hypothetical, and does not accord with our knowledge of the nature of the animal. The other assumes that the pallets lie inside the alimentary tube, or at its orifice, neither of which is the case; they are placed at the outer base of that tube, when it is protruded in search of food. Valenciennes and Quatrefages consider the posterior muscle to be that which attaches the pallet-supporting ring to the sheath. Clark "perceived in the centre of each plate a decided muscular impression." This I have not seen; but the posterior lobe or "auricle" of each valve exhibits a scar, precisely similar to that with which the corresponding portion of other bivalve shells is marked; and the muscle itself, connecting this part in *Teredo*, is very strong and conspicuous. I should be disposed to regard the muscle, which supports the pallets and is attached to that part of the sheath, as the homologue of the sinuated portion of the pallial muscle in *Pholas*. In both cases it is placed at the base of the tubes or siphons.

4. *Embryology*.—Nearly all our knowledge of this part of the natural history of *Teredo* is derived from an elaborate memoir by Quatrefages in the 'Annales des Sciences Naturelles' for 1849. The process of oviposition is successive and of long duration. During a period which varies according to the species, the female emits her eggs, which are arrested and lodged in the folds of the respiratory organs. In this singular nest they are fertilized by the spermatozoa of a male, disseminated through the mass of the surrounding water, some of which find their way into the branchial tube of the female, where they meet with the eggs and vivify them by contact. The same method of impregnation takes place in *Anodonta* or the freshwater mussel. The egg, while in the ovary, consists at first of an extremely minute globule, which is simple, homogeneous, transparent, and quite colourless. This is called "the vesicle of Purkinje." Some very fine granules soon appear in the substance of this globule; and in a short time may be seen developed in its interior a second globule called "the germinative spot of Wagner." The two globules increase together for some time before the formation of the yelk-membrane which covers the whole. In this state the egg is exactly spherical. Its volume then becomes enlarged; and after passing through other phases, it assumes the shape of a tear, and when emitted the sphere is converted into an irregular oval. The spermatozoa now attach themselves to the egg, and certain internal movements and changes ensue. These last for about two hours; the yelk-granules are distributed through-

out the substance of the egg, and ultimately separate into two nearly equal parts, one of which encroaches by degrees on the other and at last completely envelopes it. Towards the eleventh hour the yelk is transformed into an agglomerative mass, composed of two well-defined portions, and covered by a more or less folded membrane. One of these portions now separates into three lobes; and vibratory cilia make their appearance, at first short, thick, and few in number, afterwards longer, finer, and much more numerous. The cilia surround the entire body of the fry, which soon swims with great rapidity, like one of the Infusoria. This state lasts till nearly the forty-eighth hour: then the number of the cilia diminishes, and the fry falls to the bottom of the vessel, where it moves rather slowly. At the same time the yelk-membrane is divided into two equal parts. These are the rudiments of the shell, which at first is quite membranous, flexible, and irregularly oval, with a salient angle at the point corresponding with the hinge. In a short time this form is altered; the salient angle is effaced, and superseded by a re-entering angle. The shell is then symmetrical and heart-shaped, and at the same time is encrusted by calcareous salts and solidified. During the formation of the shell the mantle is developed, with delicately ciliated edges, which are destined to replace the original ciliary apparatus. The new cilia are extensible and retractile, and consist of a single row. The fry can withdraw entirely into their shells. At this stage they appear not to be sensible of noises, nor even of an agitation of the water in which they are placed. It constitutes a critical period of their lives; and a large proportion of the infantile community then perish. About the sixty-eighth hour from the production of the

egg the cilia commence growing, and become stronger.
The duration of the last period of growth is unknown.
Some of the fry survived for 130 hours. The perfect
larva swims rapidly, like a Rotifer, and has a long, narrow, and strap-shaped foot—very flexible and resembling
that of a young mussel—by means of which it creeps
with apparent ease along the bottom of the vessel. It
remains for a long time suspended in the water by a
transparent filament from the sides of the vessel. The
shell then becomes nearly globular, instead of irregularly oval; a pair of red eyes are seen in the middle of
the body; and otolites, or ear-stones, and other organs
are formed. The eyes afterwards disappear, the body is
elongated, and the animal assumes its complete form. I
have given the above description of Quatrefages nearly
in full, because it explains the embryogeny of the Conchiferous Mollusca in general. This eminent zoologist
is of opinion that the *Teredo* undergoes a true or complete
metamorphosis. In the first state of growth its integuments are membranous; it has no distinct organs; its
sole mode of locomotion is by means of cilia, which cover
the body: it is a *larva*. In the second state it has acquired
a shell; it possesses distinct organs of sense, besides a
special apparatus for swimming and a foot for creeping:
it is then a *chrysalis* or pupa. The third and last state
represents an *imago*; the transformation has been completed, and the animal thus developed enters upon a
new phase of life, with appliances peculiarly adapted
to its altered conditions. In reality, however, the evolution from a simple globule into a shell-fish endowed
with a comparatively high degree of organization, and
of a complicated structure, is not the result of sudden
changes, but is effected by a series of successive growths,
or epigenesis. The outer membrane of the egg becomes

a mantle, which at first forms the shell and afterwards the pallets and sheath; the cilia, which invest most (if not all) embryonic forms are absorbed, and a foot is produced out of the firmer tissues of the body, and substituted for the cilia; the eyes, mouth, palps, stomach, intestine, liver, heart, gills, muscles, nerves, reproductive and other organs come upon the stage and play their several parts. "Instinct" does duty as prompter. This, the inventive faculty of every creature but man, provides for its necessities of food and defence, and dictates the nature of its habits by an inscrutable kind of prescience, that is little less than divine. Laurent, Lukis, and others have also noticed the great activity of the fry in their intermediate state; and M. Kater observed them swimming freely about the piles in the dykes of Holland, and after a while attaching themselves to the wood. Like the oyster-fry, they seem capable, to a certain extent, of selecting their habitat, and they probably use their eyes for that purpose; but this can only be the case when the sea is unusually calm, their puny force being quite unequal to contend with any agitation of the water. I have just re-examined a piece of wood to which some of the fry of *T. navalis* still adhere. Each is no bigger than the smallest pin's head, and is enclosed in a pair of somewhat oval, close-fitting, semimembranous, and yellowish valves, the only opening in which serves as a passage for the foot or point of attachment. It bears some resemblance to a minute *Cythere* or crustacean of the Entomostracan kind, as well as to the pupa or last larval state of a Cirripede. The original or rudimentary valves are persistent, and form the umbonal portion of the perfect ones; they are easily recognizable in young specimens by their different shape, consistency, and colour. A similar retention of embryonic parts occurs in the case of beetles, the grubs of which do not part with their horny jaws when they attain an adult state. It is otherwise with the Lepidoptera, which exchange their larval mandibles for a suctorial proboscis. The metamorphosis of *Teredo* is not less wonderful than that which takes place in the frog, insect, or polype.

5. *Structure of Shell.*—The sculpture of the shell is excessively complicated and delicate. Harting counted 4000 denticles in the anterior portion, and nearly 10,000 in the middle division of a single valve of *T. navalis*. Dr. Carpenter kindly examined, at my request, the microscopical structure of the valves and sheath of *T. Stutchburii*. He informs me that the valves are extremely hard in texture, and that their substance has a very peculiar arrangement, corresponding generally with that of the shells of the bivalves most nearly allied to it, but having so special an adaptation to produce a fine file-like disposition of the surface, that he cannot help surmising there is more in the mechanical theory than I am disposed to admit. The sheath is destitute of anything like true structure, but has all the characters of a mere exudation shell, formed of minute calcareous particles, agglutinated together, very much like some egg-shells. He adds that the difference in texture between the two is nearly the same as that between the half chalky substance of a crab's carapace, and the almost ivory-like consistence of the black tips of its claws. I would observe that the sheath of *Kuphus arenarius* is remarkably solid and compact, with a radiating structure, and that the surface of the shells in some of the *Pholadidæ*, and even in species of *Tellina* and other genera, exhibit a file-like arrangement.

6. *Origin.*—An erroneous notion was formerly preva-

lent that the *Teredo* had been originally introduced into Europe from foreign parts—"calamitas navium ex Indiis in Europam propagata," Linné,—which seemed to be in some measure confirmed by its sudden appearance in particular years. Even Mr. Osler, so late as 1826, took for granted that *T. Norvegica* was not a native of the British seas; and he expressed his belief that, until the general use of copper sheathing, it was probably preserved only by occasional importations. But we now find that each kind of *Teredo* has its own special area of habitat. Tropical species will not live in the temperate zone, and *vice versâ*. That the *Teredo* is not of modern origin in Europe is evident from the fact that *T. Norvegica*, which at present is distributed over the North Atlantic from Finmark to Sicily and Algiers, is also found in both old and new deposits of our upper Tertiary formation. *T. megotara* inhabits the coasts of Shetland, and more northern latitudes in both hemispheres; and it occurs in a fossil state at Belfast and Uddevalla. Deshayes first noticed the same fact with regard to *T. Norvegica* being a fossil of the Italian tertiaries, as well as of the Crag; and it appears to be conclusive.

7. *Distribution in the British seas.*—Its distribution along the British coast appears to be somewhat capricious. Seaports, in which the admixture of fresh water is considerable, such as Hull and Liverpool, are exempt from the *Teredo*. But this rule has its exceptions. The Medway is infested with the Dutch shipworm (*T. navalis*), especially the upper reaches of the river, where the water becomes less salt. I extracted living specimens from the keel of a "watch boat," kept at anchor off Queensborough in that river for the purposes of the lobster trade in the Billingsgate market. Milford Haven has the Norway shipworm (*T. Norvegica*)

plentiful and of a large size. None of the other ports in the Bristol Channel are troubled with that or any other species. The dispersion of mollusca is so wonderfully rapid, that in all probability a vessel wrecked anywhere on our coast, but not driven ashore, or a newly erected submarine woodwork, will sooner or later attract the wandering fry of some *Teredo*, which must have a suitable nidus or prematurely perish. Or, as the whole ocean teems with life in various states of development, the germs of invertebrate animals (like the seeds of some plants) may remain dormant for a long period, and only become vivified when placed in favourable circumstances.

8. *Economical relations to man.*—The new Salvage Act has somewhat interfered with the liberty of conchologists in searching the shore for *Teredines*. Mr. Dennis was more than once baulked in his hopes of examining a promising piece of driftwood, seen floating towards Beachy Head, by the coastguard marking it with the broad arrow directly it reached the shore. A douceur is consequently necessary to secure the prize of a honeycombed log. If Crabbe were a living poet, he could not now say of the naturalist,

"His is untaxed and undisputed game."

The destructive nature of the *Teredo* is notorious; but we can hardly realize the extent of the damage which these obscure miners perpetrate, by their stealthy and incessant operations, when they attack our piers and other submarine wooden structures. Quatrefages asks us to imagine what would become of our trees and furniture, and of the beams, joists, and rafters of our houses, if they were to be gnawed by grubs measuring a foot or more in length. However, no evil is unmixed or without compensation. Smeathman, in his "Account of the Termites" (Phil. Trans. 1781), remarked that the seaworms appear to have the same scavenger office allotted to them in the waters which the white ants have on the land. It was also suggested by Laurent and others that the *Teredo* might be occasionally serviceable to us by assisting in the removal of wrecks, sunk at the entrance of harbours, which would otherwise obstruct navigation. The celebrated Redi describes it, in a letter to his friend Megalotti, as being not only eatable, but excelling all shell-fish, the oyster not excepted, in its exquisite flavour. Nardo likewise praises the *Teredo*, although in less rapturous terms: he wonders why the Venetians, who call it " bisse del legno," do not eat it. I should, for my own part, be surprised that any person having a stomach could venture to try the experiment; for the smell of even a fresh shipworm is almost enough to turn one sick. Ducks, however, seem to relish it, and not less when it is in a half putrid state. As regards man, its chief mission may be

"To fill with worm-holes stately monuments"

of his workmanship. Perhaps it is one of the creatures made not so much for our use as for our punishment. Southey tells us that Bellarmine allowed mosquitos and other small deer free right of pasture upon his corporal domains, being more indulgent to them than to heretics. He thought they were created to afford exercise for our patience, and moreover that it is unjust for us to interrupt them in their enjoyment here, when we consider that they have no other paradise to expect. Yet when the cardinal controversialist gave breakfast, dinner, or supper of this kind, he was far from partaking any sympathetic pleasure in the happiness which he imparted; for it is related of him that at one time he was so terribly bitten "a bestiolis quibusdam nequam ac damnificis" (it is not necessary to inquire of what species), as earnestly to pray that if there were any torments in Hell itself so dreadful as what he was then enduring, the Lord would be pleased not to send him there, for he should not be able to bear it. Patience, however, is not one of the cardinal virtues that we practise; and we therefore feel no compunction such as Bellarmine had, but wage an incessant war of extermination against the poor, not harmless *Teredines*.

9. *Remedies.*—Although our good neighbours the Dutch have been the principal sufferers from this maritime plague, we have not been spared. In 1826 Mr. Osler believed that the *Teredo*, as a British animal, was nearly and probably quite extinct. We should not be sorry to find that this case of "dying out" had a better foundation than many of those which have been assumed by theoretical naturalists with respect to certain harmless mollusca. Unfortunately the ravages still committed by this noxious mollusk in our harbours and naval arsenals tell a different tale. In 1860 it was proposed by a Committee of the British Association (of which Committee I was chairman) to have certain experiments made in the dockyard at Plymouth, with a view to prevent the further destruction by the *Teredo* of Government timber, which had cost the country a considerable sum every year. A small grant had been voted by the Association for such purposes. We find in 'Household Words' for 1857 the following statement: "It has been estimated that at Plymouth and Devonport alone the boring worms have in one year destroyed Government works to the amount of £8000." Permission to have these experiments made was asked

through the Port-Admiral, Sir Thomas Pasley, who expressed his entire approval, but forwarded the application to the Admiralty. It is scarcely credible that no answer was received for nearly a month, and that then came a simple refusal without any reason given for it! In France and Holland special commissions have been issued in the hope of discovering an efficacious remedy against the attacks of the shipworm; and experiments on an extensive scale are still being carried on in the last mentioned country. The preliminary reports which have appeared (especially those of the Dutch Commission in 1860, 1861, 1862, and 1864) show the great pains taken to ascertain as well the extent of the injury as the various modes already devised to prevent it. Great Britain, unlike other States, does not count a single naturalist in her national assembly; and the Government will not, unless urged by popular pressure, take the initiative, or even forward any plan of public improvement which is out of the regular groove of routine. Few persons know what a *Teredo* is; and the general ignorance of such subjects is too great for any except zoologists to distinguish this animal from wood-gnawing crustaceans, the *Limnoria* and *Chelura*. We therefore ought not to laugh at the ancients for confounding the shipworm with the grub of an insect. With all of us the material predominates over the intellectual. Wealth and its companion luxury constitute our *summum bonum*; and knowledge is ignored.

> "The world is too much with us; late and soon,
>   Getting and spending we lay waste our powers;
>   Little we see in nature that is ours;
>   We have given our hearts away, a sordid boon!"

It will of course be answered that there are other things to be learnt besides the history of shipworms.

But is anything learnt now-a-days, save only the arts of money-making and pleasure-seeking?

In all probability the constitution of a shipworm is poison-proof. Most of the remedies proposed in the last century were of this nature, and they signally failed. Quatrefages, indeed, has suggested that the production of the *Teredo* might be checked by dissolving in the water at the proper season a trifling quantity of corrosive sublimate or acetate of lead, so as to destroy the spermatozoa or fertilizing agent. He tried some experiments of this kind on a small scale in the harbour of St. Sebastian. Quatrefages is an excellent naturalist, and especially conversant with the natural history of the *Teredo*; but I fear his plan is not a practical one. The *Teredo* attacks wood in the open sea, or in harbours which the tide enters twice a day, and never in floating harbours or wet docks, to which the tide has only occasional access. Now, in order to prevent the birth of the *Teredo*, which is always going on during the summer months, it would be necessary that the tidal harbour should be enclosed; otherwise the poison must be continually applied in prodigious quantities, and at an enormous expense, or else it would be diluted to such an extent by the action of the tide and waves (to say nothing of the river which is generally indispensable as a scouring power, and therefore flows through nearly all such harbours), that it would become too weak to produce the desired effect. An eminent civil engineer, Mr. Hartley, of Liverpool, recommended green-heart timber to be used in harbours; the costliness, however, of that kind of wood is a serious objection to this remedy. Copper-sheathing and scupper-nailing are often and successfully employed to protect piles in exposed situations. The former is also expensive; and the crust of iron formed by the nails in the interstices between them (unless they are very closely driven in so as to completely cover the piles) is superficial and liable to scale off. I have known the *Teredo* bore through a pile which was supposed to be protected by large broadheaded nails in the usual way. At Christiania, in April 1863, I found that *Teredo navalis* was very destructive to the woodwork in the harbour, and to boats lying at anchor in the fiord. The chief engineer told me that all the piles had been thoroughly creosoted (10 lbs. to the square foot) before they were driven in, but not to much purpose. Some were taken up while I was there, and proved the correctness of his statement. They had evidently been well saturated with creosote, and yet were full of the shipworm. It seems that these piles had been fixed only two years previously. Another remedy that had been tried at Christiania consisted in covering the outside face of the piles with fascines of brushwood. This may partially succeed, by excluding the light and warmth of the sun, and consequently preventing the production or development of the organisms on which the *Teredo* feeds. It certainly does not love the cold shade. The maxim "obsta principiis" is particularly applicable to the present case. If we can succeed in preventing the young *Teredo* from commencing its burrow, the wood is impregnable to its attack. It is not difficult to bar its entrance when the whole body is not the size of the smallest pin's head, the foot almost microscopical, and the shell a mere film. In this state it insinuates itself between the fibres of the wood on the outside; and having once gained a footing, it works its way, slowly but surely, into the interior, where it becomes snugly lodged and irremovable. It is indeed a most troublesome guest; and a line from

Ovid's 'Tristium,' with the alteration of a single word, will tersely express the difficulty of getting rid of it.

"Ægrius ejicitur, quam non admittitur hospes."

A very slight coating of any kind, applied to the surface of the wood, will suffice to keep out the infant burglar. Tar would answer the purpose; but this is liable to be accidentally rubbed off, or removed by the continued agitation of the waves. Sir Gardner Wilkinson informs us that the ancient Egyptians glazed some of their inscriptions on stone, by covering them with a vitrifiable composition, which was exposed to a certain degree of heat, until properly melted and diffused over the surface. Perhaps wood cannot be treated in the same way; but a liquid mixture, containing the requisite ingredients, and capable of penetrating its pores or fibrous texture, might be invented and applied to a pile or the hull of a vessel. Any mineral preparation that shall adhere firmly and permanently to the wood, and not be subject to external influences, must be efficacious. Such may be the silicate of lime, invented by the late Mr. Ransome, and used for coating stone-work. Every chemist knows that this is a manifest improvement on Kuhlmann's process, which consists of liquid silicate of potash or "water-glass." Szerelmey proposed an additional wash of a soluble bitumen, and called the preparation "Silicat-Zopissa"; but his experiment has not yet been adequately tested. Zopissa appears to have been a mixture of pitch and wax, first used by the Phœnicians and Egyptians, and afterwards by the Greeks and Romans, to preserve their merchant vessels and men of war. The preparations of Ransome and Szerelmey were tried in 1860 on part of the stone facing of our Houses of Parliament, which had suffered considerable decay from being exposed to the corrosive action of the London atmosphere, as well as from an inherent defect in the material; and time will show which of these preparations is the best preventive. I recommended Ransome's process in the discussion of the *Teredo* question at the Oxford Meeting of the British Association in 1860. Messrs. Peacock and Buchan about the same time invented and patented a composition for protecting wood-bottomed vessels from injury by marine animals. This is said to form by a chemical combination with sea-water an unctuous or slimy pellicle, and to succeed admirably in preventing the growth of barnacles and similar incrustations by which ships become fouled; but I am not aware of its utility with regard to the present question. The popular notion is that the barnacle and shipworm are the same animal, the one being the part outside, and the other that which is inside the wood. Another remedy which has been proposed, is to infiltrate the wood with silicate of lime; but I fear this would be too expensive for harbour piles. Mr. William Hutton, of Hartlepool, has taken out a patent of this nature. Although it was principally intended to prevent the ravages of *Limnoria lignorum* (a small crustacean belonging to the class Isopoda, which I have before mentioned), it would also serve as a safeguard against the *Teredo*. Mr. Hutton's plan is to harden the wood by forcing it into a solution of silex with muriate of lime. Perhaps the cost of his process, but not its efficacy, might be lessened by applying the solution in the form of a wash with a brush, instead of infiltrating the wood by means of mechanical power. The pores of the outer layer would probably be thus penetrated to a sufficient depth, and the remedy be equally complete.

10. *Classification.*—The mistakes made by some of the older naturalists, and even by Linné, as to the organization and zoological position of *Teredo*, are scarcely less remarkable than the object of which they treated. In the first edition of the 'Fauna Suecica,' published in 1746, it was placed in *Dentalium*, along with that shell and *Serpula*, the tube only being regarded. In the tenth edition of the 'Systema Naturæ' (1760), it was correctly named *Teredo*; but it was classed among the "Vermes. Intestina," and described as having a mouth with two jaws, inside which was a ciliated foreskin ("præputium"), a siphon within the latter, and tubercles round the mouth. In the twelfth and perfected edition (1767) it is called a *Terebella*, and arranged between *Serpula* and *Sabella*. These were unpardonable blunders on the part of the great systematist, because in all his works above cited he especially referred to the celebrated monograph of Sellius, who had clearly shown the affinity of *Teredo* and *Pholas* as testaceous mollusks. Nearly a quarter of a century after the appearance of that monograph, Adanson made the same observation; and his 'Histoire naturelle du Sénégal' bears date three years before the tenth, and ten years before the twelfth edition of the 'Systema.' It is possible that Linné had no opportunity of becoming acquainted with Adanson's work on Senegal for many years after it was published. The communication between Sweden and France in their time could not have been so intimate as it afterwards became. No such excuse however can be offered for Lamarck's ignorance of the writings of his distinguished countryman, seeing also that, at the date of the 'Histoire Naturelle des Animaux sans Vertèbres,' more than half a century had elapsed since the publication of Adanson's second memoir on *Teredo* in the

'Mémoires de l'Académie Royale.' Lamarck described the valves as containing a muscle which is protruded at the posterior end, and the pallets as apparently branchial! Both O. F. Müller and Fabricius had long previously adopted the views entertained by Sellius and Adanson as to the natural position of this mollusk; each in fact gave the only species known to him the name of *Pholas teredo*. The familiar and appropriate name of this genus has not escaped the experimental handling of systematists. It is the *Siphonium* of Browne, *Xylophagus* of Gronovius, and *Teredarius* of Duméril; and it has been divided by other writers into minor and more or less equivalent genera.

11. *Indigenous species.*—I propose to admit into the list of British Mollusca only such species as inhabit fixed and submerged wood on our coasts, and which of course are really indigenous; but I consider those found in floating wood, and brought from distant parts of the world, as no more entitled to be classed with native productions than *Hyalæa (Cavolina) tridentata*, several species of *Ianthina*, or *Spirula australis*, none of which live in the British seas, although they are occasionally drifted hither by the Gulf stream. Some of the *Teredines* which pay us a visit in this way, reach our shores in a fresher state than others; *T. megotara* frequently, and *T. malleolus*, *T. excavata*, *T. bipinnata*, and *T. cucullata* now and then, have the animal entire, although dead or scarcely alive, according to the length of the voyage.

TEREDO NORVE'GICA*, Spengler.

T, norvagicus, Spengl. Skr. Nat. Selsk. ii. (1) p. 102, t. ii. f. 4–6 B, & 7.
T. norvagica, F. & H. i. p. 66, pl. iv. f. 1–5.

BODY whitish, or of a light-greyish tint, semitransparent: *tubes* separated for about one half of their extent; orifices encircled with fine cirri, which are longer and more numerous in the incurrent or alimentary tube than in the other, and are often of various colours, or edged with brown, red, rose, or yellow.

SHELL convex, solid and opaque, scarcely glossy; it is parted in the middle by a slight longitudinal crest, with a broad but shallow furrow on the posterior side: *sculpture* divided into three distinct portions, viz. anterior, middle, and posterior: the anterior consists of sharp, narrow, and fine transverse plates, from 60 to 80 in number, which are more remote at first, and become closer in subsequent stages of growth; the edges of these plates are microscopically notched across in an oblique direction; this portion represents a triangle having an acute apex at the beak of the valve, and a broad and somewhat curved base: the middle portion extends the whole length of the shell, and is strap-like; the upper part lies between the inner line of the anterior area and the crest which separates one side from the other; the lower part is open outside, and bounded by the crest on the inner side; the broadest part is at the point of the angle where the anterior and middle portions join; this middle portion consists of numerous extremely delicate and nearly equal striæ, the edges of which are exquisitely beaded; these striæ are longitudinal, with an oblique tendency towards the posterior end, and they diverge from the transverse plates at a right angle: the posterior portion is always smooth, or only marked with concentric and slightly raised lines of growth: *colour* whitish, with often a tinge or stain of brown on the anterior side, especially the separating line: *epidermis* membranous, yellowish-brown, sometimes of a very dark hue: *margins* obtusely angular on the upper part of the anterior side, with a large triangular excision on the lower part, so that when the valves are united in their natural position, the opening or gape is broadly heart-shaped; bluntly pointed or rounded in front; and incurved on

* Inhabiting Norway.

the posterior side, which is terminated by a semicircular expansion, usually termed an "auricle;" in younger specimens this auricle is entire, and has a high shoulder above, on a level with the umbo, but in aged specimens the shoulder is worn down by the continual attrition of that part, and a notch is formed above; dorsal margins sloping abruptly and equally on each side: *beaks* much incurved, situate near the anterior end, at about one-third the length of the dorsal line; umbones or rostral portion prominent: *hinge-line* angular and irregular, considerably projecting in the middle: *hinge-plate* very broad, and extremely thick, folded over the anterior dorsal area, and abruptly truncated and flattened, or occasionally excavated, on the other side; the centre is furnished with a callous protuberance, as well as with a short peg-like tooth or process, which is stronger and more conspicuous in the right than in the left valve: *apophyses* very broad, and often jagged at the edges: *inside* glossy, furnished in front with a rather large and solid pear-shaped excrescence, and having the posterior auricle separated by a strong ridge, which forms a shelf or ledge in aged specimens: *muscular scars* large but not strongly marked: *pallets* large; blades oval, wedge-shaped and truncated or squarish in front, somewhat convex outside and concave inside, of a laminated structure, and more or less covered (especially at the outer end) with the same kind of epidermis as invests the shell; stalks cylindrical, of a much more solid substance than the blades, varying in length, being usually about one-third the length of the blades; the stalk occasionally extends into the blade at its narrower or inner end, and appears like the midrib or nerve of a leaf: *sheath* thick, sometimes indistinctly annulated; septa or plates in the neck of the sheath broad and imbricated outwards; they are divided near the opening of the sheath by a sharp ridge on each side, which separates the branchial and excretal tubes of the animal, and is continuous in perfect specimens, so as to form two distinct holes. Valves, L. 0·6, B. 0·65; pallets, L. 0·8, B. 0·3; sheath, L. 12, B. 0·75.

Var. *divaricata*. Shell stunted, distorted, and thicker, having the anterior area much more developed than usual, and scarcely any posterior auricle. *T. divaricata* (Deshayes, MS.), Fischer, in Journ. Conch. v. p. 137, pl. vii. f. 7-9.

HABITAT: In oak, fir, and birch wood composing the timbers of sunken vessels, piers, shipping-stages, and

gates of harbours and docks, as well as occasionally the
stakes of fishing-weirs, and submerged trees, all around
our coasts from Alderney (Lukis) to Shetland (J. G. J.).
It is, however, a local species. The variety is sometimes
met with. Fossil valves have been found in blue clay at
Belfast (Hyndman), and in an oak tree dug up in excavating a deep sewer there (Thompson); in a piece of wood,
more than twenty feet below the surface, at Ayr (Landsborough) : and sheaths in a fossil state have been found by
Mr. Grainger in the Belfast clay-beds, by Mr. Maw at
Strethill, and by Mr. S. Wood in the Red and Coralline
Crag. Newer Italian tertiaries (Soldani and Brocchi).
The foreign distribution of this species extends from
Finmark (Sars, M'Andrew, and Danielssen) to Algiers
(Deshayes). It inhabits the boughs of trees laid down
in Kiel bay for the mussel-fishing (Meyer); and the
variety destroys, in conjunction with *T. minima*, the
fixed stages for shipping marble from the quarries at
Marola on the coast of Piedmont (Capellini).

Olaf Worm first recorded it, in his 'Museum Wormianum' (1655), from Bergen. The pallets bear some
resemblance to battledores or to the bats of French
washerwomen; they are not unfrequently distorted.
Montagu fancied that the imbricated plates which line
the neck of the sheath might be intended to ensnare
the animalcula on which this *Teredo* feeds. He does
not say what kind of a trap they make. According to
Deshayes, Algerian specimens are much smaller than
those of Europe. Some sheaths at Port Patrick were
said by Mr. Thompson to have attained the extraordinary
length of 2½ feet. I am not aware that this species has
ever been found in floating wood; the specimens mentioned in the 'British Mollusca' from this source, as if
on my authority, were the young of *T. megotara*.

It is the *T. navium* of Sellius, *T. navalis* of Gmelin and of almost every subsequent writer until Lovén identified that species with the *T. marina* of the first-named author, *T. nigra* of De Blainville, *T. communis* of Osler, *T. Bruguierii* of Delle Chiaje, *T. fatalis* and *T. Deshaii* of Quatrefages, and *T. Senegalensis* of Laurent but not of De Blainville. The sheath appears to be the *Fistulana corniformis* of Lamarck ; and I suspect that, in one of the earliest stages of growth, it is the *Dentalium bifissum* of Searles Wood from the Coralline Crag, the smaller opening of which exhibits the same internal ridge or partition between the pallial tubes that is so characteristic of this part of the sheath in *T. Norvegica*. No *Dentalium* has any such process.

2. T. NAVA'LIS*, Linné.

T. navalis, Linn. S. N. p. 1267; F. & H. i. p. 74, pl. iv. f. 7, 8, and xviii. f. 3, 4.

SHELL resembling that of *T. Norvegica*, except in being of a much smaller size, and having a thinner texture and finer sculpture: the posterior auricle in the present species is proportionately larger, not placed so high up, more compressed, and better defined both outside and inside (especially the latter) by means of a thin overlapping plate, which separates the auricle from the rest of the valve ; the colour also is fresh, although occasionally deepened by an extraneous stain ; and the epidermis is slighter : the *pallets*, however, exhibit the most remarkable and characteristic difference ; the blade is oval and forked or deeply indented and excavated in the middle at its outer edge : the outside is slightly gibbous and glossy or prismatic, and the inside is flat and of a dull chalky hue and cellular substance : the stalk never extends into the blade: and the pallets in this species are altogether more compact, and not laminar as in the other species: *sheath* usually less solid in proportion to its size, and more tortuous ; it is irregularly annulated in young specimens ; septa or internal plates arranged

* Infesting ships.

close together, slight, and scarcely raised, but existing in all perfect specimens; siphonal or longitudinal ridge perceptible only in the young; aperture obliquely truncated in front, and sometimes also at the back, making that part similar to the slit end of a *Dentalium*. Valves, L. 0·3, B. 0·3; pallets, L. 0·2, B. 0·1; sheath, L. 6·0, B. 0·3.

Var. *occlusa*. Shell like the analogous variety of *T. Norvegica*.

HABITAT : (both the typical form and variety) in fir wood or deal, composing the harbour piles at Sheerness (Sir Everard Home), Herne Bay (Hanley), Yarmouth pier or jetty (Rev. H. R. Nevill), Ramsgate pier (Rev. Sir Charles Macgregor, Bart.); in elm stakes used by fishermen for fastening their nets at Broadstairs (Metcalfe); boats left long at anchor, and shipping-stages in the lower reaches of the Thames and Medway (Baxter). It swarms along the European coasts from Christiania (Asbjörnsen) to Sicily (Delle Chiaje and Philippi), as well as in the Black Sea (Pallas and Heinrich) and Oran in Algeria (coll. Deshayes); with *T. Norvegica* in the boughs of trees, placed in Kiel bay to collect the fry of the common mussel (Meyer); "Hell-gate, New York, in a British frigate sunk during the revolutionary war" (Tryon).

This is the Dutchman's pest; and he does not seem to be troubled with any other kind, at least of the mollusk tribe. It is extraordinary that the animal of such a common species has never been described by any author, except in a general way by Home and Vrolik. Mr. Hanley procured some remarkably fine sheaths from the pier at Herne Bay (supposed by him to belong to *T. megotara*), which measure upwards of a foot in length: for a couple of them I am indebted to his kindness. They are much more solid than those taken from honeycombed pieces of wood, and have almost the polish of ivory. Sometimes the pallets are distorted, and the stalks are now and then double. The stalk passes through the pallet; but the upper part of it is seldom visible, being covered by an accretion of the less compact substance which forms the plate or main body of this appendage.

It was first identified by Lovén, and afterwards recognized by Thompson and the authors of the ' British Mollusca,' as the *T. navalis* of Linné. His description was taken from the sheath only, and is so vague that it may fit any species. Hanley remarked, in his ' Ipsa Linnæi Conchylia,' as follows : " It is impossible to determine, from the language of Linné, to what particular species of shipworm the very comprehensive term *navalis* should be restricted. Our author has not indicated the possession of examples ; consequently his cabinet affords no assistance in the investigation." I was inclined at one time to adopt the specific name *marina*, given by Sellius, which is prior to *navalis* ; but I now believe that the word " marina " was used by him only as an epithet, in an opposite sense to " terrestris." Linné, in the first edition of his ' Fauna Suecica,' described the sheath as a *Dentalium* (in the index as *Teredo navis*); and he adds that it is the *T. navalis* of Sellius, and inhabits ships and submarine piles or stakes. In the last edition of the ' Systema Naturæ ' the ' Fauna Suecica' is quoted, and then Vallisnieri, Sellius, and Plancus. The first and last of these authors intended *T. Norvegica*. That species, as well as the present, still inhabits the coasts of Sweden, as they probably did in Linné's time ; and since the name *Norvegica* is free from any doubt, and it is therefore advisable to retain it under the circumstances, there seems to be no alternative between rejecting altogether the time-honoured name *navalis*, and applying it to the species now described. Da Costa called it *Serpula Teredo*, Spengler *T. batavus*, Lamarck *T. vulgaris*, and Van der Hoeven *T. Sellii*.

3. T. PE′DICELLA′TA*, Quatrefages.

T. pedicellatus, Quatref. in Ann. Sc. Nat. 3ᵉ sér. (Zool.) t. xi. p. 26, pl. i. f. 2.

BODY not so long as that of *T. navalis*, and of a thinner texture: *tubes* rather short, separated half way (Quatrefages).

SHELL scarcely distinguishable from that of *T. navalis*. It is always smaller; the striæ which cover the anterior area are usually fewer, and consequently more remote, and the auricle of the posterior area (especially in the young) is placed somewhat higher up. The *pallets* however are unmistakeably distinct. They are to a certain extent compound, and consist of three separate portions. The stalk is very long and cylindrical: the blade or middle portion is roundish-oval, not much raised, and flat below; the upper part of the blade on each side is dark brown or chocolate, and forms a strongly marked band; it is laminated on the under side: the third or outer portion is square, and is often notched or bifurcated like the outer part of the pallet-blade in *T. navalis*, but never so deeply nor excavated; this third portion is sometimes ivory-like, as well as the stalk and blade, and at other times yellowish-brown, or of a horny substance. The *sheath* is thinner and more decidedly jointed; and it is always shorter and narrower than in *T. navalis*, showing that the animal of the present species does not burrow so deeply. Valves, L. 0.2, B. 0·2; pallets, L. 0·175, B. 0·05; sheath, L. 0·25, B. 0·2.

Var. *truncata*. Corresponding with the varieties of the preceding two species.

HABITAT: Fir and oak used in submarine and fixed woodwork at Guernsey, Alderney, and Sark (Lukis). It was originally discovered by Quatrefages in the Bay

---

* From the long pallet-stalks.

of Passages (province of Guipuscoa) on the north coast of Spain; Toulon (Eydoux and Gay); Provence (Martin, *fide* Petit) ; Algeria (coll. Deshayes).

Some valves which I received from the late Dr. Lukis are of a greenish-brown colour; these he found in oak. He also sent me a piece of a deal plank, which had formed part of a shipping-stage at Alderney, and had been under water for twenty years : the outside was fretted by *Chelura terebrans*; the interior was full of *T. pedicellata*; and through their crowded galleries a huge *T. Norvegica* pursued its solitary course, but without interference on either side. The present species produces at an early age. Its sheath is a beautiful object, the points being imbricated like the segments of the stalk of an *Equisetum*; the orifice in very young specimens resembles a key-hole. Dr. Lukis assured me that this kind caused great destruction in the Government works and new pier at Alderney : no endeavour was made to prevent or stop it.

This is not a satisfactory species, because its sole distinction depends on size and the pallets, and it has never been seen in company with *T. navalis*. The last reason has, of course, a limited value, although it is by no means unimportant when considered in connexion with other circumstances and analogous cases. The pallets are hoe-shaped, with a long handle, and a separate shelly process or membranous fringe at the other extremity. Fischer conjectured that *T. pedicellata* might be the young of *T. Norvegica* or of *T. navalis* ; but the pallets of each species, when first formed, exhibit exactly the same relative characters as in subsequent stages of growth.

4. T. MEGO'TARA*, Hanley.

T. megotara F. & H. i. p. 77, pl. iv. f. 6, and xvii. f. 1, 2.

BODY pale bluish-white: *mantle* not very thin: *foot* muscular and coriaceous, attached by a thick and powerful cylindrical stalk (Clark).

SHELL convex, solid, opaque, and rather glossy, parted in the middle by a slight longitudinal crest, with a very broad but shallow furrow on the posterior side: *sculpture* divided into four distinct portions, viz. anterior, middle, furrowed, and posterior: the anterior consists of sharp, narrow, and fine transverse plates from 25 to 30 in number, which are more remote at first and become closer at advanced periods of growth; the edges of these plates are microscopically notched across; this portion represents a triangle having an acute apex at the back of the valve and a broad and nearly straight base: the middle portion extends the whole length of the shell and is strip-like, the upper part lying between the inner line of the anterior area and the crest which separates one side from the other, and the lower part being open outside and bounded by the crest on the inner side; the broadest part is at the point of the angle where the anterior and middle portions join; this middle portion consists of 15–20 extremely delicate and nearly equal striæ, the outermost of which are exquisitely beaded, and the inner rows strongly but closely notched across; these striæ are longitudinal, with an oblique tendency towards the posterior side, and they diverge from the transverse striæ at a right angle: the furrowed portion is marked with curved but not much raised transverse steps, which gradually widen as they approach the front or ventral edge: and the posterior portion is almost smooth or only marked near the furrow by indistinct lines which form a continuation of the steps above mentioned: *colour* milk-white: *epidermis* membranous, creamcolour, more persistent on the anterior area: *margins* acutely angular on the upper part of the anterior side, with a large triangular excision on the lower part, so that when the valves are united the opening or gape is broadly heart-shaped; they are bluntly pointed or rounded in front, and incurved on the posterior side, which is terminated by a large compressed and rounded ear-shaped expan-

* Great-eared.

sion, occupying at least one-half of that side, and raised above the rest of the shell: *beaks* much incurved, situate near the anterior end, at about one-third the length of the dorsal line; umbones prominent: *hinge-line* very irregular: *hinge-plate* very broad and extremely solid, folded over the anterior dorsal area, which represents a thickened sinuosity; it is deeply notched on the other side, in consequence of which the auricle rises more abruptly; the centre is furnished with a large callous protuberance or knob, as well as with a short peg-like tooth or prong, which is stronger and more conspicuous in the right than in the left valve: *apophyses* rather narrow and regular, not much curved, but occasionally twisted: *inside* glossy, furnished in front with a rather large and solid pear-shaped excrescence; the auricle is separated by a slight and indistinct rib, but there is no shelf or ledge such as is observable in all the other species before described: *muscular scars* distinct; the muscles themselves adhere very closely, and can be easily seen in living specimens; anterior narrow and placed obliquely across the centre of the hinge-plate; posterior broad and large, occupying about one-half of the auricle: *pallets* large and leaf-like; blade oval, squarish in front, slightly convex outside and concave inside, covered with a glossy white epidermis; the outside front is wedge-like and partly excavated by a semicircular impression (exposing the laminated structure of the blade), which extends inwards over one-third or more of the blade; stalk short, stake-like, more solid than the blade; it is continued on both sides far into the blade, and on the under side may be traced the whole way from one end to the other, like a midrib; the upper surface of the blade near the insertion of the stalk is sharply excavated on each side, but not to any great distance: *sheath* usually thin, except at the neck, which is lined with imbricated plates, and these latter are crossed by a sharp siphonal ridge on either side. Valves, L. 0·4, B. 0·4; pallets, L. 0·4, B. 0·15; sheath, L. 3–6, B. 0·45.

Var. 1. *excisa*. Shell similar to the stunted variety of each of the foregoing species.

Var. 2. *striatior*. Shell more convex and not so solid; anterior area larger, and more closely and finely striated; hinge callosity not so prominent.

Var. 3. *mionota*. Shell smaller, with the auricle less developed and not reaching so far down; pallets shorter, having the semicircular part in front more deeply excavated.

HABITAT: Submerged woodwork at Wick (Peach); fir wood at Lerwick and the Whalsey Skerries, Shetland, in the first case composing the timbers of a sunken vessel, in the other the supports of a shipping-stage used in one of the fishing-stations there; and also in the hull of a small craft, lying at anchor in the Skerries Sound, and employed by the Commissioners of Northern Lighthouses on service between that place and Lerwick (J. G. J.). These are the only cases in which, to my knowledge, the present species of *Teredo* can be said to be a true native of the British seas. It is not unfrequently found in floating trees and pieces of fir cast ashore on the east and north of the Shetland Isles, after a continuance of easterly winds (having been drifted from the opposite coasts of Norway); in pieces of Canada timber, which apparently have been transported by the Gulf-stream, aided by a succession of westerly gales, especially during each equinox, on various parts of our shores including the Channel Isles, Sussex, Devon, Dorset, Cornwall, Bristol Channel, Galway, Waterford, Dublin, Antrim, Arran (in Scotland), Scarborough, and Aberdeenshire; in a piece of oak thrown ashore in Cornwall (Couch); in the knee-timber of a vessel stranded at Lulworth (J. G. J.); and in teak, as well as in deal, at Guernsey (Lukis). The first variety only occurs in drift wood; Mr. Dennis found some of a much smaller size than usual in a bamboo on the Sussex coast. The second variety is also imported from distant shores, and can scarcely be considered British. The third may be referred to the same category. Dr. Lukis noticed it at Guernsey, and Mr. Dennis on the Sussex coast, in fir timber; and a remarkably stunted and minute form, in pieces of cork (having been evidently once the net-floats of fishermen), has been taken at Plymouth by Mr. Webster, at Falmouth by Mr. Norman, in Swansea and Carmarthen Bays by myself, and at Aberdeen by Professor Macgillivray. This last variety was described by me as *T. subericola* in the 'Annals and Magazine of Natural History' for August 1860, under the impression that it was a distinct species. The typical form and first two varieties were detected by Mr. Hyndman in pieces of drift wood, that were dug up in making a public sewer at Belfast—thus showing the existence, at a period antecedent to our own, of oceanic currents and other conditions similar to those which still prevail. Malm discovered a valve in the Uddevalla deposits. This species is widely distributed over the North Atlantic. Torell found it on the west coast of Spitzbergen in drift fir wood of two kinds, one from Norway or Siberia, and the other probably from Canada; Fabricius has recorded it from Greenland, Mohr from Iceland (spoiling valuable pieces of drift timber), and Müller from Norway and Denmark; Lilljeborg found it at Mangesund, Upper Norway, in the timbers of a sunken vessel, and also at Bergen; Deyenburg at Lysekihl, Bohuslän (about 12 Swedish miles north of Gottenburg), with *T. Norvegica* and *T. navalis*; D'Orbigny (père) at Rochelle, Cailliaud at Croisic, and M'Andrew (var. *mionota*) in the North Atlantic, in floating timber; Stimpson has described it (under the name of *T. dilatata*) as infesting harbour buoys and fixed woodwork at Lynn, New England; and Tryon states that the range of this species extends from Massachusetts to South Carolina. The last-named locality affords some clue to a fact which puzzled me not a little, viz. the occurrence in drift wood of *T. malleolus* (a native of the West Indies) together with the present species, which I received from Dr. Lukis and Mr. Dennis. The proximity of South Carolina to the Gulf of Mexico, and the course of the great "river in the ocean" along the Atlantic coasts of North America, indicated by Captain Maury in his 'Physical Geography of the Sea,' may account for this commixture of different kinds of *Teredo* in the same piece of floating timber.

T. megotara is intermediate in size between *T. Norvegica* and *T. navalis*, from both of which it may easily be known by the large auricle on the posterior side and by the strong and projecting hinge; the pallets are more like those of *T. Norvegica*, but they are flatter and of a more delicate texture, with a semicircular impression in front, and shorter stalks; the sheath is of variable thickness, and is sometimes altogether wanting, except at the neck, which is regularly laminated with a siphonal ridge down the middle of each side. The mouth of the sheath in very young specimens is crossed by a slight and curved rib, that separates the tube and resembles the handle of a basket. A specimen which I took out of a piece of Canada pine measured 21 inches from the valves to the pallets.

I concur with the authors of the 'British Mollusca' in rejecting the specific name *nana*, given by Dr. Turton to this species; not only because it is inapplicable, but also because his description was insufficient and taken from immature and imperfect specimens. At the same time I regret that the name which they substituted for it is open to objection as pleonastic or redundant, being compounded of two Greek words signifying greatly and large-eared; *megalota* would be more correct. It is the *Bruma dell' oceano* of Vallisnieri, *T. oceani* of Sellius, *Pholas Teredo* of Müller and Fabricius, *T. navalis* of Möller, *T. dilatata* of Stimpson, and (according to Fischer) *T. denticulata* of Gray; the young is probably *Pholas Teredula* of Pallas, from the coasts of Belgium.

Among the species brought hither by the Gulf-stream from the shores of Northern and Central America, those most commonly met with are

T. MALLEOLUS, Turton.

Valves white, elongated, and tapering towards the front; the auricle is narrow and wing-like, higher than the beak, and projecting from the upper part of the posterior side: *pallets* short, with a broad blade, which in the young is transversely oval, giving a mallet-shaped appearance to these appendages: *sheath* not long, but rapidly increasing in size; it is thin, and has delicately imbricated plates. Size of the valves nearly the same as in *T. Norvegica*.

HABITAT: Drift wood, Guernsey (Lukis); Torbay (Turton); Exmouth (Clark); Sussex (Dennis); Swansea and Carmarthen bays (J. G. J.); Miltown-Malbay (Harvey); Belfast (Thompson); young, in cork, Plymouth (Webster); Falmouth (Norman): Cailliaud found it also in drift wood at Croisic, Loire-Inférieure. Specimens sent to me by Dr. Philip Carpenter for examination came from St. Vincents. I therefore infer that the West Indies (and not Sumatra, as stated by Forbes and Hanley) is its native place.

The valves (but not the pallets) of *T. bipinnata*, Turton, apparently belong to the present species. As more than one kind of *Teredo* often inhabit the same piece of wood, mistakes are liable to be made in extracting the valves and pallets; such may account in a great measure for the confusion that exists in public and private collections, and which has found its way into systematic works. A specimen in the British Museum, named

"*T. carinata*, Gray," is composed of the valves of *T. malleolus* and the pallets of *T. Stutchburii*, De Blainville.

T. BIPINNATA, (*bipennata*) Turton.

Valves resembling those of *T. megotara*, but more convex and of a thinner texture: the striated strip is longer; the furrow is reddish-brown, delicately and closely marked across with curved lines, and divided down the middle by a slight groove; the auricle is equally large and prominent, but does not reach quite so far down as in that species, and it is separated inside by a well defined shelf or ledge: *pallets* five times the length of the valves; blades composed of from 40 to 50 narrow funnel-shaped joints, set one within another, with feathered edges which are fringed on each side: stalk varying in length (being sometimes only as long as the blade, and at other times three times as long), quill-shaped, cylindrical, and slender, minutely tuberculated, and often closely annular or tracheiform towards the blade: *sheath* thick and solid, increasing rapidly; neck finely and closely wrinkled but not laminated. Size of the valves about the same as in *T. megotara*.

HABITAT: Drift wood at Guernsey (Lukis); Exmouth (Turton); Beachy Head (Dennis); British Channel (Bulwer); Scarborough (Bean); Roundstone, Connemara (Walpole); Miltown-Malbay, Clare (Harvey); Youghal (Ball); Waterford (Humphreys). On the French coast it has been noticed at Cherbourg and in the Gulf of Gascony by Fischer, at Pouliquen by Petit, and at Croisic by Cailliaud. Dr. Philip Carpenter has also recorded it from Vancouver's Isle and California, and I received specimens from him as West-Indian: there seems to be no good reason for considering it Sumatran. It occurs with *T. cucullata*.

Dr. Turton stated that the feathered pallets could be ejected and retracted at pleasure, and that they were probably "instruments of absorption, as the animal is furnished with a single terminal tube, whose office may perhaps be the discharge or deposit of its eggs or spat!" He may have been, like Bellario, " a learned doctor," each in his own profession; and we will charitably think that the physician understood the constitution of his patients better than that of the *Teredo*.

This species is the *T. navalis* of Spengler, *T. bipinnata* of Fleming, and *T. pennatifera* of De Blainville. The type examples of Spengler in the Royal Museum of Copenhagen are composed of the valves of *T. bipinnata* and the pallets of *T. Stutchburii*.

It is very difficult to say what the *T. palmulatus* of Lamarck may have been. He described the pallets only, which are apparently the same as those of the " Taret de Pondicheri," figured by Adanson in the 'Mém. de l'Acad. Roy.' for 1759. The habitat given by Lamarck is " L'océan de grandes Indes, les mers des pays chauds."

The less-known visitants are *T. excavata* from drift fir, Guernsey (Lukis) and Sussex (Dennis); *T. bipartita* from West-Indian cedar, Guernsey (Lukis); *T. spatha*, with the last; *T. fusticulus* from the same kind of wood, at Leith (J. G. J.) These have simple pallets. *T. cucullata* from drift fir, Guernsey (Lukis), and Sussex (Dennis), and from teak, with the next species, Belfast (Thompson); and *T. fimbriata* (*T. palmulata*, F. & H. i. p. 86, pl. ii. f. 9-11, but not of Lamarck or Philippi) from teak ship-timber, Belfast (Thompson); Exmouth (Clark); and Leith (J. G. J.). These last have compound pallets. All the above (except *T. fimbriata*) were fully described by me in the 'Annals and Magazine of Natural History' for August 1860. *T. spatha* and *T. cucullata* are probably West-Indian, because I received from Dr. Philip Carpenter for identification specimens of both, which were found by the late Professor Adams at Jamaica. *T. fimbriata* is said by Dr. P. Carpenter to be a native of Vancouver's Isle.

T. minima of De Blainville is common in the Mediterranean, but has not been noticed on our shores. It has rather long and large close-jointed pallets with plain edges; the valves are very much smaller than those of any British species, and somewhat resemble the stunted form of T. *navalis*. The pallets of this species and *T. fimbriata* may be taken for miniature ears of barley with long stalks. *T. minima* is the *T. bipalmata* and *T. bipalmulata* of Delle Chiaje, *T. palmulata* of Philippi, *T. Philippii* of Fischer, and *T. serratus* of Deshayes's MS.

Having disposed of the headless mollusks, which are represented by the classes Brachiopoda and Conchifera, we next proceed to consider such as have a head. These exhibit a greater diversity of shape and a more complicated structure; their organs and functions are more specialized. Thus creation moves, step by step, higher and higher, until at length that mental pinnacle is reached, which is attainable only by the chiefest among our own kind. In the suggestive language of Tennyson,

> "All nature widens upward. Evermore
> The simpler essence lower lies;
> More complex is more perfect, owning more
> Discourse, more widely wise."

The first in order among the Cephalic Mollusks is a peculiar class, partaking somewhat of the nature of the Acephala, and forming a link between the two. It is the

SOLE'NOCONCHIA*, (SOLÉNO-CONCHES) Lacaze-Duthiers.

BODY cylindrical, gradually tapering to a rather fine point: *mantle* sheath-like, contractile, thickened in front, where it forms a circular collar, thin and membranous in the middle, constricted behind and terminating in a short tubular process: *head* small and indistinct, not visible outside, furnished with a pair of horny jaws and a spinous tongue: *mouth* internal, surrounded by labial palps: *tentacles* thread-shaped, long and numerous, arranged in two bunches, one on each side of the mouth; they are contractile and ciliated: *gills* rudimentary and obscure, placed above the liver: *foot* remarkably flexible, and divided into three lobes, the middle one of which is conical and extensile; it occupies the front and issues from the collar of the mantle: *posterior tube* serving the purposes of a branchial and excretory duct, as well as assisting in the work of reproduction.

SHELL tubular and resembling an elongated funnel, more or less curved, and open throughout, with the broader end in front; the narrower or posterior end is channelled and sometimes slit.

This small eccentric class comprises the "tooth shells," so called from their resemblance to the tusks or canine teeth of some animals. Their nature in a zoological point of view was but little understood until of late years. Linné placed them in his "Vermes. Testacea;" Lamarck and Cuvier considered them Annelids; De Blainville and Deshayes restored them to the rank of Mollusca. But the skilful and patient investigations of Lacaze-Duthiers have at last solved a problem the interest

---

* From the tube-like shell.

of which, in the estimation of a conchologist, surpasses that of the still sought-for discovery of the sources of the Nile. His "Histoire de l'Organisation et du Développement du Dentale" appeared in the 'Annales des Sciences Naturelles' for 1856 and 1857, and is worthy of his academical fame. His researches were prosecuted at St. Malo; *D. Tarentinum* was the subject. He killed and prepared the animals for anatomical dissection, either with prussic acid, or by drowning them in seawater, particularly in that which contained the putrid corpses of their late companions. In the delightful 'Sea-side Studies' of G. H. Lewes will be found a thoughtful discussion of the very difficult question whether the simpler animals feel pain. He answers it in the negative; and I agree with him to a certain extent. A predaceous beetle with a pin through it will eat up other insects confined in the same collecting-box; and every part of a polype cut in pieces will flourish. At all events the Invertebrata appear to be exempt from that sense of apprehension, or anticipation, which we regard as the worst pain. The *Dentalium* burrows in sand by means of its conical foot in a slanting direction; the narrow end is of course uppermost, and is kept in communication with the water or air for the purpose of respiration. It feeds on Foraminifera and other minute organisms, which it catches with its thread-like tentacles. These are of all lengths and sizes, and are insinuated among the grains of sand on every side; they are covered with cilia, especially at the points, which resemble suckers. They are thrown off by the *Dentalium* under certain conditions, and may occasionally be seen detached and wriggling like taper hair-worms. *Terebella* and other tubular annelids have similar organs. Being highly contractile, these tentacles convey the food to the funnel-shaped mouth, in which, by the aid of labial and ciliated palps, the animalcula are quickly engulfed: then the masticatory apparatus comes into play. This consists of a tongue or lingual riband, armed with five rows of sharp spines, one in the middle, and two on each side. The central tooth is usually called a "rachis," and the side teeth "pleuræ;" they are arranged thus, 2 . 1 . 2. The front set of pleuræ are armed with crochets or "uncini." The apparatus now described seems to have an office analogous to that of the tongue in many cephalophorous mollusks, and it is certainly not a gizzard as Mr. Clark supposed. The shelled Foraminifera found in the stomach of a *Dentalium* are perfect, and the sarcode must be extracted from them by some secretion answering to the gastric juice of the Vertebrata. *Dentalium* has no eyes; they would be useless to an animal always buried in sand. They have otolites or ear-stones, which serve as organs of hearing; these are extremely numerous, calcareous and globular, and are enclosed in two nearly spherical pouches lined with vibratile cilia, which are in constant action, and agitate the otolites by an incessant tremulous movement. The organs of circulation and respiration are of a rudimentary kind; there is no heart. The sexes are separate. There are no external organs of generation; but impregnation is effected by the male emitting his spermatozoa, and the female her eggs at the same time, in the water. The process may be partly compared to the chance shedding of pollen in the air by diœcious plants. Lacaze-Duthiers noticed that the spermatozoa lived six hours after performing the act of fecundation. The egg is at first oval, afterwards pear-shaped, and ultimately divided into segments like those of an Annelid. Such eggs as do not arrive at maturity speedily decompose, and are cleared out by swarms of Iufusoria, which appear to be generated from the corruption. In the first stage of development the germ is motionless; in the second stage it is propelled by vibratile cilia, which are set round a large lobe in front, similar to that observable in the larvæ of many mollusca, and it swims rapidly; in the third stage it crawls by means of a disk-like foot. In swimming it does not come to the surface of the water, as do the fry of the oyster and other mollusca. The shell is formed during the third period, but is only detected by its iridescent lustre, being exceedingly thin and transparent, a mere film. This state continues till the fifth and occasionally the sixth day after birth. The embryonic period lasts from thirty-five to forty days. If any of the fry die, *Paramecia* and *Plœsconiæ* (Infusoria) are bred from the decaying matter, and, entering into the shells of living individuals, soon destroy them. Lacaze-Duthiers observed a current of water passing through the shell from the opening at the smaller end. He discovered the *Dentalium* at low-water mark, where its presence was betrayed by a small groove in the sand; and he seems to have got a knack of finding them, for he says he easily procured 200 live specimens at the recess of a single high spring tide. They prefer certain spots, especially patches of coarse sand mixed with broken shells and interspersed with *Zostera*. In this part of his researches he derived much assistance from the hydrographical survey of France, the minute accuracy of which he greatly praises, not merely as regards zoology, but as subservient to the navigation of the coast. I fear we cannot say so much for ourselves on this side of the Channel, when we reflect on the shameful delay that takes place in the publication of our charts, and even now find that the hydrographical survey on the west of Scotland has been stopped. All we can boast of is a long annual list of wrecks. We are a people that have had losses; like Dogberry, we can afford them: but a superabundance of wealth will not restore drowned mariners to life. The *Dentalium* is hardy, and apparently abstemious. Lacaze-Duthiers kept some alive in a flask of sea-water with a little sand for more than eighteen months. It is much more active at night, and sensible of light. A ray of the sun or the flame of a candle will cause it to withdraw its foot. This organ acts as a piston in expelling at the other end the eggs and seminal fluid, as well as perhaps the fæces and exhausted water. The point of the young shell is pear-shaped, and bears some resemblance to a baby's feeding-bottle with the hole at one end instead of in the middle. It is broken off when too small to contain the terminal tube or process of the mantle; and this part of the shell is continually rubbed away as the animal increases in size, until at last it becomes truncated, and a short pipe is formed with an oblique slit in front to accommodate the terminal tube. The slit is extended in certain species, although this distinctive character is confined to adult specimens. The inside of the shell is white as porcelain, and brilliant as varnish. The epidermis is slight and easily abraded. The microscopical texture of the shell is scarcely different from that of *Patella*. It is most complicated, being composed in a great measure of prisms, interlacing fibres, and anastomosing canals—not of cellular elements. The quantity of animal matter which it contains is next to nothing.

From the above account, which I have mainly derived from the memoirs of Professor Lacaze-Duthiers, it is evident that *Dentalium* is an object well deserving the study of conchologists. Thanks to him, its position among the Mollusca may now be considered settled. Its symmetrical organization and habits connect it with the Acephala; its spinous tongue, indicative of a head, allies it to the Gasteropoda. Its shell, although univalve, is tubular and pervious, never conical or spiral; in all these respects it differs from the shell of *Patella*, which is never tubular or pervious, but always conical and when young exhibits a distinct spire. Its relation to the adult *Fissurella* is merely one of analogy. For all these reasons I see no alternative but to adopt the opinion of the learned French academician by making it the type of a separate class. Argenville, in his 'Zoomorphose' (1757), gave the first idea of the animal. De Blainville called them 'Cirrobranches,' mistaking the tentacles for gills. Deshayes and Clark unfortunately tripped after him; and both appear to have made several mistakes, although of a contradictory nature, with regard to the anatomy of the animal.

> " Velut silvis, ubi passim
> Palantes error certo de tramite pellit;
> Ille sinistrorsum, hic dextrorsum abit; unus utrique
> Error, sed variis illudit partibus."

It is unnecessary to notice the attempts of other systematists, who, so far from contributing anything to our former minimum of knowledge, did their little best to lead us also astray. I may add that the views of Lacaze-Duthiers have been most satisfactorily confirmed by an elaborate essay of Sars on his *Siphonodentalium vitreum*, which is perhaps the type of a new family of the present class.

Family DENTALI'IDÆ, H. & A. Adams.

Genus DENTA'LIUM*, Linné. Pl. V. f. I.

See the account of the class for the characters of the family and genus.

We find in Aldrovandus that, according to Brasavolus, the generic name was anciently "antale" or "dentale," the two names signifying a difference of size only. They were not considered Conchæ, being neither bivalves nor univalves. Valerius Cordus called the larger sort an "Enthalium," and the smaller a "Dentalium." Some persons ate them raw as well as cooked; and druggists sold the shells for medicinal purposes, believing them to be of a mineral nature. Nicodemus Myropous put the names into a Greek dress, viz. ἄνταλι and τένταλι. Martini distinguishes the "Antales" as being smooth, and the "Dentales" as fluted and angular.

1. DENTALIUM EN'TALIS†, Linné.

D. entalis, Linn. Syst. Nat. p. 1263; F. & H. ii. p. 449, pl. lvii. f. 11.

BODY milk-white: *tentacles* slender and extensile, with oval tips: *foot* flanked on each side by an irregularly scalloped lobe.

SHELL tapering, not much curved, often irregularly divided into segments by the successive accretions of growth; it is solid, opaque, and glossy: *sculpture*, slight concentric lines of growth, and occasionally a few indistinct and extremely fine longitudinal striæ towards the narrower end; these striæ, when they occur, are not very numerous, and are only visible with the aid of a magnifier: *colour* ivory-white, with sometimes an ochreous stain on the narrower part, caused by

* Tooth shell.
† Corrupted from *Enthalium*, an ancient name of the genus.

an admixture of mud with the sand in which this species burrows: *margin* at the anterior or broader end more or less jagged, owing to that part of the shell being newly formed and consequently much thinner than other parts; at the posterior or narrower end it is usually truncated in adult specimens, and furnished with a very short sloping and oblique pipe or tubular appendage having a pear-shaped orifice; there is also occasionally at the point on the convex side a notch or groove, in a line with the front or smaller part of the tubular appendage, and this notch is rarely extended into a short and narrow slit or channel. L. 1·5. B. 0·185.

Var. *anulata*. Narrower and more regularly cylindrical, ornamented with white ring-like marks of growth.

HABITAT: Gregarious in sand, from 3 f. to the greatest depth explored on our coasts. Captain Beechey dredged it alive in 145 f. off the Mull of Galloway. It is much more common in the north than in the south. The variety occurs in Shetland at a depth of from 85 to 90 f. Its annulated appearance reminds one of the testaceous sheath in certain species of *Teredo*. As an upper tertiary fossil, *D. entalis* is generally diffused both in time and space, from the glacial "drift" to the red Crag at home, and from the newer deposits in the Christiania district, at a height of 100-150 feet above the sea-level (Sars) to the miocene formation in the Vienna basin (Hörnes). Its foreign distribution in a recent state is also very extensive, although it is probable that *D. Tarentinum* has been mistaken for it in compiling many local lists. Steenstrup collected it in Iceland; Lovén, Sars, and others in Scandinavia (4–200 f.); Macé at Cherbourg; Cailliaud in the Loire-Inférieure; and H. Martin in the Gulf of Lyons; Olivi has recorded it from the Adriatic, Maravigna and Scacchi from Naples, Forbes from the Ægean, Mighels from the State of Maine, and P. Carpenter from Northwest America.

This *Dentalium*, if placed in a vessel of sea-water without sand, is evidently uneasy: it contrives to jerk about slowly and clumsily, by attaching the central point of its foot like the sucker of a leech; and then, spreading out the side lobes to their full extent triangle-wise, it doubles up the foot, and twists itself round with a sort of flapping movement. If placed in a bed of sand, deep enough to cover the shell at a moderately inclined angle, the foot becomes conical and elongated, and soon effects a passage for the whole body, leaving only the top uncovered, to keep the gills supplied with water or air. The 'Proceedings of the Zoological Society' for 1864 contain some interesting particulars of the use and mode of capture in Vancouver's Isle and British Columbia of *D. pretiosum* (Nuttall), which appears to be identical with our species. Mr. Lord says that these shells were employed as money by the Indians of North-west America before the introduction, by the Hudson's Bay Company, of blankets, which to a great extent superseded the tooth-shells as a medium of purchase. "A slave, a canoe, or a squaw, is worth in these days as many blankets; but it used to be so many strings of *Dentalia*." The value of a *Dentalium* depends upon its length. Twenty-five long shells, strung together end to end, make a fathom, and are called a "Hi-qua." At one time such a string would have been worth about £50 sterling. The shells inhabit the soft sand, in the snug bays and harbours that abound along the west coast of Vancouver's Island, at a depth of from 3 to 5 f. The habit of the *Dentalium* is to bury itself in the sand, one end of the shell being invariably downwards, and the other end close to the surface. " This position the wily savage turns to good account, and has adopted a most ingenious mode of capturing the much-prized shell. He arms himself with a long spear, the haft made of light deal, to the end of which is fastened a strip of wood placed transversely, but driven full of teeth made of bone, resembling exactly a long comb with the teeth very wide apart. A squaw sits in the stern of the canoe, and paddles it slowly along, whilst the Indian with the spear stands in the bow. He now stabs the comb-like implement into the sand at the bottom of the water, and after giving two or three such stabs draws it up to look at it; if he has been successful, perhaps four or five *Dentalia* have been impaled on the teeth of the spear." At one period, perhaps a remote one in the history of the inland tribes of Indians, *Dentalia* were worn as ornaments; they are found in old graves, quite 1000 miles from the sea, mixed with stone beads and small bits of the nacre of *Haliotis*, of an irregular shape, but with a small hole drilled through each piece. Rows of these tooth-shells may be seen in the ethnological cases at the British Museum. Sometimes the top of the shell is excavated instead of truncated, and in such case the pipe does not project beyond the edge. The lip of the pipe is expanded and reflected in some of my specimens. The fry are very slender, and are marked with a few slight concentric ribs; the point forms an oval bulb, and has a minute circular orifice.

It is the *Tubulus antalis* of Martini, and *D. Indianorum* of P. Carpenter. In Gmelin's compilation the description is made up of this species and *D. Tarentinum*. The same confusion exists in works of the older writers on European and British shells.

2. D. TARENTI'NUM *, Lamarck.

D. tarentinum, Lam. An. sans Vert. v. p. 345. *D. Tarentinum*, F. & H. ii. p. 451, pl. lvii. f. 12.

BODY yellowish-white: *tentacles* very long, ringed like worms, with sucker-shaped tips : *palps* usually eight in number, four on each side of the mouth, but difficult to make out : they are of different sizes, and covered with vibratile cilia : *foot* flanked on either side by a sinuated symmetrical lobe or flap.

SHELL less slender and rather more curved than *D. entalis*, not so apt to be segmented, very solid and opaque, mostly dull and lustreless: *sculpture*, fine and regular longitudinal striæ towards the point; and the entire surface appears, under a good magnifying power, covered with extremely numerous and delicate impressed lines in the same direction ; there are also the usual marks of growth : *colour* creamy, with sometimes a reddish-brown tinge, or clouded rings denoting the periodical lines of growth, and occasionally a pinkish hue near the point: *margin* at the anterior end jagged, as in the other species ; at the posterior end it is abruptly truncated, and furnished with a very short and small straight pipe, placed in the middle and having a circular orifice ; it has no notch, groove, slit, or channel. L. 1·3. B. 0·2.

HABITAT : From low-water mark at spring tides (Oxwich Bay, near Swansea, J. G. J.) to 25 f., in the Channel Isles, South of England, Bristol Channel, Cardigan Bay (J. G. J.), Bantry Bay (Mrs. Puxley and J. D. Humphreys), and Arran Isle, co. Galway (Barlee). At the latter place it was dredged with *D. entalis*, but in a larger numerical proportion. Fossil in the Subapennine tertiaries (Brocchi), and Sicily (Philippi, as *D. entalis*). The present species has a southern range from the north of France to Gibraltar, both sides of the Mediterranean, and the Adriatic, in 3–40 f. It has usually been regarded as *D. entalis*.

* From its having been found at Tarento, in Italy.

The present species does not generally attain the same size as the last, although I received from Lady Wilkinson a specimen two inches long and only half-grown, which she picked up on the sands in Oxwich Bay. The shell differs from *D. entalis* in being shorter, broader, thicker, not glossy, and having distinct and regular striæ; in the posterior end being abruptly cut off, and the terminal pipe being round with a circular orifice, and in never having any notch or slit; it is also sometimes of a pinkish hue at the point. In the adult the striæ cover the whole surface, and not merely the narrower part; in the young these are fine ribs.

Lister first noticed this shell as British, from Barnstaple Bay. Da Costa described and figured it as *D. vulgare*, a name which ought in justice to be preferred, because that given by Lamarck was not only long subsequent in point of date, but unsupported by a proper description. He says *D. Tarentinum* is slender, somewhat curved, and smooth, with a reddish base. However, I suppose we must accept the proposition made by the late Mr. G. B. Sowerby in the 'Zoological Journal' for 1829, and use the latter name as the one best known to conchologists. It is not the *D. dentalis* of Linné, as supposed by Montagu and his followers. The young is the *D. striatum* of the last-named author, although not of his predecessor, Born. In a worn state it is Turton's *D. labiatum*, and *D. politum*, afterwards changed to *D. læve*.

The collection of Mr. J. D. Humphreys contains a specimen of *D. dentalis*, from Bantry, mixed with the last species. *D. dentalis* is common on the western shores of France, from the mouth of the Loire southwards, as well as in Portugal and Spain, the Mediterranean, Adriatic, Ægean, Madeira, and Canary Isles. Fossil in the

Red and Coralline Crag (S. Wood). It has nine longitudinal ribs, besides frequently a stria between each rib, but no fine impressed lines as in *D. Tarentinum*; and it is more angulated. This may have been the shell of which Miss Pocock found several specimens "on the sandy coast of Cornwall, near Lelant, in the year 1802," but which Donovan mistook for another species and named *D. octangulatum*. Perhaps *D. dentalis* may hereafter be discovered on our southern or Irish coasts. It is the *D. novemcostatum* of Lamarck, and *D. vulgare* of H. and A. Adams.

D. abyssorum of Sars once lived, and possibly survives, in our northern seas, I dredged two or three young specimens in Shetland on different occasions; but they had a semifossilized look. This species inhabits the western coasts of Sweden and Norway, at depths varying from 40 to 150 f. Sars has identified it with *D. striolatum* of Stimpson from the east coast of North America; and it is most likely the *D. attenuatum* of Say from Massachusetts. *D. abyssorum* is one of our glacial relics; it occurs in the boulder-clay at Bridlington (S. Wood, as "*D. entale*") and Wick (Peach); Moel Tryfaen (Darbishire); Banff (Forbes, as *D. dentalis*); Preston (J. Smith, as *D. striatum*); newer and older deposits at Christiania (Sars), in the former at 100–120 feet, and in the latter at 460 feet above the sea-level. It is longer and thinner than *D. dentalis*, and has more ribs; it is not so finely striated as *D. Tarentinum*, and wants the impressed lines. The terminal process is like that of *D. entalis*.

D. striatum of Born (*D. octangulatum*, Donovan, *D. octogonum*, Lamarck, and *D. striatulum*, Turton) is a tropical shell, and has been wrongly considered British on very suspicious authority. Turton's collection contained specimens; and I have likewise one which Dr. Leach sent to Mr. Dillwyn, under the name of *D. octohedra*, as found in Kent.

D. eburneum, afterwards *D. album* of Turton (*D. variabile*, Deshayes), is another un-English or spurious species; its native country is said to be the East Indies.

D. semistriatum of Turton must be, provisionally at least, placed in the same category, although specimens were taken by Mr. Humphreys from the stomach of a red gurnard at Cork. I believe Turton's specimens came from the same quarter, notwithstanding that Dublin Bay is the locality given by him. It may be the *D. semipolitum* of Broderip and Sowerby, or *D. semistriolatum* of Guilding: if the former, the habitat is unknown; if the latter, it is West-Indian.

D. clausum of Turton is certainly not a *Dentalium*, nor even a shell; it seems to be the lower part of the quill of a sea-bird's wing feather.

The cases of British species of *Ditrupa* (a genus of testaceous Annelids) may easily be distinguished from the shells of any species of *Dentalium* by their being constricted near the front, and never having the tubular appendage at the smaller end. They are thicker, and of a crystalline structure. Such are *Ditrupa arietina*, Müller (*Dentalium subulatum*, Deshayes), and *Ditrupa gadus*, Montagu (*Dentalium coarctatum*, Desh.).

Class GASTERO'PODA.

(See Vol. I. p. 51.)

In considering the natural distribution of this group, it will be found that the systems of classification which have been propounded by naturalists since the post-Linnean revolution are so numerous, that the student is apt to be lost in the perplexing labyrinth into which they lead him. That of the great Cuvier, however, seems to have stood its ground better than any other, and is commended by its greater simplicity. It is founded on differences in the nature and position of the gills or respiratory organs. Some modification has been rendered necessary by the investigations of later physiologists; and I will submit a scheme, which appears to me sufficient to classify the Gasteropoda, without making any pretence to novelty or perfection. I would adopt the following eight orders.

1. CYCLOBRANCHIATA, (*Cyclobranches*) Cuvier.

GILLS arranged in two separate rows, one on each side of the body, and covered by the mantle. *Chitonidæ.*

2. PECTINIBRANCHIATA, (*Pectinibranches*) Cuvier.

GILLS consisting of one or two plumes (usually a single plume), placed above the head, or on either side of it, and covered by the mantle. *Patellidæ, Trochidæ*, and many other families, having (invariably in the young state) a spiral or turbinated shell with an entire mouth.

3. SIPHONOBRANCHIATA, Goldfuss.

GILLS consisting of one or two plumes, placed obliquely on the anterior part of the back, and contained in a cavity of the mantle, which is prolonged into a tubular canal. *Muricidæ*,

Cypræidæ, and other families, having a spiral shell with a channelled mouth.

4. PULMONOBRANCHIATA, Sowerby.

Respiratory apparatus consisting principally of an internal cavity or pouch, formed by a fold of the mantle, and lined with a network of vessels. *Limacidæ*, *Helicidæ*, and other land and freshwater univalves, besides a few marine kinds, some of which are naked and others provided with shells.

5. PLEUROBRANCHIATA, Gray.

GILLS forming a single row, placed on the right side of the body, and covered by the mantle. *Bullidæ*.

6. NUDIBRANCHIATA, (*Nudibranches*) Cuvier.

GILLS exposed, and forming a tuft on the back. *Dorididæ* and most Sea-slugs.

7. PELLIBRANCHIATA, Alder & Hancock.

Respiratory apparatus consisting of a net-work of vessels diffused over the outer surface of the mantle. *Limapontiidæ*, and small Sea-slugs of an inferior type.

8. NUCLEOBRANCHIATA, De Blainville.

Respiratory apparatus consisting of symmetrical filaments associated with the digestive organs in a nucleus placed on the back. *Carinaria* and a few other pelagic mollusca of a peculiar kind (Heteropoda), none of which are British.

In the Prosobranches of Milne-Edwards (which constitute the first three orders) the gills are almost always enclosed in a vaulted chamber or cavity, which is placed on the *front* part of the body; the sexes are separate; and the shell is complete in all stages of growth. In his Opisthobranches (which constitute the fifth and sixth orders) the gills are never enclosed in a special cavity or receptacle, but are more or less exposed at the back or sides on the *hinder* part of the body; they are hermaphrodite; and the shell is completely formed in the fry, but often disappears in the adult or is incomplete. According to Lacaze-Duthiers the Gasteropoda are formed on an unsymmetrical plan; the organs of digestion are placed on one side, instead of in the middle as in the Acephala; and the organs of sense are more developed, and usually lodged in a head.

Our knowledge of the plan of arrangement, so far as regards the teeth on the lingual membrane of such Gasteropoda as possess this curious apparatus, is too imperfect to make it form part of any scheme of classification. Lovén, Troschel, Gray, and Macdonald have to a certain extent pursued the subject, and attach much importance to it. Dr. Gray separated on this ground his Ctenobranchiata into two suborders—Proboscidifera and Rostrifera—treating the one as zoophagous, and the other as phytophagous: but we find in the latter division *Conus, Cyprœa, Aporrhais, Fusus, Vermetus, Cœcum, Capulus, Calyptrœa,* and many other genera which are not vegetable-eaters, *Pleurotomatidœ* placed among the Proboscidifera, and *Conidœ* among the Rostrifera (both of these families having precisely the same kind and disposition of teeth), besides many other like incongruities. At the same time it is evident that this spinous organ of deglutition affords a useful character to distinguish certain genera and even higher groups; and I trust that a further examination of the subject will enable us to make it available for that purpose.

The embryology, or history of the development of the Gasteropoda, has been carefully investigated by a host of able physiologists from the time of Stiebel (1815) to this day. Grant, Quatrefages, Dumortier, Leuckart, F. Müller, Laurent, Sars, Van Beneden, Rathke, Lovén, Milne-Edwards, Nordmann, Kölliker, Gegenbaur, Krohn, Clarapède, Vogt, and Lacaze-Duthiers are some of those who have distinguished themselves by such researches. All their observations show that the Gasteropoda pass through a series of metamorphoses before attaining their perfect state, and that the duration of the larval state is often considerable, compared with the whole period of their existence.

Their shells appear to have a more uniform structure than those of the Acephala. Dr. Carpenter says "There is not by any means the same amount of diversity in the structure of the shell in the different subdivisions of this group as that which we have met with among the Conchiferous Acephala. There is a certain typical plan of construction that seems common to by far the greater number of them; and any considerable departures from it are uncommon. The small proportion of animal matter contained in most of these shells is a very marked feature in their character, and it serves to render other features indistinct." A univalve shell consists of three layers of cellular plates, each of the upper two layers lying unconformably on the one immediately below it, and every plate being composed of a single series of elongated prismatic cells, which cohere lengthwise. He dissents from the idea of Dr. Gray that the structural arrangement is the result of crystalline action. The shells of mollusca were formerly regarded as a mere exudation of calcareous matter, the particles of which were held together by a sort of animal glue. Carpenter is of opinion that the appearance of prismatic crystallization in certain shells is entirely due to the moulding of the calcareous matter within their cells. He agrees with Dr. Bowerbank in his account of the composition of univalve shells, as evincing a definite organic arrangement and not a simple crystallization.

Order I. CYCLOBRANCHIATA.

Family CHITO'NIDÆ, Guilding.

BODY oval, oblong, or elongated, semicylindrical, rounded at each end: *mantle* thick, covering the back, and encircling the sides with a girdle which is free at its edges: *head* sessile, surmounted by a membranous veil or hood, and containing a pair of horny jaws and the front of a long and slender tongue bristling with numerous teeth, which extends into the interior of the body, and is folded up within it: no *tentacles* or *eyes*: *gills* forming a row of small pyramids on each side, which meet behind the head, lying between the mantle and the foot, and extending from behind to the front: *foot* muscular, occupying the whole of the under surface: *vent* or excretory duct placed opposite to the head at the end of the foot.

SHELL composed of separate arched plates, which are inserted in the mantle along the back breadthwise; they are usually external.

I am not surprised at Lamarck calling this a singular and strange group, nor that there has been such difficulty in assigning to it a definite place among the Invertebrata. In the larval state they resemble Isopodous Crustaceans, or they might even be mistaken for tiny Trilobites; and the adult may be compared to *Onisci* deprived of antennæ, eyes, and feet. They are also not unlike species of *Aphrodita*. When a boy I was cruelly deceived in thinking that I had found a huge and new *Chiton*, having got hold of a Sea-mouse in the sand at low water. De Blainville believed that their natural affinities lie with the Annelids, and he raised them to a tribal rank under the name of Polyplaxiphora. The circulatory system is complicated; Cuvier ascertained that each auricle opened into the heart by two distinct orifices, a disposition of which he had not detected another instance in the animal kingdom. Milne-Edwards considered them a satellite group of the Mollusca, fancifully comparing the organization of the Invertebrata to the sidereal system. But the general plan of their structure is that of the limpet; the only differences of any importance consist in the latter having tentacles and eyes, which are wanting in the *Chitonidæ*, and in the shell of the one being a single piece, while the other is composed of several pieces which form together an elongated buckler. In the genus *Cylichna* we find one species (*C. truncata*) with tentacles, and another (*C. cylindracea*) without tentacles; and in each of the genera *Eulima, Mangelia,* and *Amphisphyra*, similar discrepancies occur with respect to the presence or absence of eyes in certain species. The most obvious distinction between *Chiton* and *Patella* consists in the arrangement of the gills and the multivalve or univalve character of the shell. It seems sufficient to group them in two families, separate but not widely apart. Adanson and Ström pointed out the affinity of *Chiton* to *Patella*; and Poli showed that their spinous tongues were exactly similar. The Rev. Lansdown Guilding, in a valuable monograph of the present family (Zool. Journ. 1830), called this apparatus "trachyderma."

Genus CHITON*, Linné. Pl. V. f. 2.

BODY oval or oblong: *girdle* scaly, bristly, tufted, or membranous, and fringed with short spines.

SHELL usually boat-like, composed of eight plates, which are

---

* Coat of mail.

external and overlap one another in an imbricated or tile-like fashion; the last or hindmost plate has a small overhanging boss in the middle.

These " punaises de mer," as Vallisnieri calls them— Petiver has a prettier name, " Oscabrions "—move very slowly, creeping or rather gliding, onwards, backwards, or sideways, with an imperceptible and stealthy pace. Mr. Guilding says of the West-Indian kinds (and his remarks will in most particulars apply to the British species), "They seem to feed entirely by night. Though they remain stationary during the day, when disturbed they will often creep away with a slow and equal pace, often sliding sideways, and creeping under the rocks and stones for concealment. If accidentally reversed, they soon recover their position by violently contorting and undulating the zone; and for defence they sometimes (when detached) roll themselves up like wood-lice. Some of the larger kinds, especially of *Acanthopleura*, are eagerly devoured by the lower orders in the West Indies, who have the folly to call them ' beef; ' the thick fleshy foot is cut away from the living animal and swallowed raw, while the viscera are rejected. We have here a large pale *Chiton*, which is said to be poisonous." Ladies who are not good sailors, and are fond of trying new preventives against sea-sickness, may (if they can) swallow raw *Chitons*, and so imitate the Iceland fishermen, who pretend that the " hav-bœggeluus " (sea-bugs) are an effectual remedy against this malady, and also that they quench thirst. One kind is easily procured at low water on most of our beaches by turning over loose stones. Such an occupation just before encountering a voyage might beguile the tedious interval—or perchance the deglutition of these strange boluses might by anticipating the evil rob the passage of its horrors.

Poli called the animal *Lophyrus*, and he has given some particulars of its anatomy. Neither Cuvier nor Leach found any male organ in the individuals they examined; and little seems to be known of their sexual relations. Their embryogeny, however, is no longer a mystery. In the 'Transactions of the Royal Academy of Sciences at Stockholm' for 1855 will be found a most interesting account by Professor Lovén of his observations on the development of *C. marginatus*. He says that some individuals, kept in confinement, laid their eggs, loosely united in clusters of from 7 to 16, upon small stones. Each egg has a thick envelope. The embryo, which is exactly of an oval shape, and without any trace of shell, is divided by a circular indentation into two nearly equal parts. The upper half is fringed with cirri, by means of which the embryo swims; and each side of the line of indentation is furnished with a tuft of very fine filaments. Close to this line on either side are perceptible two dark points, which are the eyes. When freed from the egg, the embryo assumes a more lengthened shape; the lower half soon afterwards exhibits transverse furrows and joints, of which seven (besides the front lobe) are distinguishable; and some granulations now make their appearance as the first rudiments of the shell. The animal bends itself frequently; it is still quite soft, and can only swim. Subsequently it begins to crawl. The eyes are then more conspicuous; the joints become separated, and acquire a shelly consistence; the cirri and tufts disappear; and the head is perfectly formed with its membranous hood. The embryo at this stage sometimes swims and sometimes crawls. The eyes are placed on distinct protuberances, and consist of pigment-spots and lenses; and the foot is rather enlarged, although some time elapses before this part attains its full size in proportion to the head. Lovén justly remarks that, if we compare the development of *Chiton* with that of other Mollusca, it is evident that the circle of cirri, by means of which the animal moves in its first or swimming stage, corresponds with the cirri of the front lobe in the young of other Gasteropoda and of the Acephala. Mr. Clark recorded some important remarks on the reproduction of *C. marginatus* in the 'Annals and Magazine of Natural History' for December 1855, being the same year as that in which Lovén's were published. One of several individuals, placed in a vessel of sea-water on the 23rd July 1855, poured out for several minutes a continuous stream of flaky-white viscous matter, like a fleecy cloud, and then discharged ova—not in volleys, but one or two at every second for at least fifteen minutes, forming a batch of from 1300 to 1500; a thousand or more remained in the ovary, perhaps not sufficiently matured for parturition. The fluid and ova were emitted "from under the centre of the coriaceous integument of the posterior terminal valve," in the same way as the author had described it to take place from the posterior extremity of *Dentalium*. Each egg was enveloped in a pale yellow membrane, and was of a somewhat globular shape, being a little compressed or oblate at what may be termed the axis; it appeared to be about the 100th of an inch in diameter. The ova were entangled in the tenacious fluid which had been previously poured out—this being seemingly a provision for preventing their being washed away until the fry were prepared to emerge. In about 24 hours afterwards the fry became disengaged from their common nidus, and swam about with great vivacity in every direction, crossing a large breakfast saucer in 30 or 40 seconds. They had by that time lost the subglobular figure, and taken that of a subelongated oval, approaching the shape of an adult *Chiton*. When the swimming-action commenced, only half the animal was liberated from the capsule or membranous integument, the other half being still enclosed, with the empty portion of the capsule folded over it. With a power of 300 linear Mr. Clark saw the elements of the four anterior valves, as well as the buccal depression and head; at this stage of development he could not perceive any metamorphosis. In the course of the next five days the animal had altogether cast off the embryonic covering, when it exhibited the complete form of a *Chiton*, and adhered to the bottom of the vessel. He apparently did not at any period detect the eyes which Lovén had noticed. Mr. Clark further remarked that the fry during its phase of rapid movement often rolled itself into a ball. The slight discrepancy between these observations of the Swedish and English naturalists may be accounted for by those of the former being more complete, and perhaps having been made under more favourable circumstances. Twenty years ago Milne-Edwards published, in conjunction with Quatrefages and Blanchard, the result of anatomical and zoological researches made on the shores of Sicily and France. *Chiton* was one of the subjects of their investigation; but I am not aware that any details were given. Milne-Edwards was induced, however, by these researches to declare that he had arrived at a different conclusion from that which was hazarded in the 'Vestiges of Creation,' viz. that the embryo of the higher animals, including man himself, presented in succession modes of organization analogous to the permanent state of the principal lower types of the animal kingdom. On the contrary, he was of opinion that the embryos of the Mollusca and the Mammalia had their own respective modes of organization, and that the theory above mentioned was by no means justified by the facts. Each plate of a *Chiton* has its sides diagonally parted, and is divided into three triangular areas. The base of the central area is covered by the edges of the preceding plate, and the base of each lateral area is inserted in the girdle or marginal band of the mantle. The front plate, being that which protects the head, is semicircular; the hindmost plate is oval, and is furnished with a boss or point, which overhangs the rear and corresponds with the apex of the cone in the shell of *Patella*. In many species the plates are inserted more firmly in the girdle by means of marginal notches. These were first noticed by Fabricius in his description of *C. marmoreus*. They vary in number and fineness according to the species. The spines and valves which cover the girdle in most species are calcareous. The structure of the shell agrees with that of *Patella*, although the details are somewhat different. Carpenter says, " The external layer, which is usually impregnated by colouring matter, does not exhibit the laminations which are seen in *Patella*, but in their stead presents everywhere a delicate fibrous structure, the fibres being arranged parallel to the surface. The superficial part of this layer is perforated by large canals, which pass down obliquely into its substance, without penetrating so far as the middle layer. The purpose of these canals, which remind us of the perforations of *Terebratula*, is by no means apparent. In the deeper part of this coloured external layer, which is of great toughness, there is a layer of minute cells which seem to lie between the fibres; and below this, again, is a layer entirely composed of large flat pavement-like cells, as in *Patella*. The internal layer seems to have the same nearly homogeneous texture as the external." The tubular structure of the outer layer appears to be accompanied by the absence of an epidermis, respecting which I offered an explanation in my account of the Brachiopoda at p. 6 of Vol. II. The second volume of the 'Zoological Journal' (1825) contains an accurate description, by the Rev. R. T. Lowe, of some Scotch *Chitons*; and Baron Middendorff has given an elaborate essay on the Russian kinds, with details of their anatomy. The genus abounds in species, which are all more or less gregarious. Reeve has lately enumerated 189, and this list is not complete. The British *Chitons* live attached to rocks, stones, and old shells; they inhabit various depths of water, and many live between tide-marks. Some of their shelly plates occur in upper tertiary strata; others of extinct form have been found in older and even ancient formations. Gray has made twenty genera out of the one so familiar to us by name. I do not consider it necessary to apply this rate of multiplication to our native species: the following conspectus may suffice to distinguish them:—

A. Girdle covered with spines, and having also tufts of bristles. (*Acanthochites*, Leach, *fide* Risso.) 1. *C. fascicularis.* 2. *C. discrepans.*

B. Girdle spinous, without tufts. (*Acanthopleura*, Guilding). 3. *C. Hanleyi.*

C. Girdle covered with scales or granules. (*Lepidopleurus*, Leach, *fide* Risso.) 4. *C. cancellatus.* 5. *C. albus.* 6. *C. cinereus.* 7. *C. marginatus.* 8. *C. ruber.*

D. Girdle apparently reticulated. 9. *C. lævis.*

E. Girdle membranous. 10. *C. marmoreus.*

1. CHITON FAS'CICULA'RIS*, Linné.

C. fascicularis, Linn. S. N. p. 1106; F. & H. ii. p. 393, pl. lix. f. 5.

BODY oblong, yellowish with often a tinge of brown: *mantle* fleshy, bordered by a narrow hem of a paler and almost transparent hue: *girdle* moderately broad, more or less closely covered with short spines, which are usually tawny or greyish; besides this armature there is a thick tuft of 14 longer spines, or rather bristles, of a paler or whitish colour (occasionally greenish or golden), between each plate of the shell at the point of junction on both sides, and 4 more, close to the front or head-plate, making in all 18; margin fringed with spines of an intermediate length, and finely ciliated at its outer edges: *head* representing an arc of ⅔rds of a circle: *mouth* large, of a purplish colour, and star-shaped, being divided into a dozen lobes, each of which radiates from the centre and is defined by a black line: *gills* visible throughout, larger towards the tail, and diminishing in size towards the head: *foot* oblong, of an orange tint, broader in front, and bluntly pointed behind, thicker towards the sides than in the middle of the sole: *vent* conical and short, projecting above the tail or hinder extremity of the foot, and placed in a channel or notch.

SHELL formed of the usual number of plates, which are shield-like and somewhat compressed, solid, opaque, and of rather a dull hue; they occupy ⅘ths of the entire breadth; when separated, the notch in front of each is very large and deep, and is flanked on either side by a broad shoulder: *sculpture,* rather fine but not very numerous oval granules, like those of shagreen, on each side of a broadish central ridge or keel, which extends along the back; they are arranged lengthwise in lines converging towards the beak or point of the ridge; their tops are flattened and sometimes slightly concave; the central or dorsal ridge is closely striated longitudinally or divided by lines, and sometimes punctured, exposing the tubular structure; it has usually a rubbed and somewhat polished appearance: *colour* brown, chocolate, orange, yellow, pinkish, or red, now and then mottled or streaked with white, pale green, or brown: *beaks* small and rather promi-

* Covered with small bundles or tufts.

nent: *inside* smooth and polished, of a greenish cast: *notches* slight, 5 on the head-plate, 1 on each side of every middle plate, and 2 on the tail-plate, making altogether 19. L. 0·75. B. 0·375.

Var. 1. *attenuata*. Much longer and narrower in proportion to the breadth.

Var. 2. *gracilis*. Longer than usual, with finer sculpture: *girdle* broader and membranous, sparsely set with spines, and mostly having an extra tuft (occasionally two) at the tail. *C. gracilis*, Jeffreys, in Ann. Nat. Hist. Feb. 1859, p. 106.

HABITAT: Rocks, stones, and oyster-shells, on every part of our coast from low-water mark to 25 f.; off Mull of Galloway in 145 f., as *C. discrepans* (Beechey). Var. 1. Oban (Barlee). Var. 2. Weymouth (Metcalfe and Damon); Lulworth (J. G. J.); Grouville Bay, Jersey, with *C. discrepans* (Norman); Milford Haven (M'Andrew and Jordan); Lough Strangford (Adair). A specimen from the last mentioned locality measures nearly an inch and a half in length, while the largest that I have of the typical form (from Unst) is scarcely an inch long. Fossil in the Coralline Crag, Sutton (S. Wood); South Italian tertiaries (Philippi). The foreign distribution of this species extends from Finmark (Sars) to the Ægean (Forbes), Barbary (Brander, *fide* Linné), Morocco (M'Andrew), Algeria (Weinkauff), and Canary Isles (M'Andrew), at depths ranging to 20 f.; but some of the southern localities which have been published probably belong to *C. discrepans*. Malm found it attached to *Laminaria saccharina* on the coast of Bohuslän at a depth of 12 f. The variety *gracilis* occurs at Etretat, in Normandy (J. G. J.), and in the Loire-Inférieure (Cailliaud).

This handsome species crawls backwards as well as forwards. Mr. Jordan remarks that it appears much more sensible of cold than the *Littorinæ*, and that even about the middle of November it was difficult for him to find two or three specimens in an hour's search at Tenby, in a spot where he could during the month of August get more than as many dozen in the same time. The fleshy part of the girdle must be porous or vascular, because it becomes swollen and puffed up if confined by a ligature; it is often raised in folds or puckered, to admit water to the gills. The dorsal ridge is formed by the wearing away of the granulated surface, showing that this part of the shell is never renewed. The plates are frequently encrusted by small spiral *Serpulæ* and Foraminifera. In young shells the triangular compartments are to be seen, as in other species of *Chiton*.

It may be the "Kalison" of Adanson. The short description by Linné of *C. fascicularis*, and the habitat (Barbary), are rather more applicable to *C. discrepans* than to the present species. Writers on the Mediterranean shells have evidently mistaken one for the other. Pennant says his *C. crinitus* has only seven valves; but his figure shows eight and the usual number of tufts. I am also disposed to refer to *C. fascicularis* the *Acanthochites æneus* of Risso, and certainly the *Acanthochætes vulgaris* of Leach. I cannot maintain the distinction which at first seemed to exist between the typical form and the variety *gracilis*, and which induced me to describe the latter as a separate species. Both have every character in common, except the additional tuft; and that is not constant.

2. C. dis'crepans *, Brown.

C. discrepans, Brown, Ill. Conch. p. 65, pl. xxi. f. 20; F. & H. ii. p. 396, pl. lviii. f. 4.

Body oblong: *girdle* broad, covered with a thick pile, like velvet, which is usually of a greyish tint; tufts similar in number and arrangement to those of *C. fascicularis*, but not so large; they are whitish or tawny, with sometimes a greenish hue; spines of marginal fringe not longer than those which form the pile.

Shell more convex in the middle than the last species, occupying only one-half of the entire breadth: *plates* similar in shape: *sculpture*, very fine and numerous round granules, arranged in rows which converge in a curved direction towards the beak in each plate; their tips are flattened in adult specimens, but seldom concave; ridge prominent and rather sharp, separated from the granulated portion on each side of it, closely striated or lineated lengthwise, and having a rubbed or polished appearance: *colour* greyish, mottled with dull reddish-brown; the ridge is generally darker and sometimes marked by a black cuneiform streak: *beaks* sharp and projecting: *inside* smooth and polished, of a greenish cast: *notches* as in *C. fascicularis*, but sharper. L. 1·25. B. 0·6.

Habitat: Not uncommon on rocks and stones in the Channel Isles, from low-water mark to 25 f.; sometimes associated with *C. fascicularis*, which is much less frequently met with in this outlying part of Great Britain. The only other British locality that I am aware of is Coomb, in Lantivet Bay, Cornwall, as *C. crinitus* (Couch): I have not seen the specimens. It occurs on the coast of France from the Boulonnais to Nice; Corsica (Payraudeau); Sicily, as *C. fascicularis*, var. *major* (Philippi); Balearic Isles and Mogador (M'Andrew); and Lovén has enumerated it among the Scandinavian mollusca as *C. crinitus* ("Boh-Norv."); but I fear he mistook a variety of *C. fascicularis* for the present species.

* Different, *i.e.*, from *C. fascicularis*.

Mr. Dennis, as well as Mr. Jordan, observed that specimens found between tide-marks in Herm and Jersey were very much finer than those dredged in deep water off the last-mentioned island. This species differs from *C. fascicularis* in being larger, and usually longer in proportion to the breadth; the central ridge is more prominent; the granules are much smaller and more numerous, and they are invariably round instead of oval; the girdle is broader, and clothed with a thick pile; the tufts are not so large or conspicuous; and the notches are deeper. The young have a remarkably elongated shape.

The locality (Tenby) assigned by Brown to *C. discrepans* belongs to *C. fascicularis*; but his statement that the "papillæ" are round can only apply to the former species. Sowerby considered it (but erroneously) the *C. crinitus* of Pennant, which is nothing more than *C. fascicularis*. I believe *Acanthochites communis* and *A. carinatus* of Risso may be referred to *C. discrepans*.

3. C. HANLE'YI *, Bean.

C. Hanleyi (Bean), Suppl. Thorpe's Brit. Mar. Conch. p. 263; F. &. H. ii. p. 398, pl. lxii. f. 2.

BODY oblong: *girdle* rather narrow, tough, covered with numerous short whitish spines; those at the posterior side of each plate, issuing from the corner where it overlaps the next plate, are a little longer than the rest, and assume a somewhat tufted form.

SHELL convex: *plates* shield-like, with a wide and deep notch in front, moderately solid and opaque, not glossy: *sculpture*, numerous but not crowded bead-like tubercles, arranged in longitudinal rows, which appear in some specimens chain-like; these tubercles are smaller, finer, and closer on the crest or back of each plate, and become coarser and irregular at the

* Named in honour of Mr. Silvanus Hanley, one of the authors of 'British Mollusca' and other works on conchology.

sides; there is no distinct ridge: *colour* dirty brown or ashy: *beaks* small and moderately pointed: *inside* porcellanous; the margin has no notches, but is indistinctly and microscopically crenulated. L. 0·4. B. 0·2.

HABITAT: Stones and old shells, from 20 to 80 f., in the following localities:—Plymouth, in trawl refuse, with *Odostomia truncatula* and other south of England shells (Jordan); Scarborough (Bean); Cullercoats (Alder); Co. Galway (Barlee); Co. Antrim (J. G. J.); Oban and Hebrides (Barlee, M'Andrew, and J. G. J.); Moray Firth (Gordon); Shetland (Barlee and J. G. J.): it is not common. Coralline Crag, Sutton (S. Wood). It inhabits every part of the Scandinavian coast, from the south of Sweden to Finmark, at depths varying from 35 to 120 f.; Malm noticed it on *Lophelia* (*Oculina*) *prolifera*. I dredged in the Gulf of Spezzia a young shell which I considered to be the present species; and M. Petit states that Mr. Shuttleworth found two specimens on a *Cardium* peculiar to the Caribbean Sea, which he received among some West-Indian shells. These southern localities, however, want confirmation.

The lingual membrane is armed with numerous teeth arranged in rows, two of which are more prominent than the rest and are furnished with black hooks. Specimens from the North Sea attain a considerable size. I have one from Shetland fully three-quarters of an inch long, and a plate which must have belonged to a specimen twice that size.

It is the *C. strigillatus* of S. Wood. The *C. Nagelfar* of Lovén is *C. Hanleyi* of an extraordinary large size; and so is the *C. abyssorum* of Sars.

4. C. CANCELLA'TUS * (Leach?), G. B. Sowerby, Jun.

C. cancellatus, Sow. Descr. Cat. Brit. Chit. p. 4, f. 104, 104 *a. b*, and 105;
F. & H. ii. p. 410, pl. lix. f. 3.

BODY oblong: *girdle* narrow, irregularly coated with small rather shiny yellowish-white granules; margin closely fringed with short spines.

SHELL semicylindrical, very convex: *plates* transversely oblong and narrow, moderately solid and opaque, and slightly glossy; each of the middle plates is divided into three distinct compartments (as described in the account of the genus), the lateral compartments in this species being elevated considerably above the middle portion, but together scarcely equalling it in superficial area: *sculpture*, extremely minute round, compressed, and close-set granules, arranged in numerous chain-like rows, which are longitudinal on the first and last plates and on the middle compartment of the other six, and converge to the centre or apex of the triangle in the side compartments, so as to present a somewhat divaricating appearance; there is no central ridge: *colour* yellowish-white: *beaks* inconspicuous, except on the tail-plate: *inside* glossy, exhibiting some of the chain-like sculpture, beside sharp semicircular leaves at each side of all but the head-plate, which form the shoulders of those plates; margin not notched, but indistinctly and microscopically crenulated. L. 0·225. B. 0·125.

HABITAT: Stones, old shells, and occasionally *Ulvæ* and small sea-weeds in the laminarian zone, Channel Isles, south of England, Isle of Man, north and west of Ireland, Hebrides, and Shetland, at depths between 5 and 40 f.; it is rather local, but not uncommon. Its foreign distribution is wide, and embraces the Norwegian and Swedish coasts from 50 to 150 f., and those of France from Etretat to the Gulf of Lyons.

Malm found it on *Lophelia prolifera*. The links of the chain-like rows of granules on this small and pretty

* Latticed.

species resemble punctures, and produce a latticed appearance.

It is the *C. albus* of Pulteney and Montagu (but not of Linné), *C. alveolus* of Sars, and probably *C. tuberculatus* of Leach's 'Mollusca of Great Britain.'

5. C. cine′reus*, Linné.

C. cinerea, Linn. S. N. p. 1107. *C. asellus*, F. & H. ii. p. 407, pl. lix. f. 1, 2, and (animal) pl. A A. f. 5.

Body broadly oval, brownish-yellow, orange, or of a somewhat tawny fleshcolour: *mantle* thin: *girdle* rather narrow, covered with small oval, rather shiny, yellowish-white or darker-coloured granules, which lie one upon another in a thick heap; margin closely fringed with sharp whitish spines: *head* semicircular, surrounded by a narrow hood: *mouth* forming when at rest a transverse and concentrically wrinkled slit; but when open and showing the teeth, it becomes circular: *gills* pale brownish-yellow; only from 6 to 10 of the plumes or leaflets nearest the tail on either side are visible, the others being convoluted and withdrawn: *foot* oval, broader in front, and margined by a pinkish line: *vent* short and tubular.

Shell compressed: *plates* as in *C. cancellatus*, but less solid in proportion to the size; lateral compartments indistinct: *sculpture* similar to that of the last species, although much finer and never exhibiting a punctured or cancellated appearance: *ridge* slight, more or less conspicuous: *colour* pale yellowish, often irregularly streaked lengthwise with dark lines, and sometimes having a transverse mark of the same hue on the lateral compartment near the beak in each plate: *beaks* small: *inside* porcellanous, streaked in the middle like the outside, displaying the leaf-like shoulders described in *C. cancellatus*; margin not notched, but crenulated in the same way as in the last two species. L. 0·5. B. 0·35.

Var. *Rissoi*. Shell of a uniform pale yellowish colour. *C. Rissoi*, Payraudeau, Moll. Cors. p. 87, pl. iii. f. 3, 4.

Habitat: Stones, and old shells (especially oysters), everywhere in the laminarian, coralline, and deep-sea

---
* Ash-coloured.

zones; occasionally between tide-marks at high springs; off Mull of Galloway, 145 f. (Beechey). Macgillivray says he found it at Aberdeen on a starfish! The variety is from the west of Scotland in deep water. " Glacial " bed at Fort William (J. G. J.); Coralline Crag, Sutton (S. Wood). Greenland (Fabricius and Eschricht); Iceland (Steenstrup and Torell); Scandinavia, 1–130 f. (Müller and others); north of France (De Gerville); Vigo Bay (M'Andrew); and along the coasts of the Mediterranean to the Ægean, 5–10 f. (Forbes, as *C. Rissoi*).

Chemnitz called the specimens in Spengler's cabinet " the negress," owing to their swarthy complexion. When this *Chiton* opens its mouth and shows its teeth, a double row of black glistening points, separated by a central column, is suddenly unfolded, and as rapidly withdrawn; this operation is repeated several times in the course of a minute. Is it caused by the blind cravings of hunger, or is it a process like that of rumination, or merely for the purpose of keeping the teeth clean? Mr. Dennis says that all the specimens which he dredged in 17 f., seven or eight miles off Blatchington, on the Sussex coast, are small and light-coloured in comparison with those procured by him at low water. The largest specimen I have came from Oban, and measures $\frac{8}{10}$ths of an inch in length by $\frac{1}{2}$ an inch in breadth; the smallest is not much more than $\frac{1}{20}$th of an inch long. The fry are broader than the adult, and their granules are tubercular, few in number, and apparently analogous to the external bulbs of the tubular perforations in shells of Brachiopoda. *C. cinereus* may be distinguished from *C. cancellatus* by its larger size, expanded and compressed shape, finer sculpture, the lateral compartments being inconspicuous, and by its central ridge, beaks, and thicker coating of granules on the girdle, which is broader than in that species.

It is the *C. asellus* in Spengler's monograph of the genus (Skr. Nat. Selsk. 1797), *C. islandicus* of Gmelin (from Schröter's 'Einleitung'), *C. fuscatus* of Leach (but not of Brown), and *C. scoticus* of the same author; the variety is *C. onyx* of Spengler.

6. C. ALBUS *, Linné.

C. albus, Linn. S. N. p. 1107; F. & H. ii. p. 405, pl. lxii. f. 2.

BODY narrowly oval, brownish yellow: *girdle* rather broad, regularly and closely beaded with glittering equal-sized oval granules, which have their smaller points towards the beaks of the shell; margin fringed with short spines.

SHELL rather convex: *plates* narrowish, solid and opaque, somewhat glossy; lateral compartments slightly raised: *sculpture*, numerous and small granules, arranged in irregular and wavy lines which converge towards the beaks; there are also in adult specimens a few darker marks of growth in each plate: *ridge* sharp and conspicuous: *colour* yellowish-white: *beaks* small, prominent: *inside* porcellanous, with sometimes a bluish tinge, displaying broad leaf-like shoulders on all the plates except that which covers the head: *notches* slight but distinct, 13 on the head-plate, 11 on the tail-plate, and 2 on each of the other plates (one on either side), making altogether 36. L. 0·35. B. 0·2.

HABITAT: Stones, old shells, and sea-weeds, from low-water mark to 30 f.; Ballaugh, Isle of Man (Forbes, from whom I received a specimen in 1841, with a note of this locality, and named " Chiton, new sp."); west coast of Scotland (R. T. Lowe and others); Burghead, Moray Firth (Murray, *fide* Gordon); Buchan, Aberdeenshire (Dawson); Wick (Peach); Orkneys (Thomas); Lerwick and other parts of Shetland (J. G. J.): it is a

* White.

local species. Fossil at Fort William (J. G. J.). Its foreign distribution is entirely northern, viz. Spitzbergen, Greenland, Iceland, the Faroe Isles, Norway, Sweden, and Denmark, in 10–150 f. (Torell, König, and others); the coast of Russian Lapland, on the White Sea (Middendorff); Massachusetts (Gould); New England (Stimpson); and State of Maine, in the stomachs of fishes caught in Casco Bay (Mighels).

This approaches *C. cinereus* nearer than any other species: but it is narrower and higher, and of a uniform yellowish-white colour; it has a rather prominent ridge and beaks; the sculpture is finer, and not chain-like, but irregularly disposed in a radiating and wavy manner; its margin is notched; and the granulation of the girdle resembles bead-work. Spiral Foraminifera (*Discorbina rosacea* and *Truncatulina lobatula*) seem fond of attaching themselves to the girdle. The fry have disproportionately large beaks. My finest specimen is from Scalloway, and measures $\frac{6}{10}$ths of an inch in length, and half as much in breadth.

It is the *C. oryza* of Spengler, *C. aselloides* of Lowe, and *C. sagrinatus* of Couthouy.

7. C. MARGINA'TUS*, Pennant.

C. marginatus. Penn. Brit. Zool. iv. p. 71, tab. xxxvi. f. 2. *C. cinereus*, F. & H. ii. p. 402, pl. lviii. f. 1 (as *C. marginatus*).

BODY oval, pale fleshcolour: *mantle* thin, edged with a narrow border of light brown: *girdle* of moderate breadth, usually puckered on the inner side (owing to the contraction of the mantle), covered with minute close-set roundish granules, which lie evenly on the surface; it is of different colours, and often variegated by alternate patches of reddish-brown and yellow; margin thickly fringed with short but conspicuous spines of a yellowish tint: *head* thick, transversely oval: *mouth* round and plaited: *gills* from 15 to 20 on each side, triangular, apparently not continued behind the head:

* Bordered.

foot lanceolate, truncated in front, and broader towards the tail, which is bluntly pointed.

SHELL somewhat convex: *plates* broad, rather solid and opaque, without lustre; lateral compartments scarcely (if at all) raised, but marked by a slight ridge which extends on each side from the beak to the front corner: *sculpture* like shagreen, composed of not very small oval flattened granules, which are arranged in two indistinct sets of rows, one lengthwise on the middle compartment, and the other nearly at a right angle on the lateral compartments from each side to the beak: *ridge* distinct and prominent: *colour* various, forming different combinations of yellow, reddish-brown, and green, often mottled, or the plates are party-coloured, seldom of the same hue throughout: *beaks* strong, prominent, and conspicuous: *inside* porcellanous, with frequently a greenish tinge in the middle, displaying broad leaf-like shoulders on all the plates except that which covers the head; the terminal plates often exhibit white lines which radiate outwards, and represent so many segments: *notches* deep, 8 on the head-plate, 10 on the tail-plate, and 2 on each of the other plates, making altogether 30, besides occasionally an intermediate and slighter notch. L. 0·6. B. 0·4.

HABITAT: Under stones below high water of neap tides on all our coasts; common. It is diffused everywhere throughout the North Atlantic from Faroe (Landt) and the Loffoden Isles (Sars) to Mogador (M'Andrew). In the North Sea it seems to frequent deeper water; Asbjörnsen and other writers on the Scandinavian Mollusca give depths varying from 2 to 40 f. According to Gould, a single specimen was found on the coast of Massachusetts; and the *C. dentiens* of that author, from Vancouver's Island, appears to be undistinguishable from our shell.

This *Chiton* uses the under side of the head, as well as the foot, in crawling. From one specimen that I was observing on the 3rd of June 1864, a thin stream of milky fluid issued, immediately beneath the anal tube, at short intervals for about two minutes; the discharge was so copious that the water in the vessel became turbid. This was probably a seminal secretion. The colour of the shell is extremely variable. Out of more than five hundred specimens Bouchard-Chantereaux was unable to find two marked exactly in the same way. He describes the tongue as horny, bristling with six longitudinal rows of small tricuspid teeth, those of the two central rows being blackish and much stronger than the others. *C. marginatus* differs from *C. cinereus* in being usually of a larger size, narrower, and more convex or arched; the plates are broader; the colour is variegated, not streaked; the sculpture is much coarser, and not chain-like; the granulation of the girdle is finer, more minute, and even; the marginal spines are stronger and more conspicuous; and the edges of the plates are deeply notched, instead of being slightly and indistinctly crenulated. The habitat of the two species is also different; this is littoral, while the other prefers deeper water. In the fry of the present species the front of each plate is curved.

Two specimens of *C. marginatus* in Turton's collection, affixed to separate cards, are named in the Doctor's handwriting " *Chiton ruber* "; one from " Dublin Bay," and the other from " Portmarnock." They correspond with his description of *C. punctatus*. Both have been painted red! A daughter of Dr. Turton told me that when her father went out shell-hunting, some young ladies would occasionally go before him on the beach, and drop here and there shells which they had taken with them, in order to play him a merry trick. Let us suppose that these were the artists who so ingeniously beautified the specimens above noticed, finding such perhaps an easy feat compared with that which Shenstone's Laura could not accomplish—

" With fresh vermilion paint the rose."

A specimen was described by Captain Brown as having only five plates, under the name of *C. quinquevalvis*. Other synonyms of the ordinary form appear to be *C. cimex*, Chemnitz, *C. cimicinus*, Landt, *C. cinereus*, Laskey and Lowe (but not of Linné), *C. fuscatus*, Brown and Macgillivray, *C. variegatus*, Leach (but not of Philippi), and *Lepidopleurus carinatus* of the same author. It may be partly the *C. punctatus* of Linné.

8. C. RUBER* (Linné), Lowe.

C. ruber, Lowe, in Zool. Journ. ii. p. 101, pl. v. f. 2; F. & H. ii. p. 399, pl. lix. f. 6, and (animal) pl. A A. f. 6.

Body oval, inclining to oblong, yellow or creamcolour, and apparently of a granular texture: *mantle* thin : *girdle* rather broad, of a mealy aspect, covered with numerous minute spherical granules which lie evenly on the surface, as in *C. marginatus*; it is chequered with alternate patches of red and white; margin thickly fringed with very short spines of the same colour as the patch to which they belong : *head* semioval, edged with a narrow band of brown, which is surmounted by a line of darker hue : *mouth* when closed forming an arched slit, also surrounded by a darker line, and concentrically wrinkled : *gills* more exposed than in *C. cinereus* : *foot* elliptical, bordered by a light-brown band, which is much narrower than the one round the head, and likewise surmounted by a dark line : *vent* or excretal duct broad and wedge-shaped.

Shell convex : *plates* broad, solid, opaque, and glossy; lateral compartments indistinct : *sculpture*, parallel lines of growth, which are sometimes remarkably strong and conspicuous; with a lens of moderate power it appears otherwise to be quite smooth; but if a Coddington or Stanhope be used, the whole surface is found to be very finely and closely reticulated : *ridge* more or less prominent, but seldom distinct : *colour* reddish-brown of different shades, mottled or streaked with white or pale yellow : *beaks* strong and projecting : *inside* rosecolour in the middle of each plate, with a greenish hue on the edges and sides, shouldered as in the foregoing species :

* Red.

notches deep, 9 on the head-plate, 8 on the tail-plate, and 2 on each of the other plates, making altogether 29. L. 0·5. B. 0·275.

Var. *oblonga*. Larger, longer, and more arched. L. 0·65, B. 0·35.

HABITAT: On rocks, stones, old shells, and the "roots" of *Laminaria saccharina*, between low-water mark and 20 f., from South Devon to Shetland; it is common in the west of Scotland and Lerwick Sound, where also the variety occurs. Fossil at Fort William (J. G. J.). The only southern locality that I can find recorded is the Adriatic, according to Olivi; but its northern range is very extensive, and comprises Spitzbergen (Phipps, *fide* Scoresby); Godhaab, E. Greenland, 50–150 f. (Wallich); S. Greenland (Eschricht); Iceland (Mohr, Steenstrup, and Torell); Norway, Sweden, and Denmark, 1–150 f. (Spengler and many others); White Sea (Baer, *fide* Middendorff); Northeast America, from Cape Cod northwards (Gould, Mighels, and Stimpson).

This pretty species was first noticed as British by Professor Jamieson, in the first volume of the 'Memoirs of the Wernerian Society' (1811), from rocks in the island of Unst, and almost simultaneously, in the same volume, by Captain Laskey, from Dunbar. It may always be distinguished from *C. marginatus* by its reddish-brown colour, narrower and more arched shape, broader girdle, and especially by its smooth and glossy appearance. Young shells are longitudinally veined, showing the internal tubular structure.

It is extremely probable that the *C. ruber* of Linné may have been the species which Fabricius afterwards described with greater precision as *C. marmoreus*; and there can be no doubt that the *C. lævis* of Pennant was the species which we are accustomed to call *C. ruber*. But although Lovén and his scientific countrymen have adopted the correct names, I must confess a want of moral courage in not following their example, believing that the perpetuation of such trifling errors may cause less inconvenience to conchologists in general than the changes necessary to rectify the nomenclature of so many species. Spengler described this *Chiton* as *minimus*, and Leach (but not Lowe) as *latus*.

9. C. LÆVIS* (Pennant), Montagu.

C. lævis, Mont. Test. Brit. p. 2; F. & H. ii. p. 411, pl. lviii. f. 3.

BODY oval, inclining to oblong, reddish brown: *girdle* broad, resembling hair-cloth, covered with numerous minute and closely packed lozenge-shaped scales or spines, which are set horizontally with their points towards the outer margin; it is of a dark brick-colour irregularly flecked with white; margin fringed with a few scattered and caducous short pinkish spines, which are apparently a continuation of those which cover the girdle.

SHELL convex: *plates* broad, solid, opaque, and glossy; lateral compartments more or less distinct: *sculpture* smooth to the naked eye or examined with a lens, but exhibiting under a higher magnifying-power a series of extremely delicate striæ, running lengthwise on the middle compartment of each plate, and towards the beak on the side compartments; the surface is also covered (especially the terminal plates and the side compartments of the other plates) with small tubercles, which are very little raised and scarcely perceptible; these are the bulbs or extremities of the canals that permeate the fabric of the shell, like the tubular apparatus observable in most of the Brachiopoda; in young specimens the tubercles are perforated or open; there are likewise slight parallel lines of growth: *ridge* more or less prominent, but seldom conspicuous: *colour* reddish-brown, marbled or veined with white, and sometimes variegated with green, red, pink, or brown, rarely of a uniform dark brick-colour: *beaks* strong and projecting: *inside*

* Smooth.

fleshcolour, more or less tinged with green, slightly shouldered: *notches* deep, 16-20 on the head-plate, 15 on the tail-plate, and 2 on each of the other plates, being altogether about 45. L. 0·75. B. 0·4.

Var. *navicula*. Smaller, narrower, and more arched.

HABITAT : On rocks, stones, and old shells, from Unst to Sark, between low-water mark of spring tides and 70 f. ; apparently not gregarious, nor so common as some other species. The variety inhabits the west of Scotland. *C. lævis* has been noticed by foreign writers as far north as Vadsoe, East Finmark, in 30-60 f. (Danielssen), southward to the Ægean, in 31-80 f. (Forbes), and Algeria (M'Andrew and Weinkauff), and also in various intermediate places, at depths varying from 8 to 50 f.

According to Philippi, Sicilian specimens are much smaller than the British. The largest in my collection came from Oban, and are upwards of an inch and a quarter long. The proportion of length to breadth is variable.

It is the *C. corallinus* of Risso, *C. achatinus* of Brown, *C. Cranchianus* and *Lepidopleurus punctulatus* of Leach, and *C. Doriæ* of Capellini. Montagu described a seven-plated specimen as *C. septemvalvis*, a name which Maton and Rackett changed to *C. discors*.

10. C. MARMO'REUS*, Fabricius.

C. marmoreus, Fabr. Faun. Grœnl. p. 420 ; F. & H. ii. p. 414, pl. lviii. f. 2, and lix. f. 4.

BODY oval, inclining to oblong, yellowish or reddish-brown : *girdle* rather broad, membranous and thin, apparently smooth, but microscopically pustulated ; it is dusky-brown, sometimes

* Marbled.

barred with dark orange; margin fringed with extremely short yellowish spines.

SHELL convex: *plates* broad, solid and opaque, somewhat glossy; lateral compartments distinct, not much raised, but defined by a blunt ridge which extends from the beak on either side to the front corner of each plate: *sculpture* nearly smooth to the naked eye, exhibiting under a magnifying-power numerous minute and slight tubercles, which usually are more conspicuous on the terminal plates and side compartments, as in *C. lævis*; the parallel lines of growth are strongly marked in adult specimens: *ridge* indistinct: *colour* reddish-brown, variegated or speckled with white or yellow, sometimes in a zigzag or lightning fashion: *beaks* very strong and prominent: *inside* yellowish, tinged with pink, showing the under side or hollow of the ridge to be striated across; shoulders long and narrow: *notches* deep, 8 on the head-plate, 9 on the tail-plate, and 2 on each of the other plates, making 29 in all, besides some intermediate denticles. L. 1. B. 0·6.

HABITAT: Stones, shells, and sea-weed in the Laminarian zone, from Shetland to Scarborough (Bean); eastern shores of Ireland, as far south as Dublin Bay (Kinaghan). Fossil at Fort William (J. G. J.); Uddevalla (Malm). It inhabits every part of the Atlantic, north of Great Britain, from Spitzbergen to Zealand, and the coasts of North-east America, at depths of from 7 to 100 f. M'Andrew has recorded it as dredged at Carthagena in 5–10 f.; this appears to be the only instance of a southern locality.

According to Brown, Mr. Hancock discovered this king of the British *Chitons* below Tynemouth Castle in 1809. Laskey indicated it from Dunbar in 1811. Fabricius says that it is often found in the crops of the Eider-Duck and *Anas spectabilis*. His description of the animal and shell is most admirable; and he particularly noticed the notches on the margin of each plate or valve, as characteristic of this and other species of *Chiton*. It is stated by Middendorff that the epidermis of the girdle in *C. marmoreus* displays under the microscrope a coverlet ornamented with erect spinules. I have not succeeded in detecting any such armature in British specimens; the margin of the girdle is fringed in this way, but the surface is merely pustulated. Specimens taken by Captain Bedford in Mull are more than an inch and a half long.

It is the *C. punctatus* of Ström, who nearly a century ago showed the resemblance between the animal of *Chiton* and that of *Patella*; perhaps in strictness the specific name given by him, being the more ancient, ought to be preferred to *marmoreus*. Fleming called it *C. lævigatus*, Lowe *C. latus*, Bean *C. pictus*, Couthouy *C. fulminatus*, and Leach *C. Flemingius*.

Order II. PECTINIBRANCHIATA.

Family I. PATEL'LIDÆ, (*Patelladæ*) Guilding.

BODY semioval, more or less raised above and flat beneath: *mantle* thin, covering the back and sides: *head* snout-like, furnished with a pair of horny jaws and a long and slender tongue, which bristles with numerous teeth and is folded up within the body: *tentacles* spike-shaped: *eyes* on protuberances at the outer bases of the tentacles, wanting in certain kinds: *gills* forming a single row or plume of leaf-like plates, which issue from behind the neck on the right-hand side: *foot* very large and rounded, occupying the whole of the under side.

SHELL conical or cap-shaped; apex turned towards one end, spiral, and slightly twisted on one side, or curved, in the young state: *mouth* extremely wide, forming the entire base of the cone: *central scar* inside shaped like an amphora.

This family constitutes the vanguard of the innumerable limpet tribe. Their shells are never symmetrical, as has been stated by some writers. When first formed they are either spiral or else eccentrically twisted; the spire or twist is worn away in the course of growth. There being no communication between the mantle and the apex of the shell, the latter cannot be absorbed by the animal. The sexes are separate. Those kinds which inhabit the littoral and laminarian zones are phytophagous; and the others, which inhabit the coralline and deep-sea zones, are probably zoophagous. In Lovén's scheme of the dentition in univalve mollusca the rhachis or central plate in *Patella* and *Helcion* has six teeth, and each of the pleuræ or side plates three teeth; in *Tectura* the rhachis has from four to six teeth, and the pleuræ have none; in *Lepeta* the rhachis has only a single tooth, and each of the pleuræ two. Such differences may indicate the nature of the food; the first three genera are known to live on sea-weeds, while the last (as well as *Propilidium*) cannot derive their subsistence from any vegetable matter except diatoms.

Genus I. PATE′LLA*, Lister. Pl. V. f. 3.

BODY convex: *mantle* fringed at its edge with cirri of irregular lengths: *tentacles* rather short: *eyes* prominent: *gills* numerous and closely packed, lying between the mantle and the foot, and only interrupted on the right-hand side: *foot* thick and muscular.

SHELL conical, more or less convex, furnished with ribs that radiate from the crown, having in its embryonic state a completely spiral apex; crown prominent, eccentric but not very much on one side; the attachment of the mantle to the shell is exhibited in the middle (between the crown and the margin) as a ring-like scar.

The λεπάς of the Greeks, with whom it appears to have been rather a favourite article of food. In the

---

* A small pan.

'Δειπνοσοφισταὶ' of Athenæus, Icesius says that it is even more appetizing than the oyster, although not so digestible; Diphilus does not hold it in such esteem. The tenacity with which it adheres to the rock was well known to ancient writers. This is compared by Aristophanes with the attachment of an old woman to a youth; and Ælian remarks that, when touched, it is as difficult to remove as a pomegranate was from the fist of Milo. In one of the odes of Alcæus it is apostrophized as the child of the rock and hoary sea; and Cicero refers to it (although not by name) as an example of the sedentary nature of some marine animals, " partim ad saxa nativis testis inhærentium." With his usual power of observation, exceeding that of many subsequent naturalists, Aristotle described the habits of the limpet, and showed that it leaves its place on the rock and goes out to feed. This was confirmed by Reaumur, although Borelli and others asserted that the limpet remained all its life fixed to the same spot. It uses its foot like a snail, but travels more slowly. Bouchard-Chantereaux says that he had often seen limpets crawling, especially just after the tide had gone out. The young limpet moves freely about, and shifts its quarters; but after attaining a growth of probably a few days, it affixes itself to a particular spot, which it only quits, when covered by the sea, on the return of each tide. If it settles on a hard and rugged rock, the circumference of the shell is moulded to fit the irregular surface of its abode; the base of attachment is then bleached. Should the rock be soft, it scoops out by degrees with its muscular foot a cavity of a greater or less depth. Mr. Anderson of Wick (the highly intelligent editor of the ' John O'Groat Journal ') gave me some pieces of Old Red sandstone from that coast, in which the pits made and inhabited by *P. vul-*

gata were so deep, that little more than the crown of the shell was visible outside. On the Dorsetshire coast the chalk-rocks are also excavated in the same manner, but not so deeply. Specimens are not unfrequently found on impure limestone, which are constricted or indented at the edges, in consequence of the excavation having been hindered by the greater hardness of one side of the spot occupied by these limpets. The animal feeds on small delicate sea-weeds of a foliaceous kind, as well as on *Melobesia polymorpha*, that encrust the rocks at low water, by means of its long tongue, which is coiled spirally, like the mainspring of a watch set round with spring-cogs. This instrument is thrust out from side to side; and when charged with food, it is withdrawn into the stomach, unloaded, and again put forth. The mark left on the face of a rock, coated with a film of the fine sea-weed mentioned above, by a limpet after grazing resembles the track of a sea-worm: indeed a late eminent geologist had a large slab thus marked cut out of the rock, and sent to him with great care, in order to publish the supposed discovery of a new Annelidan ichnolite in the old red sandstone; fortunately the mistake was pointed out to him before proceeding further. Each limpet appears to have its own feeding-ground or pasturage; its tracks are sometimes numerous, and deviate in different directions. Mr. Peach has ascertained that it does not retire in the winter to deeper water on the coast of Caithness, and that it always returns home before the ebbing tide leaves it dry. Its firm adhesion to the rock is extraordinary. In order to test the strength of its tenacity, Reaumur suspended a weight of from 28 to 30 lbs. from the shell of a limpet attached to a stone; this weight it sustained for some seconds: less weights failed to overcome its resistance. He attributed the adhesive force not to muscular action, but to an invisible glue which exudes from the granulated base or sole of the foot. It may be also caused by an adaptation of the surface of this part of the animal to the frequent, although often minute, inequalities of the stone; although the glutinous and viscous fluid, which is secreted by numerous glands in the foot, appears to be the principal agent. It is said that death does not destroy the cohesion; but I do not see how such an experiment could be tried. Dr. Johnston, in his 'Introduction to Conchology,' likewise states that if, after having detached a *Patella*, one's finger be applied to the foot of the animal, or to the spot on which it rested, the finger will be held there by a very sensible attraction; and that if the spot be then moistened with a little water, no further adhesion will occur, the glue having become dissolved or weakened. When the limpet wishes to leave its abode, it has only to raise gently the edges of the foot to admit the sea and loosen the cement. Adanson believed that the adhesion was owing to the action of numerous hemispherical suckers on the under surface of the foot, aided by a viscous secretion; he observed that when the animal was detached from the rock, those suckers expanded or assumed a globular form. The foot is undoubtedly capable of considerable dilatation and contraction, and has a vascular structure; it is often much distended with water. This great French naturalist does not seem to have known the branchial organization of *Patella*; for he describes the gills as an appendage of the mantle. It was supposed by Cuvier that the common limpet was hermaphrodite. Adanson and Milne-Edwards, however, established the fact of its bisexuality; and Lebert and Robin published in the 'Annales des Sciences Naturelles' for 1846 further particulars of its reproductive organs. The last-named physiologists noticed that at the end of April these organs (which in each sex are placed at the left-hand side of the body) were wanting in nearly one-half of the individuals dissected by them, and that of the remainder the males were in the proportion of 3 to 8 or 10 of the females. Fischer has given us some information as to the mode of its oviposition. This takes place in the months of March and April, when all the rocks at low water, as well as the shells of old limpets, are covered with an immense quantity of the fry. He is of opinion that immediately on the eggs being excluded from the ovary, they are developed and attach themselves. Gray suspected the sexes to be distinct in the limpet; although he could not discover any external difference in the animals, except a slight variation of colour. He says that in autumn he found a white, milky, glairy fluid in some individuals, and ova in others. My late friend Dr. Lukis noticed, in taking up a limpet while in the act of crawling, that young ones were attached to the under side of the foot; and he inferred that it carried its offspring about with it for protection. But it is more probable that the fry became accidentally entangled in the gelatinous fluid which exudes from the foot, than that the phenomenon which he observed was an instance of molluscan στοργή. The shell represents part of a cone whose section is an irregular ellipse. It is composed of three layers, as in many other univalves. According to Carpenter the inner and outer layers in *Patella* are rather less compact than usual; the middle layer is " composed of tolerably regular polygonal cells, which form only a thin layer in some parts, whilst in others they are elongated into prisms."

Gaza, a Byzantine philologist who flourished in the 15th century, appears to have been the first to give this shell the name of *Patella*. It was, notwithstanding, called by the ancient name *Lepas* by other writers, and even as late as 1616 by Colonna. Aldrovandus included the genus with *Balanus*: Lister had the merit of separating and distinguishing them. Nor have all modern zoologists been uniformly successful in recognizing the natural position of *Patella* among the Mollusca. In the opinion of Lamarck it belongs to the same family as *Phyllidia*; but the gap between the Pectinibranchs and Nudibranchs seems much too wide to be bridged over by even his engineering. Most of his followers placed *Patella* alongside of *Chiton* in the order *Cyclobranchiata*. The present genus was for some time the receptacle of miscellaneous and incongruous organisms. Among these were *Patella unguis*, Linné (*Lingula*), *P. anomala*, Müller (*Crania*), *P. orbiculata*, Walker (according to Mr. Norman "the calcareous disk of the termination of a tentacle of *Echinus*"), *P. extinctorium* and *P. tricornis*, Turton (opercula of species of *Serpula*): *Ancylus fluviatilis* and *A. lacustris* were also placed in the same genus. *Patella*, as now restricted, is very rich in species, although their tendency to vary is so great that the number of those described by authors is evidently excessive. All of them inhabit rocks and shingly beaches, and are strictly littoral. The distribution of the genus is world-wide. As to its fossil ancestry, Searles Wood says, "Shells of this form have early made their appearance, and several have been figured from the secondary formations." De Montfort, perhaps for the sake of variety, changed the generic name to *Patellus*.

PATELLA VULGATA*, Linné.

P. vulgata, Linn. S. N. p. 1258; F. & H. ii. p. 421, pl. lxi. f. 5, 6.

BODY brownish-yellow or dusky, with a bluish tinge: *mantle* fringed with slender cirri or filaments of different lengths and sizes, which correspond with the ribs and striæ of the shell; some of these cirri above the head are much longer than the rest, and are in the proportion of 1 to 4 or 5 of the latter; the mantle is often edged with a narrow band of a darker colour: *head* short, bulging, and strong: *mouth* provided with two lips, which are placed laterally: *tentacles* awl-shaped, not retractile, darker at their tips; they curl towards each other and lie flat on the head, when the animal is at rest: *eyes* small, on slight eminences outside the swollen bases of the tentacles: *gills* of a drab or yellowish colour; branchial artery transparent, thicker, and funnel-shaped at its origin, and having smaller veins issuing from it during its course, at a right angle: *foot* attached to the rest of the body by a series of powerful but short interlacing muscles; the sole is lead-coloured, or more or less deeply tinged with yellow; margin thin with a pale border.

SHELL forming usually a regular and somewhat raised cone, solid, opaque, and of a dull hue: *sculpture*, numerous ribs, which radiate from the apex and become stronger and broader at the lower part or margin; between each rib are 2 or 3 (sometimes more) parallel striæ or finer ribs; in some specimens the ribs are irregularly granulated or studded with knob-like tubercles; the surface is also covered in fresh and less rubbed specimens with close-set microscopical longitudinal lines, and with numerous but irregular concentric lines of growth: *colour* greyish or pale brownish-yellow, with often purplish longitudinal rays arranged in duplicate; it is rarely speckled with white, or of a uniform dusky hue: *beak* or apex blunt, often worn so as to expose the crown, which is of a reddish or orange tint; it is sometimes nearly central: *mouth* or aperture roundish-oval, with the broader part behind: *margin* scalloped or indented by the ribs and intermediate striæ: *inside* nacreous and glossy, often yellow or exhibiting the coloured rays (especially at the margin); it is minutely but irregularly lineated in a concentric direction from the margin to that part which is always covered by the edge of the mantle, and micro-

* Common.

scopically fretted in the last mentioned part ; margin bevelled ; *annular scar* broad ; *central scar* of every colour from white to dark brown. L. 1·75. B. 1·5.

Var. 1. *elevata.* Shell much smaller, rounder, and higher.

Var. 2. *picta.* Shell smaller and thinner; with alternate rays of reddish and dark blue.

Var. 3. *intermedia.* "Animal black or dark-coloured" (Knapp). Shell rather smaller, flatter, and oval, with finer ribs, and an orange crown; inside golden-yellow, or tinged with fleshcolour (occasionally creamcolour) in the centre, and beautifully rayed towards the margin.

Var. 4. *depressa.* Body creamcolour ; mantle fringed with yellowish-white or pale drab cirri ; foot light orangecolour. Shell usually much depressed, and more oblong than the ordinary form ; ribs finer but sharp ; beak nearer to the anterior end ; inside porcellanous, with a pale orange head-scar or spatula. *P. depressa*, Pennant, Brit. Zool. iv. p. 142, tab. lxxxix. f. 146 (not *P. depressa* of Gmelin) ; *P. athletica*, F. & H. ii. p. 425, pl. lxi. f. 7, 8.

Var. 5. *cærulea.* Shell depressed, roundish-oval; ribs more delicate, and less regular : *inside* dark blue. *P. cærulea*, Linné, S. N. p. 1259.

HABITAT : Between tide-marks, on rocks and stones, everywhere, most plentiful. The first variety occurs in North Devon, the opposite coasts of South Wales, and at Sark. The second is not uncommon in the Channel Isles. The third is also from the Channel Isles, where the late Dr. Knapp first noticed it ; and Mr. J. D. Humphreys found it at Cork. The fourth frequents rocks only at low water; it is the *P. Tarentina* of Lamarck, *P. Bonnardi* of Payraudeau, and *P. athletica* of Bean. It is in most cases easy to separate this from the ordinary form ; but the variety *intermedia* connects the two, and I cannot find a single permanent character which will serve to distinguish any of them. The colour of the animal is of every hue and shade ; nor is the shell less variable, taking into account the shape, height, position of the apex, sculpture, and inside lining. I once considered myself an adept at picking out the variety *depressa* (or "China limpet," as it has been called) by merely seeing the outside; but I have since failed, and a recent examination and comparison of a great many living individuals of each form has quite convinced me that they are not separate species. The fifth variety inhabits flat stones and slabs of rock at low water, often in places where streams empty themselves into the sea; in its younger state it is the *P. aspera* of Philippi. The common limpet is fossil in raised beaches, including that near Macclesfield at a height of 500–600 ft. (Darbishire), Moel Tryfaen, 1300–1400 ft. (Capt. Lowe), Fort William 10 ft.(J.G.J.), and the Red Crag (S. Wood); Uddevalla (Hisinger and Malm); newer glacial formation near Christiania,120 ft. (Sars); Palermo (Philippi). Its distribution in a recent state comprises every coast between the Loffoden Isles (Sars) and the Ægean (Forbes); and Weinkauff has enumerated it as an Algerian species. The variety *intermedia* has been found in Brittany by Cailliaud, and in Spain by M'Andrew. Philippi noticed the variety *depressa* as fossil in an ossiferous cavern at Mardolce, in Sicily; and it inhabits the shores of France, Spain, Italy, Greece, and North Africa.

The limpet appears to have formed a considerable part of the food of the primitive inhabitants of North Britain, where heaps of their shells are continually being turned up. In the ruins of a so-called Pictish fort near Lerwick the shells are partially calcined; and those of the common periwinkle, which are also found there, must have been subjected to the action of fire in order to extract the animals. Roasted limpets are capital eating. A few years ago I was a guest at a dinner-party in the little island of Herm. The hour was unfashionable, one o'clock; and the meal was served on the turf in the open air. This consisted of fine limpets, laid in their usual position, and cooked by being covered with a heap of straw, which had been set on fire about twenty minutes before dinner; there was also bread and butter. The company were a farmer, two labourers, a sheep-dog, the late Dr. Lukis, and myself. We squatted round the smouldering heap, and left on the board a couple of hundred empty shells. The limpet used to be eaten by the Faroese; and in Ireland and the north of England the consumption was prodigious between twenty and thirty years ago, according to the accounts furnished by Mr. Patterson and the late Dr. Johnston. The former estimated that 11¼ tons of boiled limpets were sold in one season about Larne, co. Antrim; and the latter states that nearly twelve millions had been collected yearly on the coast of Berwickshire, until the supply was almost exhausted. These quantities were exclusive of what were collected to feed the pigs and poultry. The Shetlanders are either more fastidious, or prefer real fish; they will not even eat an oyster. Some of the Orkneymen seem to be imbued with a similar prejudice; for we find in the life of Sir Walter Scott, that "the inhabitants of the rest of the Orcades despise those of Swona for eating limpets, as being the last of human meannesses." The limpet is not omitted in the old pharmacopœia; and Rondeletius prescribes it eaten raw as a gentle purgative. It is a most taking bait for coal-fish. In Shetland it is chewed by the fishermen, and spat into the sea to attract a shoal; this they call "sowing." The yellow or "ware-limpet" (var. *depressa*) is preferred by them as bait; but according to Bean and Alder it is rejected by the fishermen in the north of England. Sea-fowl of all kinds are also fond of the limpet. The bill of the oyster-catcher is said to be admirably adapted for forcing it from the rock; and the pious Derham tells us that "the Author of Nature seems to have framed it purely for that use." Something must now be said for the limpet itself, as well as about its persecutors. It appears from the experiments of Beudant to have an unusual capability of living in fresh water. This may be accounted for by its littoral habit, which exposes it to rain and the efflux of streams into the sea, as well as to the continual percolation of fresh water which takes place on shingly beaches. The animal is occasionally monstrous. Fischer noticed a limpet on the French coast which had the left tentacle branched or double, with two eyes at its base. The shell is as much entitled to the name *polymorpha* as to that of *vulgata*. In the 'Zoologist' for October 1860 will be found an excellent remark by Mr. Norman, as to the variation of its form resulting from habitat; and I cannot do better than give it in his own words. "It will be found to be a general rule with regard to the limpet, that the nearer high-water mark the shell is taken, the higher-spired, more strongly ribbed, and smaller it will be; and that the lower down it lives, the flatter, less ribbed, and larger it becomes." In the intermediate space, and under local conditions, other forms of course occur, which partake of some of the above characteristics in a modified degree. Specimens which I found in a particular spot at Lerwick were excessively thin, and as if they were exfoliated, probably owing to a deficiency of calcareous material. One shell from Balta Sound is of an extraordinary thickness and weight: it had been inhabited by a colony of the burrowing cirriped, *Alcippe lampas*; and the poor limpet must have spent much of its time, as well as all its substance, in adding layer after layer to provide a roomy lodging for its troublesome parasites. In some specimens the crown is depressed, the rest of the cone being considerably raised. The inside of old shells is often garnished with irregular pearly excrescences. My largest specimens were taken at Lulworth, and on Uyea Island; they measure 2½ inches by 2¼. The variety *depressa* is very pretty, and especially when the interior is streaked with violet-brown rays on a porcelain ground. So is the variety *picta*. In Da Costa's time such shells were called by the English "Auriculas," and by the French "Soucis" or marigolds, from their resemblance to those flowers. The spire of the very young shell is slightly twisted on one side, with an inclination to the posterior or broader end; it has one whorl and a half. The tongue is rather longer than the shell; and, according to Forbes and Hanley, it is armed with 160 transverse bands of teeth, each band having 12 teeth, or 1920 in all. Mr. Spence Bate has examined the lingual ribbon in the variety *depressa*. This is broader and shorter than in the common kind, but offers no other distinction than that the teeth are perhaps somewhat larger.

It is the *P. vulgaris* of Belon, Petiver, Da Costa, Landt, and others. Gmelin divided it into a great many species, chiefly from the descriptions of Schröter. The local names are innumerable. De Montfort reckons no less than fifty-two. I will cite a very few only—"flither" of the English, "flia" of the Faroese, "flic" of the Normans, "œil de bouc" of the French, and "lapa" of the Portuguese.

Genus II. HEL'CION *, De Montfort. Pl. V. f. 4.

Body convex: *mantle* fringed at its edge with cirri, which are alternately long and short: *tentacles* rather long: *eyes* prominent: *tongue* shorter than in *Patella*: *gills* not so numerous as in that genus, and forming a shorter plume, which is interrupted over the head: *foot* thick, of a cellular texture.

Shell cap-shaped, having in its embryonic state a slightly twisted apex; crown never prominent, but inflected towards the anterior end, and placed near the margin—especially in the young, where it is almost terminal: *scars* slight and indistinct.

Besides the differences in the arrangement of the pallial cirri, and in the shorter branchial plume, the shell may always be known from that of *Patella* by its shape being semioval instead of resembling a peaked hat; the crown is incurved, and the apex nearly terminal in the present genus. The fry is not spiral, as in *Patella*. The shell of *Helcion* is also usually a thinner shell, with an opalescent hue; and in the only species that we possess the surface is smooth, or never distinctly ribbed. *Helcion* is found on Laminariæ and sea-weeds of a similar kind, which constitute its food; and it is therefore sublittoral in its habits. The species are few, but have an extensive range, including Europe, West and South Africa, Cape Horn, and Australia.

It is the genus *Nacella* of Schumacher, *Patina* of Leach, and partly *Calyptra* of Klein.

Helcion pellu'cidum †, Linné.

Patella pellucida, Linn. S. N. p. 1260; F. & H. ii. p. 429, pl. lxi. f. 3, 4, and (animal) pl. A A. f. 1.

Body creamcolour, with a slight tinge of brown in front: *mantle* often bordered by a grey or leadcoloured line, and fringed with from 30 to 65 fine white cirri, half of which are more than twice as long as the intermediate ones: *head* trans-

---

* A breast-collar. † Transparent.

versely triangular: *mouth* minutely scalloped or puckered: *tentacles* slender: *eyes* small: *gills* of a whitish colour: *foot* oval, equally broad at both ends; sole yellowish-white, edged with a narrow brown line.

SHELL resembling the "cap of liberty," convex, semitransparent, and glossy: *sculpture*, sometimes very slight and indistinct angular lines, which radiate on all sides from the beak, and vary in number and regularity; the surface, however, is frequently quite smooth; it is covered with numerous and close-set microscopical concentric striæ, which in old shells are raised into distinct marks of growth: *colour* yellowish-brown of different shades passing into horncolour, and adorned with from 25 to 40 narrow bright blue streaks on a brown ground: these streaks radiate from the beak, and are more or less interrupted; crown marked with a dusky spot and occasionally with a short linear ray of the same colour: *beak* sunk below the level of the crown and inconspicuous: *mouth* oval: *margin* compressed at the sides, even and smooth: *inside* shining and polished, as if highly glazed, opalescent or lilac in adult shells. L. 0·8. B. 0·6.

Var. *lævis*. Shell more or less solid, opaque, compressed, and expanded. *Patella lævis*, Pennant, Brit. Zool. iv. p. 144, pl. xc. f. 151.

HABITAT: On *Laminariæ* at low water (and as deep as 15 f., Forbes and M'Andrew), on all our coasts; the young are sometimes also found on the under side of large stones which are uncovered at spring tides. Mr. James Smith has enumerated it as fossil from Dalmuir, Ayr, Banff, and Ireland; and Sars, from the newer glacial formation near Christiania, at a height of 100 ft. Living in the Faroe Isles (Mörch); North Cape (Danielssen); Scandinavia, from the shore to 20 f. (Brander, Lovén, and others); Heligoland (Frey and Leuckart); coasts of France (De Gerville and others); Vigo, 8 f., and Cascaes Bay, 15–20 f. (M'Andrew); Mediterranean (Linné); Sicily (Maravigna); Mogador, littoral to 3 f. (M'Andrew and R. T. Lowe). The variety seems to have an equally wide distribution, although Lovén says that he had not met with it.

Lister figured both forms of the "blue-rayed limpet," or "peacock's feathers." The young attach themselves to the upper side of the fronds of the smooth tangle (*Laminaria saccharina*), and sometimes of *L. digitata*, (according to Macé, *Halymenia palmata* also,) which supply them with succulent and abundant pasturage: when it grows older, it attacks the stalks, and afterwards gets to the base of the plant, into which it eats its way until it becomes almost buried in a cup-shaped cavity; it is then fat and lazy. The best way of procuring such last mentioned specimens is to tear up by its roots the large tangle, which girdles the rocks at low water, and waves forwards and backwards like a field of ripe corn in a summer breeze. As, however, it is not an easy matter for a lady collector to do this, she may avail herself of the next storm, and hunt for the pretty prize among the sea-weeds thrown up on the beach. This remarkable habitat was first noticed by M. le Gentil, in the 'Mémoires de l'Académie' for 1788. If it had been known to English naturalists, so many of them would not have persisted in considering the ordinary form on the leaves and the variety imbedded in the roots as different species. The crown is the same in each. The animal crawls with an undulating motion. Some individuals, which I observed in a glass vessel of sea-water, now and then protruded their jaws and the front of their tongues, apparently for the purpose of cleaning their teeth; and after doing this, they ejected from the mouth a thick fluid of a brownish colour—possibly the scrapings of the lingual ribbon. The beak is almost terminal in young shells. Specimens taken from the stalks of *Laminaria* at Dover and in North Wales are fully an inch long, although very convex, thin, and beautiful. They evidently would never have assumed the shape of the variety.

It is the *Patella intorta* of Pennant, *P. minor* or Wallace, *P. cæruleata* of Da Costa, *P. cærulea* of Montagu (but not of Linné), and *P. cornea* of Michaud. The very young is Montagu's *P. bimaculata*. Couch's shell of the last name was apparently a simple Ascidian, perhaps a species of *Cynthia*.

H. pectinatum (*Patella pectinata*, Linné) was wrongly admitted into British catalogues on the authority of Laskey. Linné gives as its habitat the Mediterranean; Payraudeau, Corsica; and R. T. Lowe, Mogador and Senegal.

Genus III. TECTU′RA*, (*Tecture*) Cuvier. Pl. V. f. 5.

BODY more or less depressed: *mantle* fringed at or near its edge: *tentacles* variable in length: *eyes* prominent, wanting in some species: *gills* forming a short plume, which is free, and contained in a cavity over the neck on the right-hand side of the head; it is extensile, and sometimes protruded beyond the shell: *foot* of moderate thickness.

SHELL conical, usually depressed, furnished with striæ which radiate from the crown, having in its embryonic state a curved or semispiral apex; crown not prominent, but projecting horizontally, and placed near the front margin: *pallial scar* nearly marginal.

I do not see any reason for placing this genus in a separate family from that which includes the last two genera. The difference in the length of the branchial apparatus, on which so much stress has been laid by some conchologists, is comparatively unimportant. In each of these three genera the gills compose a single row or plume, which is elongated and attached throughout in *Patella*, and less so in *Helcion*; while in *Tectura*

* A covering over.

it is short and free, except at the base. Lovén, who is certainly not inferior to any one in his knowledge of the organization of the Mollusca, reunites all in the old genus *Patella*. Certain species are eyeless; but the genera *Eulima, Mangelia, Cylichna,* and *Amphisphyra* offer analogous cases of such a deficiency of the so-called visual organs.

The name *Tectura* has the precedence of *Acmæa* (Eschscholtz) by three years. It was originally *Tecture*; and although the termination is not Latin, I am inclined to adopt it as now spelt, in justice to Audouin and Milne-Edwards, the distinguished French zoologists, who first indicated the genus, as well as to Cuvier, who afterwards named it and defined the characters in his report to the Academy of Sciences in 1830 on their account of the Invertebrata of the French coasts. The name *Acmæa*, besides, is objectionable, being derived from an adjective. Quoy and Gaimard called this genus *Patelloidea*, and Gray *Lottia*. Forbes proposed to form another genus, with the name of *Iothia* (afterwards changed by him and Hanley to *Pilidium*), for one of the species. The *Tecturæ* inhabit both the Atlantic and Pacific Oceans; and some have been found in the newer tertiary strata. Their bathymetrical range is extensive. The littoral species have eyes, while those living in deep water have none.

1. TECTURA TES'TUDINA'LIS*, Müller.

Patella testudinalis, Müll. Prodr. Zool. Dan. p. 237. *Acmæa testudinalis*, F. & H. ii. p. 434, pl. lxii. f. 8, 9, and (animal) pl. A A. f. 2.

BODY white: *mantle* covered with vibratile cilia; margin fringed with minute white cirri: *head* large, somewhat rounded and convex: *tentacles* awl-shaped, long, and slender: *eyes* small: *gills* whitish, lanceolate, and ciliated: *foot* oval and very broad, with plain and nearly level sides.

* Like tortoiseshell.

SHELL depressed, rather thin, opaque, and devoid of lustre: *sculpture*, numerous very fine and sharp longitudinal striæ, which radiate from the beak; they are not visible by the naked eye; the surface is also covered with close-set and almost microscopical concentric striæ, varied by occasional lines of growth that are more conspicuous: *colour* greyish with dark reddish-brown longitudinal streaks, which often are confluent and forked, giving a tessellated or clouded appearance like that of tortoiseshell; sometimes the colour is reddish-brown varied by broad rays or spots of white: *beak* rather sharp, placed usually about one-third nearer the anterior end: *mouth* roundish-oval: *margin* expanded, even and smooth: *inside* shining and polished, except at the margin, chocolate-colour in the centre or dorsal scar, porcelain-white and highly glazed in the middle, and of a dull hue at the margin, which is rather broad and bevelled to a sharp edge. L. 0·85. B. 0·7.

HABITAT: On the under side of stones, at low water and as deep as 20 f. in the laminarian zone, Shetland Isles (Barlee); Orkneys (M'Andrew and Thomas); Caithness (Peach); Sutherlandshire (J. G. J.); Aberdeenshire (Macgillivray); Moray Firth (Gordon and others); west of Scotland (Brown and others); Belfast (Hyndman); Lough Strangford (Dickie); Bangor, co. Down (Clealand); Dublin Bay (Lloyd and others); Isle of Man (Forbes); Berwick Bay (Howse); Northumberland and Durham, as far south as Hartlepool (Hancock and others), and living in 40 f. (Alder). It is also common and widely distributed throughout the Arctic and North Seas from Greenland to Iceland, and from Nova Zembla to the South of Sweden, as well as Canada and the north-eastern coasts of the United States.

Forbes noticed the migratory habit of this remarkable species, in his account of a shell-bank in the Irish Sea; and the Tyneside Naturalists' Field-Club have given some curious details of its southward march. Specimens collected by Captain Bedford at Oban are nearly

1¼ inch long, and one found by Mr. Macdonald in the Moray Firth is a trifle longer; in North America it even exceeds these dimensions. Dr. Wallich procured bright-coloured specimens at depths of from 50-150 f. off Godhaab in East Greenland.

A less distinctly striated form is the *Patella testudinaria* of Müller, although not of Linné; and the young is the *P. tessulata* or *tessellata* of the first-named author. It is likewise the *P. Clealandi* of Sowerby, *P. clypeus* of Brown, *P. amœna* of Say, and *P. Clealandiana* of Leach. The *P. Clealandi* of Couch, from Gorran, in Cornwall, appears to have been only a white variety of *T. virginea*.

2. T. VIRGI'NEA*, Müller.

Patella virginea, Müll. Prodr. Z. D. p. 237. *Acmæa virginea*, F. & H. ii. p. 437, pl. lxi. f. 1, 2.

BODY milk-white or pale yellowish-white, faintly suffused with pink : *mantle* thick, fringed with unequal filaments a little within the margin, where it is banded with pink at intervals corresponding to the coloured rays on the shell : *head* having a rosecolour tinge, very short, broad, and semicircular, furnished with a lappet on each side : *tentacles* rather long, contractile, and ciliated : *eyes* small : *gill-plume* falciform, of the palest drab, coarsely pectinated, also contractile and ciliated : *foot* roundish-oval, smooth, delicately veined with white.

SHELL most commonly depressed, more or less solid, according to habitat (specimens from the laminarian zone being thinner than those found between tide-marks), opaque, somewhat glossy : *sculpture*, numerous fine thread-like striæ which radiate from the beak ; these, however, are often indistinct and apparently wanting ; concentric striæ and marks of growth as in *T. testudinalis* : *colour* yellowish-white, with a pinkish tinge and from 16 to 20 pink or brownish longitudinal rays, which are rather broad, and are occasionally interrupted or spotted with white, so as to give an appearance of coloured

* Maidenly, or graceful.

chainwork: *beak* rather sharp, placed near the anterior end, which it sometimes overhangs: *mouth* usually more round than oval, but variable in this respect: *margin* even and smooth: *inside* highly polished, porcelain-white or pinkish, and frequently exhibiting in the young near the crown two of the outside rays, which are darker than the others and assume the shape of a reversed V; margin rather broad, and bevelled to a sharp edge: *pallial scar* marked on the inner line with a row of several white dots, that probably correspond with the fringe of cirri on the mantle. L. 0·4. B. 0·3.

Var. 1. *conica*. Shell much smaller, more conical, and higher, with the crown nearly central.

Var. 2. *lactea*. Shell milk-white.

HABITAT: Common on shells and stones in the laminarian zone, and occasionally at low water, throughout the British Isles. Var. 1. occurs in deeper water. Var. 2. Scarborough (Bean). Fossil in the Scotch and Irish newer pliocene beds (J. Smith, Forbes, Jamieson, Crosskey, and J. G. J.); Uddevalla (Malm); Christiania, 120–200 ft. (Sars); Calabria and Tarento (Philippi); Red Crag (S. Wood). It is found living in every part of the North Atlantic, from Iceland (Torell) to the Canary Isles (M'Andrew) and Azores (Drouet), as well as on both sides of the Mediterranean, and in the Ægean (Forbes); perhaps also at Sitka Island as *Patella pileolus* or *P. Asmi* of Middendorff. The range of depth varies in these foreign localities from 3 to 60 f.

Specimens taken by Mr. Jordan on the shore at Guernsey are larger and thicker than any other which I have seen; their diameter exceeds half an inch. The apex of the fry is white, and has an incomplete whorl.

The little pink-rayed limpet has had many hard names given to it, besides those of Middendorff. It is the *Patella minima* of Gmelin (from Schröter), *P. parva* of Da Costa, *P. æqualis* of J. Sowerby, *Ancylus Gussonii* of Costa, *Patelloides vitrea* of Cantraine, *Patella pellucida* of Philippi, and *P. pulchella* of Forbes.

3. T. FULVA *, Müller.

Patella fulva. Müll. Prodr. Z. D. p. 237. *Pilidium fulvum*, F. & H. ii. p. 441, pl. lxii. f. 6, 7, and (animal) pl. A A. f. 3.

BODY whitish: *mantle* fringed at the margin with fine short transparent cilia: *head* prominent, furnished beneath with two triangular lappets, one on each side: *tentacles* conical and short, not bearing any tubercle or eye-stalk: *eyes* none: *foot* oval, thick, occupying a space equal to about two-thirds of the mouth or base of the shell.

SHELL cap-shaped or semioval, rather thin, semitransparent, lustreless: *sculpture*, numerous fine and sharpish ribs which radiate from the beak; their crests are minutely beaded; the concentric striæ are as close-set as in other species, but are much stronger and somewhat imbricated; the shell appears under the microscope to be permeated by exceedingly fine longitudinal lines: *colour* orange, bright reddish-brown, or yellow, sometimes diversified by white rays of various widths: *beak* sharp, placed very near the anterior end: *mouth* oval: *margin* very thin, slightly scalloped by the ribs: *inside* highly glossy, coloured like the outside: *central scar* forming a semicircular lobe in front and an oval one behind: *pallial scar* too faint to be perceptible. L. 0·25. B. 0·185.

Var. 1. *albula.* Shell white.

Var. 2. *expansa.* Shell larger, more depressed, and broader in proportion to the length.

HABITAT: Common on stony ground in 10–40 f., in many parts of the west of Scotland; Moray Firth (Dawson); twenty miles off Kinnaird's Head, Aberdeenshire, in 30 f. (Thomas); Shetland, 40–90 f. (M'Andrew and others); Cork Harbour, on *Pinna rudis* (Humphreys); Youghal, with *Crania anomala* (Miss M. Ball); off Cape Clear and Mizen Head in 50–60 f. (M'Andrew); west of Ireland, 100 f. (Hoskyns, *fide* King). The 1st variety is found occasionally with

* Deep yellow.

coloured specimens; and the 2nd is Zetlandic. I am not aware that *T. fulva* has been found fossil in this country. Mr. Searles Wood's specimens from the Coralline Crag, which he described as this species, appear to belong to *Lepeta cæca*. Sars has enumerated it among the shells from the older and newer glacial formations near Christiania; in the former at a height of 400–440 ft., and in the latter at 120 ft. Its foreign habitat is entirely Scandinavian, and comprises all that coast between Finmark and Bohuslän, its bathymetrical range extending from 10 to 160 f.

The animal swims or floats in an inverted posture, but slowly. Its lingual apparatus does not seem to differ materially from that of *T. testudinalis* or *T. virginea*. In all of them the central teeth are square, and the laterals elongated and hooked. *T. fulva*, however, has but a single row, while each of the other two species has a double row of central teeth. The shell is often encrusted by zoophytes and sessile Foraminifera. It differs from *T. virginea* in texture, colour, and sculpture. It is never so large as that shell; my finest specimen is $\frac{4}{10}$ths of an inch long.

This is the *Patella Forbesii* of James Smith.

Genus IV. LE'PETA *, Gray.   Pl. V. f. 6.

BODY depressed: *mantle* thick-edged: *tentacles* short: *eyes* wanting: *foot* thin.

SHELL conical, somewhat depressed, furnished with tuberculated striæ, which radiate from the crown, and are crossed by concentric lines: *beak* in the embryonic state curved, and always inclining towards the rear; crown nearly central: *pallial scar* placed within the margin.

Indicated and named by Gray; defined by H. and

---

* Possibly derived from Lepas, the ancient name of the limpet.

A. Adams. It depends on conchological characters. The apex of the shell turns backwards, instead of forwards or towards the head, which latter is the case with *Tectura* and the other preceding genera of the same family. The animal is blind, an infirmity that it shares with *T. fulva* and the succeeding genus *Propilidium*.

LEPETA CÆCA*, Müller.

Patella cæca, Müll. Prod. Z. D. p 237.

BODY whitish: *tentacles* setose: *foot* large: *liver* green (Müller and Stimpson).

SHELL having an oval outline, moderately solid, opaque, slightly glossy: *sculpture*, very numerous and close-set fine striæ, which radiate from the beak, and are crossed by slighter concentric and imbricated striæ, the intersection of which causes the longitudinal striæ to be granular or nodulous, especially towards the margin; marks of growth distinct: *colour* milk-white: *beak* blunt, much worn in full-grown specimens: *mouth* oval: *margin* thin and even, minutely tuberculated in immature specimens: *inside* porcelain-white, and partly iridescent: *central scar* large and conspicuous: *pallial scar* rather broad and glossy, placed between the central scar and the margin. L. 0·5. B. 0·35.

HABITAT: Off Unst, in Shetland, at a depth of from 80 to 90 f.,—Mr. Dawson having found a fine and fresh but somewhat broken specimen in sand which I dredged there last summer. I should not be so well satisfied of this evidence that it is British, if it had not been confirmed by my discovering a smaller specimen (*having the dried remains of the animal in it*) among some of *Tectura fulva* which Mr. Barlee dredged on the west coast of Scotland in 1846. He was never, as I believe, acquainted with this species, nor had any shells from Scandinavia, where it is rather common. I may

* Blind.

also mention that Mr. Dawson dredged several specimens in the Moray Firth, but in apparently a semi-fossilized state. *L. cæca* occurs in the Red and Coralline Crag; Uddevalla (Lyell); Christiania district, in the older portion of the post-pliocene or glacial formation, 400–440 ft. (Sars); Antwerp Crag (coll. Nyst). It inhabits the Arctic and North Seas, from Spitzbergen (Goodsir) and sea of Okhotsk (Middendorff) to Gottenburgh (Malm), as well as the eastern and western coasts of North America (Couthouy, Stimpson, Bell, and P. Carpenter); the depths in these localities vary from 20 to 100 f.

According to Lovén its dentition agrees with that of *Tectura fulva*.

The present species is the *P. candida* of Couthouy and *P. cerea* of Möller.

Genus V. PRO'PILI'DIUM *, Forbes and Hanley.
Pl. VI. f. 1.

BODY compressed: *mantle* finely ciliated at its edge: *tentacles* rather long and slender: *eyes* wanting: *gills*, according to Forbes and Hanley, apparently forming two short triangular plumes, which are furnished with large cilia: *foot* thick.

SHELL conical and much raised, cancellated, having in all states of growth a minute spiral apex, which is inflected towards the rear; crown central: *inside* furnished in the middle with a shelf-like triangular plate, which covers about one-half of the crown: *central scar* indistinct: *pallial scar* situate within the margin.

A singular genus, agreeing with *Lepeta* in the retrogressive inclination of the beak, but differing from that and every other genus of the *Patella* family in always

* From its affinity to the genus *Pilidium* proposed by the same authors.

having a distinctly spiral apex and a plate or septum inside the crown. The use of this last-mentioned process is not known. It is too small to contain or support the viscera, as in *Calyptræa* and allied genera; but it may be homologous with the internal process of *Puncturella*.

PROPILIDIUM ANCYLOI'DE *, Forbes.

Patella? Ancyloides, Forbes, in Ann. Nat. Hist. v. p. 108, pl. ii. f. 16.
Propilidium Ancyloide, F. & H. ii. p. 443, pl. lxii. f. 3, 5, and (animal) pl. A A. f. 4.

BODY whitish with a faint tinge of yellow: *mantle* fringed at its edge with close-set but distinct cilia, which correspond with the striæ of the shell: *head* semicircular, margined with light brown: *mantle* scalloped or puckered: *tentacles* tapering to a fine point, delicately ciliated, destitute of eye-stalks: *foot* oval, broader in front than behind.

SHELL having an oval outline, compressed at the sides, rather thin, semitransparent, glossy at the apex, but elsewhere of a dull hue: *sculpture*, very numerous and close-set fine striæ, which radiate from the beak and are exquisitely granulated in consequence of their being intersected or decussated by equal-sized concentric striæ: *colour* dirty white, occasionally diversified by a few clear longitudinal rays or lines: *beak* smooth and highly polished, styliform and slender, pinched up into a minute spire of between one and two whorls, which curls downwards at the posterior end: *mouth* oval, of nearly the same breadth throughout: *margin* thin and even, minutely tuberculated in immature specimens: *inside* nacreous, furnished in the centre with a thin laminar partition, like the half deck of a vessel, which has its opening towards the head or anterior part; *pallial scar* broad. L. 0·15. B. 0·115.

HABITAT: Not uncommon on stones and among nullipores, in co. Galway (Barlee); Strangford Lough on oysters, and on the Antrim Coast in 18–100 f. (Hyndman and others); Ballantrae, Ayrshire (Getty); Lamlash Bay (Smith, Forbes, and others); Oban, 20 f.

* Having the aspect of an *Ancylus*.

(J. G. J.) ; Mull and Skye, 30-90 f. (Forbes, M'Andrew, and Barlee) ; Moray Firth (Dawson) ; Shetland, 75-80 f. (Barlee and J. G. J.). It has not yet been noticed as fossil; and the only foreign locality is the coast of Sweden, at a depth of only 12 f., with *Mytilus Adriaticus* and *Branchiostoma lanceolatum* (Malm).

The animal is active for its size. Forbes and Hanley remarked that the tongue is very long, and the brown central spines conspicuous under the microscope, resembling bramble-thorns in miniature.

It was named by the late Mr. W. Thompson "*Patella? exigua*, Forbes."

Family II. FISSURE'LLIDÆ, (*Fissurelladæ*) Fleming.

Body conical or semioval: *mantle* folded in front, so as to form a tubular process, which occupies a slit in the margin or near the summit at that end of the shell or else a hole in the crown: *head* prominent, with a short muzzle, furnished (as in the *Patellidæ*) with jaws and a spinous tongue, which latter is shorter than in that family and scarcely convoluted: *tentacles* spike-shaped: *eyes* seated on short tubercles, one at the outer base of each tentacle: *gills* forming two symmetrical and somewhat triangular plumes, one on each side of the neck: *foot* thick, studded at the upper side or covered entirely with papillæ: *vent* anterior, placed in the middle between the gill-plumes.

Shell cap-shaped or ovately conical, with a slit in front or near the crown on that side, or else a hole in the centre ; it is ribbed lengthwise and often cancellated by concentric or transverse striæ: *beak* turned towards the hinder part, where it forms a short and complete excentric spire, always in the young and mostly in the adult: *mouth* extremely wide and occupying the entire base.

The fissure or perforation of the shell indicates a corresponding formation of the animal, a fact which to this extent enables us to dispense with the so-called science of malacology. The fewer technical words that are used, the more easy it will be for students to learn the language of this or any other branch of natural history. The tubular process of the mantle apparently serves for the admission of aërated water to the gills, as in the Siphonobranchiata; it has been also, but without reason, supposed to have a fæcal office. The outer layer of the shell is laminated, the middle one cellular, and the inner nacreous. None of the *Fissurellidæ* can properly be called littoral, although some of them are occasionally found under stones at low-water mark. They are spread over all parts of the world.

Genus I. PUNCTURELLA *, R. T. Lowe.
Pl. VI. f. 2.

BODY conical: *mantle* protruded through a slit near the top of the shell on the anterior side, outside of which it forms a short tubular process: *foot* crested with a row of papillæ.

SHELL cap-shaped, with a slit in front of the crown: *beak* always spiral: *inside* furnished with a short funnel-shaped process having its exit in the hole abovementioned.

The name *Cemoria*, proposed by Dr. Leach, was not published before Mr. Lowe described the present genus; the type of the first-named genus is the fry of *Fissurella Græca*. The *Cemoria* of Risso (from Leach's MS.) is a fossil, and apparently a species of *Calyptræa*. Some conchologists have associated Defrance's genus *Rimula* with that of which we are now treating: the latter has an internal process, and the perforation is placed close to the crown; while the other has no such process, and the perforation is placed midway between the crown and the posterior margin. *Rimula* bears the same relation to

---

* Having a small prick or puncture in the shell.

Emarginula as *Schismope* does to *Scissurella*. Other synonyms of *Puncturella* are *Diodora*, Gray (according to De Blainville), and *Sipho*, Brown.

PUNCTURELLA NOACHI'NA *, Linné.

Patella Noachina, Linn. Mant. Plant. p. 551. *Puncturella Noachina*, F. & H. ii. p. 474, pl. lxii. f. 10-12, and (animal) pl. B B. f. 4-6.

BODY milk-white with a faint tinge of brown : *mantle* thick ; tubular fold conical and short, furnished with six small papillæ in front and four behind : *head* large and broad, bilobed : *tentacles* conical and pointed, rather short, greatly diverging, and ciliated : *eyes* large and prominent : *foot* oval, broader and somewhat truncated in front, bluntly pointed behind ; upper side forming a ridge, which is studded with short white conical points (ten on each side) corresponding with the ribs of the shell ; those in front, especially the penultimate ones near the tail, are larger than the rest.

SHELL more or less raised, slightly compressed at the sides, rather thin, semitransparent, not glossy : *sculpture*, 25–30 sharpish but not much elevated ribs, which radiate from the beak to the margin, and as many smaller intermediate ones ; the surface is also covered with microscopical and close-set longitudinal striæ, and with minute white and glistening dots, which are arranged lengthwise in rows, and seem to indicate an internal tubular structure : lines of growth slight and irregular : *colour* whitish : *beak* small, ribless, incurved, and twisted to the left, forming a spire of one whorl and a half: *slit* lanceolate, extending from the crown some distance down the front, and passing obliquely in that direction : *mouth* oval, somewhat broader behind : *margin* thin, scalloped or indented by the ribs : *inside* nacreous, marked with fine concentric lines ; from the centre or crown towards the front runs a rather large vaulted sheath, occupying more than one-fourth of that side ; it covers the slit, which is continued in front of the sheath in the form of a narrow groove with thickened sides, nearly to the margin ; the sheath is strengthened at each side by a rather solid buttress. L. 0·4. B. 0·3.

Var. *princeps*. Shell higher and much narrower from being pinched up at the sides, with the mouth consequently oblong. *Cemoria princeps*, Mighels and Adams.

* So named (as a fossil) from its supposed diluvian origin.

HABITAT: Hard ground, from 25 to 90 f., in Shetland and the west of Scotland, being rather plentiful in the latter district; Aberdeen (Macgillivray); Northumberland and Durham (Alder, King, and others); Scarborough (Bean); co. Antrim (Hyndman, Waller, and J. G. J.). The specimens, however, from the last-mentioned locality are probably relics of the glacial epoch, and not recent. The variety is rare; it occurred in my Shetland deep-water dredgings. *T. Noachina* is tolerably common as a fossil in the Clyde beds; also at Fort William (J. G. J.); Bridlington (S. Wood); Kelsey Hill, Yorkshire (Darbishire); Uddevalla (Hisinger); older glacial formation at Christiania, 400–440 ft. (Sars). It inhabits every part of the sea north of Great Britain, from Gottenburgh (Malm) to Spitzbergen and North Greenland (Torell), at depths of from 4 to 150 f.; Canada (Bell); Maine (Mighels); Massachusetts, frequently in the stomachs of fishes (Gould); New England, 20–30 f. (Stimpson); and the variety has been taken also from the stomachs of fishes caught in 40–75 f., nearly 100 miles seaward from Casco bay.

Fabricius noticed the difficulty of keeping this mollusk alive when taken from its native habitat. In the young shell the slit is almost marginal, but recedes further from the edge in the course of growth.

The synonyms are somewhat numerous, viz. *Patella fissurella*, Müller; *Sipho striata*, Brown; and *Rimula Flemingii*, Macgillivray, who gives the following reason for that cognomen:—"One malacologist has named it after Noah, another after Dr. Fleming. I am unable to determine the priority, and therefore take the living godfather." Leach had called it *Cemoria Flemingiana*. The fry is the *Patella Zetlandica* of Fleming.

Genus II. EMARGINULA *, Lamarck. Pl. VI. f. 3.

BODY conical: *mantle* protruded from the slit in front of the shell, outside which it forms a short tubular process: *foot* studded at the upper side with papillæ: *verge* cirriform, on the right-hand side.

SHELL cap-shaped, with a vertical slit in front, which is partly filled up as the shell increases in size, so as to leave a furrow: *beak* always spiral: *inside* thickened on each side of the slit.

These pretty shells, commonly called "slit-limpets," inhabit Europe, Asia, Africa, America, and Australia. The very young resembles a *Scissurella*; it has a regular Trochoidal spire, and the outer edges of the slit are inflected: the fry has no slit.

1. EMARGINULA FISS'URA†, Linné.

Patella fissura, Linn. S. N. p. 1261. *E. reticulata*, F. & H. ii. p. 477, pl. lxiii. f. 1 (as *E. Mülleri*).

BODY white, sometimes faintly tinged with yellow or light brown: *mantle* open in front; margin finely ciliated; tubular fold forming an entrance into the branchial cavity, fringed outside with minute papillæ: *head* large and broad, usually protruded beyond the foot: *tentacles* contractile and therefore varying in length, flattened at the sides: *eyes* oval, placed on round peduncles one-third of the way up from the outer base of the tentacles: *foot* oval, crested above on each side: round the upper edge of this crest, and near its junction with the rest of the body, is a row of small milk-white papillæ or tentacular processes.

SHELL usually raised, so as to give a height in proportion to the length as 2 to 3, solid, opaque, not glossy: *sculpture*, 25-35 strong but narrow and cord-like ribs, which radiate from the beak to the margin, and as many smaller intermediate ones; sometimes these ribs are equal in size; they are crossed by from 20 to 30 somewhat slighter concentric ribs,

---

* Having a little notch in the margin.   † A cleft.

imparting a regularly and deeply cancellated or punctured appearance, and forming slight nodules at the points of junction; the surface is also covered with microscopical and close-set longitudinal striæ, and in the young may be observed the same white dots that have been described with reference to *Puncturella Noachina*: *colour* white, often more or less stained by extraneous matter: *beak* very small, ribless, incurved and slightly twisted to the left, forming a spire of two whorls: *slit* of equal width, extending from the margin in front about one-third of the way up, where it is closed by a subsequent formation of shell, and becomes as far as the crown a rather deep groove, which is somewhat closely laminated across: *mouth* roundish-oval, distinctly scalloped and notched by the indentation of the longitudinal ribs: *inside* nacreous, finely lineated in a concentric direction, and usually exhibiting the external larger ribs: the sides of the slit are thickened, and the outside groove is represented by a white ridge. L. 0·45. B. 0·35.

Var. 1. *subdepressa*. Somewhat larger, more depressed, and expanded at the sides.

Var. 2. *elata*. Also larger than usual, much higher, and more solid.

Var. 3. *incurva*. Smaller, more raised, and compressed at the sides, with the beak almost overhanging the posterior margin; sometimes of a pinkish colour inside.

HABITAT: Everywhere on shells and stones, from low-water mark at spring tides to 90 f.; off the Mull of Galloway, in 110–145 f. (Beechey). Var. 1. Shetland, in deep water. Var. 2. Fishguard, and Larne near Belfast (J. G. J.). This variety also occurs in the Red Crag; it is nearly as high as long. Var. 3. Oban, Skye, and Shetland (Barlee and J. G. J.). *E. fissura* is fossil in Ireland, according to Mr. James Smith; and it is rather common in the Red and Coralline Crag; Antwerp tertiaries (Nyst); Christiania, in the newer glacial formation, 150–200 ft. (Sars). Living in the North Atlantic, from Finmark and the Faroe Isles to the Canaries (where

Mr. M'Andrew noticed that it decreased in size), at various depths between 1 and 80 f.

Curious old Petiver called this shell the " crack'd Barnstaple Limpet," in consequence of Lister having figured it as found in that place. According to De Gerville it bears the name of " l'entaille " in the north of France. The inside is sometimes greenish or rose-colour, being probably stained by algæ or nullipores.

The first locality given by Linné for *Patella fissura* is England, on Lister's authority; his description answers to the present species, as well as to *E. rosea*. It is the *E. reticulata* of J. Sowerby, who however does not say that it is distinct from *E. fissura*, but gave it a new name because of the then prevalent opinion that no fossil was the same as any recent species: his reflection on the subject is somewhat hazy, though pious. It is also the *E. conica* of Sars (but not of Schumacher), *E. Mülleri* of Forbes, *E. lævis* and *E. fissurata* of Récluz, whose *E. tenuis* appears to be the young.

2. E. RO'SEA*, Bell.

E. rosea, Bell, in Zool. Journ. i. p. 52, pl. 4. f. 1; F. & H. ii. p. 479, pl. lxiii. f. 3.

BODY white: *mantle* not projecting beyond the shell, and having a scalloped margin; it is notched in front to form the tubular fold, which is bordered on each side by an angulated prominent lobe: *tentacles* of moderate length, and stout: *eyes* rather large, placed on distinct, although short, pedicles or stalks: *foot* large, strong, and very steep-sided. At its junction with the rest of the body is a circle of about 20 very short papillary cirri (F. & H.).

SHELL smaller and much narrower than *E. fissura*, and otherwise distinguishable in the following particulars:—it is proportionally broader in front than behind, and pinched up

* Rosecoloured.

at the sides; the front is more arched or convex, and the back more concave; the longitudinal ribs are more closely set, and mostly equal-sized; the cancellation is smaller, and exhibits round holes instead of square lattice-work; the colour is often pinkish: the beak quite overhangs the front margin in full-grown specimens, and it is invariably longer, and greatly incurved; the slit is much shorter; the mouth is smaller; and the inside is frequently reddish-brown, and the cancelli are marked by white spots. L. 0·225. B. 0·185.

HABITAT: Common in 7–25 f. on the coast of Dorset (Bell and others); Exmouth (Clark); Plymouth (J.G.J.); Cornwall (Peach and others); Scilly Isles (Lord Vernon); Channel Isles (Hanley and others). Coralline Crag at Sutton (coll. S. Wood); Palermo (Philippi). Its extra-British distribution in a recent state is entirely southern, but extensive; it embraces the coasts of France, Italy, Algeria, and the Hellespont, at depths varying from 8 to 95 f.

I have taken this living with *E. fissura*; otherwise I should have been disposed to consider it an aberrant form of that species. Mr. Alder lately compared the tongue of *E. rosea* with a drawing which he had previously made of the same organ in *E. fissura*; and he notices the following small points of difference. "The uncini or lateral spines are of three kinds. The large inner one appears to be longer and more produced at the point than in *E. reticulata* [*E. fissura*]; and the spines of the second kind, which are denticulated at the points, are four in number in *E. rosea*, while (if my drawing is correct) there are only three in *E. reticulata* [*E. fissura*]. Their tips appear to be more slender." The present species must be very prolific, judging from the extraordinary number of the ova produced in April; each is enclosed in a cartilaginous case. Specimens of a larger size than usual are only $3\frac{1}{2}$ lines long. Their height often exceeds the breadth. Those from the Mediterranean are much smaller than ours. The late Chevalier Vérany found one in the stomach of a flamingo that was killed in the neighbourhood of Nice.

It is the *E. conica* of Schumacher and Risso; but the description given by the former is generic only, and that by the latter is (as usual) almost enigmatical. Lamarck's *E. rubra* is probably also the same species. The Mediterranean or dwarf form is *E. pileolus*, Michaud, *E. capuliformis*, Philippi, *E. curvirostris*, Deshayes, and *E. Costæ*, Tiberi. We have here a goodly choice of specific names. I would have adopted the first and earliest (*conica*), if any modern conchologist of repute had set the example; it is besides more characteristic and appropriate than *rosea*. Montagu must have known the present shell, but considered it a variety of *E. fissura*; he sent a specimen with the latter specific name to Mr. Dillwyn.

3. E. CRASSA*, J. Sowerby.

E. crassa, Sow. Min. Conch. p. 73, t. 33, upper figures; F. & H. ii. p. 481, pl. lxiii. f. 2, and (animal) pl. C C. f. 2.

BODY white: *mantle* rather thick at its edge: *tentacles* thick and cylindrical: *eyes* apparently smaller in proportion than those of our other species: *foot* having narrow sides, which, at their junction with the rest of the body, are studded with about 30 short somewhat unequal cirri (Alder).

SHELL usually more depressed than that of either of the two former species, moderately solid, opaque, slightly glossy: *sculpture*, 40–50 broad and compressed longitudinal ribs (each of which is sometimes divided into three), with as many smaller intermediate ones; all these ribs are crossed by fine, equally numerous, and wavy concentric striæ or wrinkles, producing a delicately granulated appearance; the surface is likewise covered with minute white glistening dots arranged in longitudinal rows: *colour* white: *beak* small and somewhat

* Solid.

angular, usually less excentric than in the other species; it is twisted a little to the left, and forms a spire of between one and two whorls: *slit* rather narrower above than below, extending (in adult specimens) from the middle of the front margin between one-fourth and one-fifth of the way up, being closed in the line of its previous passage, and becoming a rather broad and shallow groove which is closely laminated transversely: *mouth* varying in shape from oval to roundish-oval, delicately scalloped and notched by the impression of the ribs: *inside* porcelain-white and nacreous, exquisitely and closely but irregularly lineated in a concentric direction; the edges of the slit and groove are thickened. L. 1.25. B. 1.

HABITAT: West coast of Scotland, and Shetland, in 20–75 f. (J. G. J., Barlee, and others); " at Oban it is found alive under loose stones, which are uncovered at the fall of high spring-tides, as well as by dredging; the tide sometimes retreats fourteen feet" (Bedford); co. Antrim, off the Copeland Isles, in 20–60 f. (Hyndman); Dublin coast (Thompson). Red and Coralline Crag (Wood); Opslö, near Christiania (Lyell); Belgian tertiaries (Nyst); Lamato, in Calabria (Philippi). The correctness of this last locality in some measure depends on the probability of *E. crassa* being identical with *E. decussata* of Philippi. Its foreign distribution, as a recent species, is entirely Scandinavian. Lovén, Malm, M'Andrew and Barrett, Asbjörnsen, and Koren have dredged it at different points between Bohuslän and Drontheim, in from 10 to 60 f.

This noble shell is never likely to become common in collections, until some plan is discovered for dredging in rocky ground. The young differs from *E. fissura* of the same size in being more depressed, and in its peculiar sculpture. In that species the ribs are strong, and the surface is coarsely cancellated; in this the ribs are fine and more numerous, and the surface is delicately granulated. The rows of small white dots are always visible in *E. crassa*; and the slit is shorter relatively to the size of the shell.

Genus III. FISSUREL'LA*, Bruguière. Pl. VI. f. 4.

BODY semioval: *mantle* protruded in front through a hole or slit in the crown of the shell: *foot* covered with papillæ.

SHELL ovately conical, perforated on the anterior side of the crown : *beak* spiral in the young only : *inside* thickened around the terminal perforation.

This is one of the genera of mollusks which Cuvier illustrated in his celebrated Mémoires on their anatomy. He considered it to be allied to *Haliotis*. Although the animal in its normal state extends beyond the shell, it can be entirely withdrawn into it, like *Vitrina*. Woodward has well remarked that its organization has certain homological affinities with that of the Lamellibranchiate bivalves, in the number and position of the gills, as well as in the pallial tube. According to Beudant, it is equally incapable with *Capulus* of existing in fresh water. The opening in the summit of the shell resembles a keyhole; in the young it is placed on the anterior side of the beak, which is distinctly spiral at that period of growth. The fry might be mistaken for that of *Puncturella*, if it had also an internal sheath or process. *Fissurella* is represented in all seas, scantily in the North Atlantic, but amply in southern latitudes, whence many fine and gaily painted species have been brought by collectors. The number of genera into which this has been divided by Gray and other English conchologists was noticed by Philippi as one of the curiosities of science.

* Having a small cleft in the shell.

FISSURELLA GRÆCA*, Linné.

Patella græca, Linn. S. N. p. 1262. *F. reticulata*, F. & H. ii. p. 469, pl. lxiii. f. 4, 5, and (animal) pl. B B. f. 7.

BODY creamcolour or yellowish, passing into deep orange: *mantle* ample, extending beyond the sides of the shell, and expanded over the head as a hood or veil; margin fringed with a row of very small and short but stout cirri, which correspond with the longitudinal ribs of the shell: *head* tumid and strong: *tentacles* extensile: *eyes* black and rather small: *gills* very thick, brownish: *foot* yellowish, dilated, with broad sides; the upper part is studded with a row of from 30 to 40 papillæ, which are usually by turns larger and smaller.

SHELL forming a cone of variable height, small and apparently stunted specimens being more raised than younger ones of a regular growth; it is solid, opaque, nearly lustreless: *sculpture*, generally about 25 strong and cord-like, but not much raised, longitudinal ribs, and an equal number of smaller intermediate ones; all these are crossed by about 30 narrower and imbricated concentric ribs, which by the decussation make the crests of the other ribs nodulous or vaulted; the surface of living, and especially immature specimens is covered with microscopical longitudinal striæ; in the fry are observable a few white dots, arranged in lines as in *Puncturella* and *Emarginula*: *colour* pale yellowish-white with a few broad rays of reddish-, greenish-, or dark-brown, which are sometimes intermingled or variegated: *beak* very small, only persistent in the young, inflected and twisted a very little to the left, and forming a spire of between one and two whorls: *slit* oblong, broader above than below, contracted at the outer sides, which project in the middle like the teeth of a saw: *mouth* oval, finely scalloped by the ribs and toothed within; these teeth are often double; when the shell is placed on its base, the outline of the mouth is more or less arched on each side, and resembles the sole of a human foot: *inside* porcellanous, exhibiting the coloured rays in young specimens; it is delicately lineated around the margin, as in the interior of all shells belonging to other genera of the same family: *pallial scar* wide and irregular, having a large central impression analogous to that of *Patella*. L. 1·25. B. 0·75.

* Inhabiting the Archipelago.

HABITAT: South-eastern, southern, and western coasts of England (including the Channel Isles), Bristol Channel, Isle of Man, Anglesea, all around Ireland, and the west of Scotland; not uncommon in oyster beds and on old shells and rocks, from low-water mark to 50 f.: Caithness coast (Gordon); Orkneys (Thomas); and Forbes gives it, in his report to the British Association for 1850, as living at 10 f. in Shetland. It occurs fossil at Moel Tryfaen (Darbishire); Red and Coralline Crag (S. Wood); as well as in the Belgian, French, and Italian tertiaries. South of Great Britain it has a wide distribution in a recent state, as far as the Ægean and Canaries, at depths ranging from the shore-line to 95 f. I am not aware of any northern locality.

Petiver called this the "thimble limpet," possibly from its being open at the top, like a tailor's thimble. The number of longitudinal ribs, and consequent compactness of the cancellation, vary greatly; in a specimen from Guernsey I counted no less than seventy-two of these ribs.

The only habitat assigned by Linné to his *Patella græca* was the Mediterranean. His description, although short, suits our shell; and his references, with the exception of Adanson (and perhaps also of Gualtieri and Regenfuss), are quite appropriate. Our shell is the *P. larva, reticulata* (in the index *P. reticulata*) of Da Costa, *F. cancellata*, Gray (but not of G. B. Sowerby), *F. Europæa*, Sowerby, *F. occitanica*, Récluz, and *F. Listeri*, Woodward; the fry is *P. apertura*, Montagu, *Sipho radiata*, Brown, *F. striata*, Récluz, and *Cemoria Montaguana*, Leach.

I have a worn specimen of *F. nubecula*, Linné, in Turton's collection, which, he states (in his 'Conchological Dictionary'), had been dredged off the Land's

End. Couch gives the same habitat; and Peach noticed this species as found by him at Gorran, in Cornwall; but he appears to have mistaken for it the young of *F. Græca*. Better evidence is wanting of *F nubecula* being British; it is not uncommon in the Mediterranean. This is the *F. nimbosa*, afterwards *F. rosea*, of Philippi (but apparently neither of Lamarck's species bearing these names), and *F. Philippii* of Requien.

Family III. CAPU′LIDÆ, Fleming.

Body conical or cap-shaped: *mantle* entire: *head* snout-like, furnished with jaws and a stout spinous tongue: *tentacles* awl-shaped, widely separated: *eyes* placed on slight bulgings or tubercles, about halfway up the tentacles at their outer bases: *gills* forming a single plume or row of slender elongated leaflets, and seated in a large cavity behind the head: *foot* fleshy and rounded.

Shell cap-shaped and tumid: *epidermis* velvety: *beak* spiral, turned towards the posterior side, curling downwards, and twisted to the left: *mouth* round or transversely oval, with an irregularly sinuated margin.

The beak or apex of the shell is turned to the rear and always spiral, as in the last family; in the *Patellidæ* it is turned to the front, and only spiral or curved in the very young state. In Gray's system the present family and the next are arranged in a different group from that which contains the *Patellidæ*, and the latter family is separated by *Dentalium* from the *Fissurellidæ*.

Genus CA′PULUS*, De Montfort. Pl. VI. f. 5.

Generic characters the same as those of the family.

These mollusks adhere to stones and old shells in the coralline and deep-water zones. They probably never

* A receptacle.

willingly change their places of abode, but subsist on animal or vegetable food brought by marine currents within reach of their extensile snouts. The female carries her egg-cases under the neck in front of the foot until the fry are hatched. According to Lovén the dental apparatus is nearly the same in *C. Hungaricus* and *Calyptræa Chinensis*; so that these genera must be closely allied. But the internal appendage in *Calyptræa* and other genera belonging to the same family indicates a peculiar structure of the animal which is wanting in *Capulus*.

D'Argenville called it Cabochon. It is the genus *Mitra Hungarica* of Klein, and *Pileopsis* of Lamarck. About a dozen other synonyms have been cited by Herrmannsen.

CAPULUS HUNGA'RICUS*, Linné.

Patella ungarica, Linn. S. N. p. 1259. *Pileopsis Hungaricus*, F. & H. p. 459, pl. lx. f. 1, 2 (as *C. Hungaricus*), and (animal) pl. C C. f. 3.

BODY whitish, with a yellowish or brown tint: *mantle* either the same colour as the rest of the body, and thickly covered with minute milk-white specks, or else pinkish-white or red with a border of bright yellow or orange; margin thickened, and fringed with fine filaments: *head* broad and thick, with produced lips so as to make the extremity of the muzzle appear cloven: *tentacles* variable in length, sometimes of a white or yellowish colour: *eyes* small: *foot* bordered in front by a puckered ruff or membrane.

SHELL not unlike a cornucopia, or an ancient fool's or jester's cap, with a roundish base, the height of the cone depending on the comparative dilatation of this latter part; it is rather thin, semitransparent, and of a dull hue beneath the epidermis: *sculpture*, numerous fine ribs which radiate from the beak towards the margin, near which they almost disappear, besides very slight and close-set minute transverse striæ between the ribs; marks of growth conspicuous but irregular:

* Hungarian.

colour varying from pale yellowish-white to dull reddish-brown, rarely milk-white: *epidermis* arranged in concentric layers, which are often fringed by a row of leaf-like or triangular points; it is easily rubbed off, and seldom remains on the upper part: *beak* in adult specimens overhanging the posterior side, and gradually becoming spiral with from two to three whorls, which are placed sideways, and separated by a distinct and rather deep suture: *mouth* extremely open, in consequence of the expansion of the base: *inside* lustrous, either porcelain-white or having a rosy and sometimes lovely pink tinge; it is concentrically and microscopically lineated from the middle to within a short distance of the margin (as in the *Fissurellidæ*); and the border of the margin is marked lengthwise with longitudinal lines, which run at a right angle to the other set of lines; margin in young specimens finely notched or scalloped: *muscular scar* horseshoe-shaped, with the opening in front and the broad ends on each side of the neck. L. 1·5. B. 1·75.

HABITAT: Attached to rocks and large shells, and especially frequenting oyster- and scallop-beds, from 7 to 85 f. on all our coasts; low-water mark to 20 f. in the Clyde district (Norman); 110–145 f. off the Mull of Galloway (Beechey). Norwich Crag at Bramerton (Woodward, *fide* coll. Wigham); Red and Coralline Crag (Wood); newer tertiaries of Belgium and France (Nyst); upper miocene bed at Antibes (Macé); Parma (Ménard de la Groye); North Italy (Brocchi); Sicily (Philippi). Its existing distribution comprises the North Atlantic sea-bed lying between Oxfjord in North Finmark (Sars), south coast of Iceland (Wallich), and Gibraltar (M'Andrew), also both sides of the Mediterranean, the Adriatic, and Ægean; the depths given by different observers range from 5 to 105 f.

Dr. Turton mentions, in the 'Zoological Journal' (ii. 566), "a thin laminar under valve," which he noticed on removing a live specimen of *C. Hungaricus* from an oyster; and he therefore conjectured that the present species might belong to the genus *Hipponyx* (*Hipponice*) of Defrance. I think the Doctor must have mistaken a lamina of the oyster shell for such under valve, since no one else appears to have made the same discovery. The supposed valve was not in his collection when purchased by Mr. Clark. This, in common with other univalve mollusks, when taken from the deeper parts of the sea, has a habit of getting out of the water in which it is kept. It is of a sedentary and inactive nature; and its power of adhesion is considerable. Mr. Bretherton says, in the 'Zoologist' for 1858, that it "could move for a few inches even on the smooth sides of a glass jar. The proboscis or rostrum (like that of *Cyprœa*, it appears to be of an intermediate character) is capable of extension, and can be produced beyond the shell." With respect to the embryology of the present species, Mr. Clark observes that "the matrix, or part thereof, is sometimes, perhaps always, detached, and deposited on the neck of the foot, for further development of the ova, for some time previous to their being committed to take care of themselves." The shell is frequently distorted. A specimen now before me has its sides so much compressed that they are nearly flat, and its mouth is narrowly elliptical; it had probably squeezed itself into the crevice of a rock. The spire is occasionally broken off or decollated, and replaced by a shelly convex plate.

This species is the *Patella Pileus Morionis major* of Da Costa; and the young is the *C. militaris* of Macgillivray and S. Wood, but not Linné's species of that name. Hungaria is cited in the dictionaries of Facciolati and Bayle as the country supposed to have been peopled by the Huns; and the adjective must of course be spelt accordingly.

C. militaris and *Cochlolepas antiquata* are West-Indian; Lister figured them as from Barbadoes. The reported British localities are incorrect, and depend chiefly on the authority of Bryer and Laskey. The latter goes so far as to say that he procured both in the Frith of Forth "by deep dredging." Weinkauff enumerates *C. militaris* as an Algerian shell: possibly he fell into the same error as Macgillivray and Wood.

Piliscus commodus of Middendorff has been dredged by Mr. Dawson in the Moray Firth, but apparently in a semifossil state. It is known living only in the sea of Okhotsk, although occurring in the older or glacial strata at Uddevalla, and (under S. Wood's name of *Capulus fallax*) in the Coralline Crag. Possibly *C. obliquus* of the last named author, from the Red Crag, may be the same species.

Family IV. CALYPTRÆIDÆ, Broderip.

BODY round or oval, more or less depressed: *mantle* entire: *head* not very prominent, terminating in a short but extensile muzzle: *tentacles, eyes,* and *gills* as in the *Capulidæ*: *foot* separate from the lower part of the body, and expanded.

SHELL shaped like a cap or slipper and depressed, partly spiral: *beak* turned towards the rear, and twisted to the left: *mouth* round and oval: *inside* furnished with a partition or diaphragm, the outer edge of which forms an incipient or rudimentary pillar.

The *Calyptræidæ* have the same habits as the *Capulidæ*; each family has only a solitary representative in our seas, although their members are numerous in warmer latitudes. Chenu says that the present family first made its appearance in the upper part of the Chalk formation.

Genus CALYPTRÆ'A*, Lamarck. Pl. VI. f. 6.

BODY round: *mantle* very thin: *head* large, cloven at the extremity: *foot* circular, somewhat thickened, especially in front.

SHELL conical, with a wide base: *beak* central: *mouth* circular: *diaphragm* incompletely spiral.

The only species we possess was included by Lamarck in the present genus; *Crucibulum extinctorium* (or the " cup and saucer limpet ") being the first named species. *Calyptræa*, or rather *Calyptra*, is now represented by the *Patella equestris* of Linné, according to the Messrs. Adams. These authors place our species in *Galerus*, a genus which was named, but not described, by Humphreys. *P. equestris* is the type of Schumacher s genus *Mitrularia*. Under all the circumstances I prefer retaining *Calyptræa* in the British list, leaving Schumacher's genus *Crucibulum* to stand as above, and rejecting *Galerus*. If any change were necessary, it ought in my opinion to be the adoption of *Trochita*, as proposed by the same learned Danish naturalist (Schumacher) for the *Patella Chinensis* of Linné, upon which that genus was founded.

CALYPTRÆA CHINEN'SIS †, Linné.

Patella chinensis, Linn. S. N. p. 1257. *C. Sinensis*, F. & H. ii. p. 463, pl. lx. f. 3–5, and (animal) pl. B B. f. 8–13.

BODY yellowish or whitish, minutely speckled with flake-white: *mantle* transparent, broader on one side than the other: *head* short, terminating in a cloven or bilobed muzzle; there is a slightly developed, semicircular plain-edged fleshy lobe on each side of the neck: *tentacles* thick, cylindrical, and rather short; *eyes* small, seated on tubercles: *foot* somewhat angulated in front.

* From καλύπτρα, a woman's cap.
† Like a Chinese Mandarin's hat.

SHELL usually much depressed and spread out (higher and more conical when attached to a pebble), thin, semitransparent, glossy at the point, but else of a dull and scabrous aspect: *sculpture*, numerous and fine minute striæ, which encircle the surface in a spiral direction, and are usually raised, so as to form irregular rows of short vaulted scales; marks of growth indistinctly spiral: *colour* white, with frequently a yellowish tinge at the point: *beak* small, nipple-shaped, nearly erect, representing the apex of the spire in turbinated univalves: *suture* slight: *mouth* greatly expanded: *inside* porcelain-white, rarely of a yellowish colour, highly glossy; margin extremely thin: *diaphragm* or rudimentary pillar occupying nearly one-half of the posterior side; its outline is obliquely triangular, with rounded sides and a spirally incurved nucleus; it is concave, marked with delicate and close-set flexuous lines parallel with the base, and has its inner margin double and consequently thickened. L. 0·75. B. 0·75.

HABITAT: Local but gregarious, in shelly and pebbly ground, on many parts of the Devon and Cornish coasts, and in the Channel Isles, from low-water mark to a few fathoms' depth; Weymouth (Thompson); Milford Haven, 10-12 f. (M'Andrew and Jordan); "Dublin Bay" (Turton); "a solitary small specimen has been found near Dunbar" (Laskey); "in Caledoniæ Borealis mari profundo" (Leach). The last two localities, and probably the Irish one also, are erroneous: such statements have considerably retarded our knowledge of the distribution of the British Mollusca. It has been recorded as fossil from the Norwich Crag (S. P. Woodward); Red and Coralline Crag (S. Wood); Antwerp (Nyst); faluns of Touraine (Cailliaud); Bordeaux (Grateloup); Subapennine tertiaries (Brocchi); South Italy and Sicily (Philippi). Recent:—Coasts of France, Portugal, Spain, Italy, islands in the Mediterranean, Crimea, Morea, North Africa, Madeira, and Canaries, from the shore to 55 f.

It would seem that this mollusk seldom, if ever, leaves

its place of abode. I found some at Sark, living attached to small pebbles, each pebble having scarcely a broader surface than the circumference of the shell, which closely fitted the sinuosities of the stone. Both shell and pebble were encrusted by nullipore, and had the same appearance. The mark of adhesion is glossy, but does not show any excavation. The animal must be occasionally zoophagous; for the authors of the 'British Mollusca' say, " A *Calyptræa*, which we kept in confinement, swallowed a *Goniodoris nodosa* preserved in the same vessel." Mediterranean specimens are smaller than ours; one taken by Mr. Jordan at Milford measured an inch in diameter. The fry resembles that of *Velutina lævigata* in shape and sculpture; and the animal at this stage of growth has large ciliated neck-lobes, as in other Gasteropoda. According to Audouin and Milne-Edwards (Hist. litt. de la France, i. p. 133), the female *C. Chinensis* hatches her eggs, and keeps the fry between her foot and the foreign body to which she adheres; her patelloid shell thus serves not only to cover and protect herself, but is also a shield for her offspring. The eggs are yellow, and enclosed in membranous capsules, which are flattened, transparent, and filled with an albuminous matter. These little sacs are from six to ten in number, connected one with another by a pedicle, and arranged like the petals of a rose; each capsule contains ten to twelve eggs.

China does not seem to have been known to the Romans; and Linné was quite as good a classic as his emendator Gmelin, who altered the name *Chinensis* into *Sinensis*. The synonyms are numerous, and comprise *Calyptra canaria*, Bonanni, *Patella albida*, Donovan, *P. squamulata*, Renier, *P. muricata*, Brocchi, *C. lævigata*,

Lamarck, *C. succinea*, Risso, *C. Polii*, Scacchi, *C. vulgaris*, Philippi, and *C. mamma*, Krynicki *fide* Middendorff.

Crepidula sinuosa of Turton was included by him in the catalogue of British shells, in consequence of Mr. Bean having taken specimens at Scarborough, " from the bottom of a ship just arrived from North America." It is the *C. plana* of Say; but I question its being identical with the *Patella crepidula* of Linné or *C. unguiformis* of Lamarck, as was supposed by Sowerby and Gould.

Family V. HALIO'TIDÆ, Fleming.

BODY oblong, depressed, partly spiral: *mantle* puckered in small folds at intervals on the right hand: *head* broad, with a short snout: *tentacles* filiform, long and tapering, separated by a fringed lobe or membrane, which forms a head-veil: *eyes* placed on cylindrical, but short, stalks at the outer bases of the tentacles: *gills* two, unequal in size: *foot* extremely large, thick and fleshy, encircled by a double row of festoons.

SHELL ear-shaped, nacreous, pierced on the right side by a series of holes, which are closed in the course of growth, after ceasing to be of use in containing the pallial folds; the hole last formed commences as an open notch: *spire* very short, placed on the left side, although almost terminal: *mouth* extremely large and open, occupying nearly the whole of the base; borders curved, that on the right being thick, and the other thin: *inside* highly iridescent.

This family has several points of relationship to the *Fissurellidæ*; but the shape is different, the orifices are numerous, and the shell is remarkably nacreous. There appears to be a homogeneity between all the genera or members of the *Haliotidæ*, making it difficult to distinguish one from another. We have only the typical genus.

Genus HALIO'TIS *, Linné. Pl. VII. f. 1.

Characters corresponding with those of the family.

Aristotle called it λεπὰς ἀγρία, adding that it was the θαλάττιον οὖς of others. It is mentioned by Athenæus as exceedingly nutritious, but indigestible. The Eolians gave it the pretty name of Venus's ear. It is the "Mother of Pearl" or "Norman shell" of old English writers, "ormier" (contracted from oreille de mer) of the French, "lapa burra" of the Portuguese, "orecchiale" of the Italians, and " patella reale " of the Sicilians. It adheres to rocks like the limpet. Its food appears, however, to be different from that of the *Patella*, according to the observations of Mr. Daniel, which will be given in the account of *H. tuberculata*. This inhabits the littoral zone; but a Japanese species lies deeper under water, and is procured only by diving. Cuvier found every individual which he examined to have an ovary; and he therefore concluded that the *Haliotides* were hermaphrodites. His view was adopted by Feider in his essay 'De Halyotidum structura.' Half a century has since elapsed; and it is full time to have more definite information on a subject which is so easy for any physiologist to determine. The arrangement of teeth on the lingual ribbon agrees generally with that of *Fissurella*; it is more complicated than the Trochidan form. According to Adanson, the maritime negroes of Senegal esteemed one species of *Haliotis* a great delicacy; other kinds are said to constitute part of the multifarious food of the Japanese and Chinese; and *H. tuberculata* is habitually eaten by the poor in the north of France and our Channel Isles, where it is occasionally cooked and served

* Sea-ear.

at the tables of the rich. It requires a good deal of beating and stewing to make it tender. The inside of the shell displays "all the colours i' the rainbow," or at least

"Orange and azure, deepening into gold."

Perhaps the latter description is not pictorially correct; but it poetically suggests the vivid hues which are so exquisitely blended in the *Haliotis*. The cause of this iridescence has been explained by Sir David Brewster, Dr. Carpenter, Professor Van der Hoeven, and very lately by Mr. C. Stewart. The hypothesis propounded by the first of these observers was that the peculiar appearance is owing to minute striæ or grooves on the surface of the nacre, which alternate with others of animal membrane. Mr. Stewart is of opinion that the colour is produced by the nature of the laminæ, which decompose the light in consequence of the interference caused by the reflection from the two sides of each film, as may be seen in soap-bubbles and the iridescent surfaces of many natural and artificial productions. He further believes that the nacreous or inner layer is only a modification of the previously formed prismatic layer, each layer being composed of particles or prisms mostly presenting an hexagonal outline. The microscopical structure of the shell has been investigated by Carpenter. He says that " calcified laminæ alternate with plates of a brown horny substance, much resembling tortoiseshell in its appearance; and when the calcareous matter has been dissolved away by dilute acid, these horny plates may be easily detached from each other, the basement membrane of the adjoining calcified lamina remaining adherent to one side of each of them. In immediate contact with the horny plates is a thin layer of large cells of a very peculiar aspect. The nacreous laminæ, when examined with a sufficiently high magnifying power, indicate a minute cellular structure, such as I have not observed in the nacre of bivalves. The cells are of a long oval form, and their short diameter is not above $\frac{1}{5000}$th of an inch. Their boundaries in many parts are very indistinct or even disappear altogether; so that every gradation can be traced, from the obviously cellular arrangement to the homogeneous appearance presented by the nacre of bivalve shells. Hence I should be disposed to draw the same inference, with respect to the nacreous structure, as in regard to other forms of apparently homogeneous shell-substance—namely, that like them, it was originally formed upon a cellular plan, but that the cells subsequently coalesced, their boundaries disappearing." Woodward gives seventy-five as the number of recent species, and four for the fossil (miocene) species. The distribution of this genus comprises every part of the ocean, from Great Britain southwards.

It is the genus *Auris* of Klein.

HALIOTIS TUBERCULA'TA [*], Linné.

H. tuberculata, Linn. S. N. p. 1256; F. & H. ii. p. 485, pl. lxiv. and (animal) pl. C C. f. 3.

BODY mottled with brown, green, and white, blending agreeably together : *foot* ornamented with two rows of most delicate thorn-like processes, which alternate with green filaments; the interspaces between these rows are covered with greenish tubercles; sole of a salmon-colour.

SHELL much depressed, solid, opaque, and of a dull aspect : *sculpture*, numerous narrow longitudinal grooves or striæ, which are more or less close-set and occasionally undulating; the surface exhibits also the edges of extremely minute transverse plates, that compose the structure of the shell; marks of growth sometimes fold-like : *colour* reddish-brown, mottled

[*] Tuberculated.

with pale green, and occasionally speckled with brown, pink, or white: *epidermis* fibrous, thin, of a light yellowish hue: *spire* small, somewhat raised: *whorls* three or four, rapidly increasing, and becoming less convex as they recede from the apex: *orifices* from six to eight, roundish; their sides are raised so as to resemble tubercles: *mouth* oval: *outer lip* bevelled to a sharp edge: *inner* (or pillar) *lip* broad, flattened, somewhat notched or emarginate at the base in front, and bordered outside by a pink line: *inside* dark towards the margin, although in other parts splendidly lustrous. L. 4. B. 3.

HABITAT: Rocks and large stones at low water in the Channel Isles; common. The Devonshire, Sussex, Scotch, and Irish localities mentioned by Pennant, Da Costa, Laskey, Turton, and Brown must have been from hearsay, and are manifestly wrong. Linné introduced this handsome and familiar shell into his 'Fauna Succica,' on apparently no better grounds. The principle of geographical distribution was not then known, and a long time elapsed before it was made a law. Fossil in the Sicilian tertiaries (Philippi). It inhabits the North Atlantic, from St. Malo to the Canary Isles and Azores, the Adriatic, and every part of the Mediterranean.

I include this among our Mollusca, because the Channel Isles are as much an integral part of Great Britain as are the Shetland Isles. The animal rivals the shell in beauty. From Beudant's experiments it appears that *H. tuberculata* cannot exist in fresh water. Mr. Daniel detected in its stomach different species of diatoms in considerable quantities, besides many crystalline substances of the same prismatic hue as its own shell. These last mentioned organisms may have been the spicula of sponges. The number of open orifices in the shell corresponds with that of the tubular folds of the mantle. As the animal grows, the orifices that were first formed become disused, and are filled up by successive layers of shell; the last or outermost pallial fold forms and occupies a notch or semicircular slit in front, which is subsequently converted into an eyelet-hole. Very young specimens are imperforate, and in that state resemble the shells of *Stomatia* and allied genera, which are placed by Messrs. Adams in the *Trochus* family. One in Mrs. Collings's collection has no orifice, although it is about an inch and a quarter in length; this, however, is an exception as regards size. Such young shells are finely striated in a longitudinal or spiral direction, and are adorned with two narrow rows of blue spots; in a more advanced stage they are spirally ridged and delicately cancellated. The Cherbourg fishwomen call it "si ieu" (six yeux), from an idea that the orifices in the shell are real eyelets or peepholes. The importation of *Meleagrinæ*, or true mother-of-pearl shells, from the South seas, has interfered with the sale of the "ormer" (or *Haliotis*) at Guernsey for button-making and inlaying, although, as Dr. Lukis informed me in 1859, one merchant at St. Peter's purchased every season from four to nine tons. At sorting-time every shell was separately examined; the best lots fetched on the spot seven shillings and sixpence per cwt. I found that in some parts of Guernsey the ormer was put to rather a novel use, viz. to frighten away small birds from the standing corn. Three or four shells are strung loosely together, and suspended from the top of a pole, so as to make a clatter when moved by the wind and knocking one against another. In Montagu's time they ornamented cottages there, the plaster on the outside being studded with them.

It is the *Auris vulgaris* of Klein and Da Costa.

Family VI. SCISSURE'LLIDÆ, Gray.

Body more globular than conical, spiral: *tentacles* long, ciliated: *eyes* at the outer bases of the tentacles: *foot* furnished with pointed lappets on each side, besides tentacular appendages.

Shell corresponding in shape with the body or animal, white, nacreous, and thin : *mouth* somewhat angulated : *outer lip* vertically fissured at the edge, or having an orifice a little behind it : *umbilicus* narrow, but conspicuous: *operculum* horny, thin, circular, and multispiral, with a central nucleus.

The recent genera (*Scissurella* and *Schismope*) which constitute this family have their analogues in the fossil genera (*Pleurotomaria* and *Trochotoma*), if indeed the two last named are not the same as the two first. Living species of *Pleurotomaria* have been lately described and figured in the 'Journal de Conchyliologie'; they do not seem to differ from those of *Scissurella*, except in their greater size and coloured markings. The nacreous inner layer of *S. crispata* is very evident when the shell is broken or has been accidentally exfoliated. The absence of nacre in *Scissurella* was regarded by Fischer and Bernardi as a distinctive character of *Pleurotomaria*. The slit or fissure probably serves the same purpose as that of *Emarginula*, *Puncturella*, or *Fissurella*, and the orifices in *Haliotis*, by admitting water to the branchial cavity. The conformation and habits of the animal may require this additional provision for aërating the gills. In most other respects the *Scissurellidæ* resemble the *Trochidæ*; in every one of them the foot has tentacular appendages, the shell is spirally conical, and the operculum is concentrically spiral. Too little, however, is known with regard to the animals of the present family to pursue the comparison to anything like a complete or satisfactory extent.

Genus SCISSUREL'LA *, D'Orbigny. Pl. VII. f. 2.

SHELL having a more or less depressed spire, and an open slit in the outer lip which is closed behind in the course of growth.

In the 'Zoological Journal' for June 1824, Mr. G. B. Sowerby suggested the possibility of this "proving to be either an *Haliotis* or a *Fissurella*, just emerged from the egg;" he supposed that the slit or notch existed only in young specimens. He was apparently led to form this strange conception by having mistaken for adult *Scissurellæ* shells belonging to the genus lately described by me as *Schismope*, which have an orifice behind the front margin, instead of an open slit at the edge. D'Orbigny especially notices this open slit as a generic character of *Scissurella*; and he compares it with that of *Pleurotoma*, *Emarginula*, and *Siliquaria*, placing *Scissurella* among his *Trochoidea*. Some conchologists have referred the present genus to *Anatomus* of De Montfort; but his description and illustrative figure (the latter copied from Soldani) show a flat-spired or discoidal shell, having a circular mouth with a slit on the *lower* side—certainly not the position of the slit in *Scissurella*. He evidently considered his *Anatomus* one of the Polythalamous or chambered Foraminifera, and he associated with it the fry of some mollusk which he found adhering to the "Sargasso" or Gulf-weed. I am therefore not inclined to substitute *Anatomus* for *Scissurella*.

SCISSURELLA CRISPA'TA †, Fleming.

S. crispata, Flem. Mem. Wern. Soc. vi. p. 385, pl. 6. f. 3; F. & H. ii. p. 544, pl. lxiii. f. 6.

BODY greyish-white: *head* prominent, with the mouth

* From a small slit in the shell.   † Curled.

placed underneath: *foot* oblong, rather elongated, rounded at each end, and somewhat broader in front, furnished with two pointed lappets on either side of the anterior part: *appendages* or pedal filaments two on each side behind the lappets, one in the middle, and the other close to the tail; these are long, slender, and serrated or cirrous.

SHELL somewhat globular, with a slope towards the middle or periphery, of a delicate texture, semitransparent, and glossy: *sculpture*, numerous extremely fine and curved longitudinal ribs, which are interrupted in the middle or circumference of each whorl by the encircling slit and canal; they are more close-set on the under than on the upper surface of the last whorl, and are to a greater or less extent decussated in the interstices by minute spiral striæ: *colour* pearl-white: *epidermis* thin and caducous, pale yellowish-brown: *spire* usually rather depressed, but variable in that respect: *whorls* 4, flattened above, and rapidly enlarging; the last is three or four times the size of all the others put together: *slit* long and narrow, nearly central; canal or groove (formed in consequence of the closure or partial filling up of the slit from time to time) deep and striated across; the edges of the slit and canal are somewhat thickened, sharp, and prominent: *mouth* roundish, placed obliquely, ending in a small corner at the upper part of the columella or pillar; peristome continuous: *outer lip* thin: *inner lip* folded back on the columella: *umbilicus* deep, but exposing only the under side of the last or body whorl: *operculum* filmy, having many apparently concentric volutions in the central part, the last being very large in proportion. L. 0·075. B. 0·1.

Var. *paucicostata*. Spire more raised, and the ribs on the upper side much fewer than usual.

HABITAT: Stony ground in Shetland, 18–75 f.; not uncommon. It has also been taken by Captain Thomas abundantly in 7 f. at Sanda Sound in the Orkney Isles; more sparingly by Mr. Peach at Wick, and in Dunnet Bay, Caithness; by Mr. Barlee at Skye and in other parts of the west of Scotland; by Mr. Hyndman in 27 f. on the Antrim coast; and by Captain Hoskyns in about 100 f. on a fishing-bank off the west of Ireland. The variety was found by Mr. Waller in Shetland. Believing the *S. aspera* of Philippi to be the same species as *S. angulata* of Lovén, and that the latter is merely a large form of *S. crispata*, I will venture to give the Calabrian tertiaries as the only known locality for this shell as fossil. Lovén and others have dredged the present species on the Norwegian coasts, at depths varying from 30 to 100 f., Möller and Torell in Greenland, and the latter at Spitzbergen also; Martin obtained it in the Gulf of Lyons, and Benoit in Sicily.

Dr. Fleming discovered in 1809 this remarkable little shell on the shore at Noss Island in Shetland after a storm; he sent specimens to Colonel Montagu, who pronounced them to be the fry of a *Trochus*. It was procured in a living state by Mr. Barlee on several occasions; but, unfortunately, he never observed the animal. This deficiency has been in some measure supplied by Professor Barrett, who in company with Mr. M'Andrew dredged a live specimen at Hammerfest. His description and figure in the 'Annals of Natural History' for February 1856, aided by the dried remains of the animal in specimens received from Mr. Barlee, have enabled me to give a short, though meagre, account of the soft parts. Barrett remarked that " no part of the animal was external to the shell. When it was placed in a glass of sea-water, it crawled up the side, and scraped the glass with its tongue. After immersion in spirit it became inky-black." Apparently the fry have no slit, a condition similar to that which exists in the *Fissurellidæ* and *Haliotidæ*.

S. angulata probably bears the same relation to *S. crispata*, as *Chiton nagelfar* or *C. abyssorum* does to *C. Hanleyi*. Sowerby named our shell (perhaps from inadvertence, or a typographical error) *S. crispa*.

Family VII. TRO'CHIDÆ, D'Orbigny.

BODY spirally twisted into a cone: *mantle* forming on each side of the head a distinct lobe or lappet: *head* proboscidiform, furnished with a dentate tongue, the extremity of which is convoluted within the visceral cavity: *tentacles* long and ciliated: *eyes* placed on short stalks or tubercles at the outer bases of the tentacles: *gills* composing a single plume: *foot* furnished on each side with from 3 to 6 vibracula or appendages resembling tentacles; operculigerous lobe occupying the middle of the upper part of the foot.

SHELL orbicular or conical, and spiral, more or less nacreous: *mouth* rounded: *umbilicus* depending in a great measure on the height of the cone, sometimes wanting: *operculum* horny, thin, circular, and multispiral, with a central nucleus.

The *Trochidæ* probably live on minute animal and vegetable organisms. From Lovén's account of the tongue it seems that the rachis is armed with many teeth, and that each of the pleuræ has extremely numerous regularly arranged uncini, which become gradually more slender and simple as they recede from the centre. In *Trochus cinerarius* there is a large heart-shaped tooth in the middle, and on each side of it five principal or front teeth and about ninety uncini. The sexes are separate. Many of the shells of the typical genus *Trochus* are extremely ornamental; and the animals of all are adorned with plumed filaments, and with flounces often of resplendent hues.

Genus I. CYCLOSTRE'MA*, Marryat. Pl. VII. f. 3.

BODY compressed: *head* bilobed at its extremity: *foot* expanded at each of the front corners into a short triangular process.

* Having a circular twist.

CYCLOSTREMA. 287

SHELL orbicular, white or of a uniform colour: *spire* more or less depressed, of few whorls: *mouth* nearly circular, with a free and continuous peristome: *umbilicus* distinct and deep.

All the British species are minute. They appear to be ovoviviparous, producing their spawn inside, and depositing it on extraneous substances to be developed; the spawn contains fry perfectly formed and having complete shells. The genus was founded by the celebrated novelist, Captain Marryat, by whom its characters were thus briefly described in the 12th volume of the 'Transactions of the Linnean Society' (1818) : " Testa depressa, perspectivo umbilicata; apertura circularis." He referred the *Helix depressa* and *H. serpuloides* of Montagu to this genus: but the animal of the former resembles that of a *Rissoa*, and is the type of Fleming's genus *Skenea*; the other is correctly assigned to the present genus. *Delphinula* of De Roissy or Lamarck has rough and angular whorls—although perhaps Philippi was right in adopting it for some of the species now under consideration. Fleming's genus *Cyclostrema* is very different, being represented by *Rissoa Zetlandica*. It is unnecessary to say of the genus *Delphionoidea* or *Delphinoidea* of Brown, which has been suggested by the Messrs. Adams, more than that it is both superfluous and heterogeneous.

1. CYCLOSTREMA CUTLERIA′NUM *, Clark.

Skenea Cutleriana, Clark, in Ann. & Mag. Nat. Hist. new ser. vol. iv. p. 424. *S.?* *Cutleriana*, F. & H. iii. p. 164, pl. lxxviii. f. 3, 4.

BODY clear-white: *pallial lobes* or neck-lappets distinct: *head* rather long, broad, and finely wrinkled across: *tentacles* flattish, lineated down the middle, exquisitely but rather

* Named in honour of Miss Cutler, a lady of scientific taste and acquirements.

sparsely ciliated: *eyes* proportionally large, black, placed on very short pedicles: *foot* somewhat rounded at each end; front corners curved, ear-shaped, broad and flat: *appendages* 3 or 4 on each side, filiform, and finely ciliated like the tentacles. (Clark).

SHELL globular, thin, transparent and glossy: *sculpture,* numerous fine spiral or revolving striæ, and occasional scratch-like and more minute lines of growth: *colour* clear white: *spire* raised, but blunt: *whorls* 3, very tumid, rapidly enlarging: *suture* deep: *mouth* slightly angular above; peristome somewhat reflected on the inner or columellar side: *umbilicus* narrow, oblique, exposing only the base of the last whorl: *operculum* having from six to eight volutions, microscopically and irregularly striated across in an oblique direction. L. 0·04. B. 0·04.

HABITAT: Coralline zone, 15–40 f. at Guernsey and Lulworth (J. G. J.); Falmouth (Webster and Hockin); Fowey, abundant (Barlee); Exmouth (Clark and Barlee); Skye, a single but characteristic specimen (J. G. J.). I noticed this exquisite little gem in Mr. M'Andrew's collection, from his Mediterranean dredgings; and Professor Lilljeborg gave me at Upsala two specimens of an extraordinary size (about a tenth of an inch in length and diameter) which he had dredged at Bergen and Christiansund.

The animal is described by its discoverer, Clark, as exceedingly active and rapid in its movements. Occasionally the shells of this and the next tiny species are found pierced by some of the smaller canaliferous mollusks.

I at one time believed that the present species was the *Trochus exilis* of Philippi; but I now doubt it. The peristome of that shell is represented in his figure as disconnected; in ours it is continuous. The two species are alike in other respects.

2. C. NITENS *, Philippi.

Delphinula nitens, Phil. Moll. Sic. ii. p. 146, tab. xxv. f. 4. *Trochus pusillus*, F. & H. ii. p. 534, pl. lxxiii. f. 3, 4.

BODY closely resembling that of *C. Cutlerianum*. The tentacles and lateral appendages of the foot, however, are not quite so long; the foot is shorter, broader, and more rounded at each end, with the front corners detached to a greater extent; and there are four tentacular appendages on each side.

SHELL not so globular, thin, or transparent as *C. Cutlerianum*, but somewhat depressed above and below, more glossy and almost iridescent: *sculpture* consisting of only a few indistinct grooves on the upper part of the umbilicus; the surface is otherwise quite smooth and polished, even under the microscope: *colour* whitish, with a faint tinge of yellow: *spire* not much raised, and blunt: *whorls* 3, convex, rapidly enlarging: *suture* rather deep: *mouth* as in the first-described species; peristome thickened and slightly reflected on the columellar side: *umbilicus* narrow, placed obliquely, not exposing any part of the middle whorl: *operculum* having 6-8 volutions, which are continued to the centre. L. 0·035. B. 0·03.

Var. *Alderi*. Shell thinner and more transparent. *Skenea? lævis*, F. & H. iii. p. 165, pl. lxxxviii. f. 5, 6.

HABITAT : Coralline zone on the coasts of Guernsey, Devon, Cornwall, Ireland (north, east, west, and south), west of Scotland and the Hebrides, Moray Firth, and Shetland. Mr. Cocks has taken it " attached to *Algæ* in the pools on rocks, Gwyllyn vase," near Falmouth. The variety was obtained by Mr. Barlee in Skye. A single fossil specimen of the typical form was found by Philippi in Calabria; and M'Andrew dredged this species alive in the Mediterranean.

It was described by me in the 'Annals of Natural History' for 1848 as *Margarita pusilla*. The name given by Philippi has precedence by four years; and it

* Shining.

is more correct than mine, which implies a comparison with species of a different genus.

3. C. SERPULOÏDES *, Montagu.

Helix Serpuloides, Mont. Test. Brit. Suppl. p. 147, tab. 21. f. 3. *Skenea? divisa*, F. & H. iii. p. 161, pl. lxxiv. f. 4–6.

Body pure hyaline-white: *pallial lobes* or neck-lappets of different shapes; that on the right hand is narrowish, flat, and partially serrated; the other is shorter, somewhat oval, and plain-edged: *head* rather long, broad, and finely wrinkled across, having a pale-red or pink disk [the colour of which is perceptible even in the dried animal]: *tentacles* flattish, marked lengthwise by a white line, symmetrically and elegantly clothed with long transparent close-set cilia: *eyes* very large and black, seated on small bulbs: *foot* somewhat truncated or bluntly rounded in front, having at each of the front corners a long curved linear ear-shaped process: *appendages* 3 or 4 on each side, equidistant, filiform, flattish, shorter and less slender than the tentacles, although equally ciliated; these filaments also issue from bulbs or tubercles: *verge* flat, pointed, and lying horizontally, not projecting beyond the mouth of the shell; sole not fringed at the edge (Clark).

Shell depressed, rather thin, transparent, and glossy: *sculpture*, numerous fine spiral striæ on the under side; the upper part is quite smooth or very rarely marked with a few indistinct and almost microscopical spiral lines: *colour* clear-white, with sometimes a light-yellowish tint, which is perhaps derived from a filmy epidermis that is not otherwise perceptible: *spire* scarcely raised: *whorls* 3–4, cylindrical, rapidly increasing in size; the last extremely large in proportion to the rest: *suture* rather deep: *mouth* slightly angular or forming a small corner above, in consequence of the last whorl impinging on that part of the circle; it is furnished inside with a narrow ledge in order to receive the operculum; peristome simple: *umbilicus* not large, but exposing the whole of the spire: *operculum* having 6–8 whorls and slightly iridescent. L. 0·02. B. 0·05.

Habitat: In the laminarian and coralline zones on

* Having the aspect of a *Serpula*.

all our coasts, from low-water mark to 25 f. Raised sea-bed at Fort William (J. G. J.); Calabria (Philippi); Väderöarna, in the south-west of Sweden, 12 f. (Malm); Croisic, Loire-Inférieure (Cailliaud); Gulf of Lyons (Martin); Mediterranean (coll. M'Andrew); Magnisi in Sicily (Philippi); State of Maine, "littoral, found occasionally clinging to the under side of wet stones, above low-water mark " (Mighels).

Clark says, " it is active, marches with quickness, not at all shy, and gave me good opportunities of observing its peculiarities." In Shetland it deposits its spawn in thick irregular clusters on some of the finer and membranous sea-weeds; each cluster contains a great number of fry, having their shells completely formed, and enveloped in a glairy matter.

It is the *Skenea divisa* of Fleming, and *Delphinula lævis* of Philippi. The authors of the 'British Mollusca' do not appear to have given a sufficient reason for preferring the later name "*divisa*" to that by which Montagu published this species. Philippi's specimens (only three in number, two recent and one fossil) may have been accidentally discoloured, as is sometimes the case.

C.? costulatum (*Margarita? costulata*, Möller) has been dredged by Mr. Barlee in Loch Fyne, by Mr. Waller on the Turbot Bank near Larne in Antrim, by Mr. Dawson in the Moray Firth, and by myself in Shetland; and Mr. Bean found a specimen in sand dredged at Lamlash in the Isle of Arran. I do not, however, consider any of the specimens thus procured recent. It occurs in a fossil state at Fort William (J. G. J.) ; Paisley (Crosskey) ; and at Uddevalla. The most southern point where it has been observed in a living state is Ireland; it inhabits the Arctic seas of both hemispheres. The operculum is calcareous, and of the same consistence as that of *Cyclostoma*; but this is multispiral and has a central nucleus. *C. ? costulatum* may therefore belong to the *Turbinidæ*. The shell is remarkably solid for its size (three-fourths of a line in breadth), and has strong and partly dichotomous transverse ribs; the peristome is continuous. The very little that we know of the animal is derived from Möller, who states that it is allied to that of *Margarita*, but differs in the foot of this mollusk being furnished in front with filaments. *Mölleria* would be a suitable name for the genus to which the shell in question may hereafter be assigned. Möller was the Danish governor of East or old Greenland; and, without neglecting his duties, he did much to elucidate the history of the glacial epoch, by investigating the existing mollusca of the far north.

Genus II. TROCHUS*, Rondeletius. Pl. VII. f. 4.

BODY of various sizes, but not minute: *head* prominent and stout: *foot* ridged on the upper part of each side by a digitated or fringed membrane.

SHELL conical, with an angular periphery, highly nacreous: *spire* more or less raised: *mouth* placed obliquely; lips or edges disunited on the columellar side: *umbilicus* (if present) variable in extent, even in the same species.

Rondeletius called this kind of shell a *Trochus*, because of its similarity to a Roman boy's plaything of that name. His comparison would be correct if "trochus" meant a top; but the word (derived from the Greek τροχὸς) is rendered in all the best dictionaries "a trundling-hoop for children." "Turbo" is the ancient name of a playing-top. The shells now about to be described were (and per-

* Top-shell.

haps are still) called in some parts of the north of Italy "trottola," by the fishermen at Spezzia "narnai," and by the French "culs-de-lampe." Adanson's species of *Trochus* belong to *Littorina*. Dr. Leach's posthumous 'Synopsis of the Mollusca of Great Britain' contains an extremely inaccurate account of the anatomical structure of the animal. The following are extracts: "The eggs (ova) are pedunculated; the peduncle is situated at the sides of the tentacles of the young animal:" we are also told that *Trochus* has "four tentacles." Surely the publication of such a work was not "an act of justice" to the memory of this once celebrated zoologist. In the 'Zoologia Adriatica' of Olivi will be found some curious lucubrations as to the cause of the internal lustre, resembling silver or mother of pearl, which decorates the shells of this genus. Finding that the shell of *Trochus* was composed of different layers, he at first supposed that the iridescence could only be the effect of light reflected or refracted at different angles from the distinct surfaces which resulted from the relative superposition of these layers. In consequence, however, of the experiments made by Hérissant with respect to the heterogeneous nature of shell-matter, and of Bouvier having detected by analysis a considerable proportion of magnesium in *Corallina officinalis*, Olivi hazarded another conjecture, viz. that the iridescence might arise from the admixture of some other mineral with carbonate of lime, such as is seen in mica schist. Nacre composes the inner layer of every species, and the entire substance of some; and Carpenter was able to distinguish in this nacreous composition the same minutely cellular arrangement which he had described as presenting itself so distinctly in *Haliotis*. The genus comprises a multitude of species, recent and fossil. It is evidently of great antiquity, although palæontologists are not agreed as to its origin. Sowerby assigns this probably to the Lias, Woodward to the Devonian formation, and Searles Wood to the Protozoic rocks. The distribution of existing species corresponds in extent with their number; none of the typical form appear to inhabit North-east America—only those of the section *Margarita*.

For the same reasons which I gave in the preceding volume for not dismembering *Venus* as regards the British species, I will preserve *Trochus* in its integrity, at the same time dividing it into as many sections as the gradual nature of the differences between the species may seem to warrant. It is true that all the species comprised in the so-called genus *Margarita* are quite pearly and that some of them are low-spired and umbilicate; but it must be observed that *Trochus occidentalis* (which is placed by Lovén in that genus), although pearly, is high-spired and has no umbilicus, and that *T. Vahlii* and *T. amabilis* are decidedly conical. The shells of *Gibbula* are usually low-spired and deeply umbilicate; but varieties of *T. tumidus*, *T. umbilicatus*, and *T. cinerarius* (referred to this genus by the Messrs. Adams) have the spire raised, and the base is not even perforated. Searles Wood says that in Crag specimens of *T. tumidus* (which connects *Gibbula* with *Margarita*) the umbilicus is very variable; "in some it is open, while in others it is quite covered, depending upon the elevation or depression of the spire, and also on the extension of the left lip." Again, *T. lineatus* is our only representative of Klein's genus *Trochocochlea*, in which the spire is raised, the base imperforate, and the pillar lip furnished with a blunt tubercle or notch; the last two characters are common, however, to several species of *Gibbula* and the typical section *Ziziphinus*, which last has a pyramidal spire. It is also not generally known, but not less the fact, that young shells of *T. lineatus* (the type of *Trochocochlea*) are always deeply umbilicate.

A. Small, pearly, and umbilicate. *Margarita*, Leach.

1. TROCHUS HELICI′NUS *, Fabricius.

T. helicinus, Fabr. Fn. Grœnl. p. 393; F. & H. ii. p. 531, pl. lxviii. f. 4, 5, lxxiv. f. 10, and (animal) pl. C C. f. 4.

BODY orangecolour, the upper part marked with close-set longitudinal purple lines or streaks: *pallial lappets* oval, small, purplish-grey: *head* short and rounded, semicircular in front, and divided into 12–15 lobes, which form a sort of fringe; some of these lobes in front are cloven: *tentacles* slender, flexible and contractile, ringed or annulated, and thickly covered with short cilia giving a bristly appearance; tips blunt: *eyes* rather large, and there is a supplementary pair of a smaller size on the inner base of the tentacles: *foot* thick and gibbous, lanceolate, rounded in front and bluntly pointed behind, with a pale line in the middle at the posterior end towards the tail; sole plain-edged when fully expanded, at other times minutely and irregularly scalloped or jagged at the edges: *appendages* 6 on each side, annulated and setose like the tentacles; each filament has a dark eye-like tubercle at its base; there are sometimes two of these filaments between each of the penultimate and caudal pairs.

SHELL somewhat globular, rather thin, semitransparent and lustrous: *sculpture*, several slight spiral striæ on the under side, and occasionally some faint and indistinct spiral lines on the upper side, and a few puckers near the suture: otherwise the surface is quite smooth and highly polished: *colour* orange or reddish-brown, sometimes variegated by purplish or azure tints on the upper parts: *spire* more raised in female than in male individuals; apex blunt: *whorls* 5, convex and gradually enlarging in the former, compressed and rapidly increasing in the other sex: *suture* distinct but not deep: *mouth*

* Like a *Helix*.

inclined to be angular above: *outer lip* plain in the female, and spread outwards in the male: *inner lip* folded back a little on the umbilical cavity: *umbilicus* narrow but deep, exposing the base of the penultimate whorl: *inside* iridescent: *operculum* having about a dozen volutions, becoming slightly concave towards the centre; the nucleus forms a boss or projecting point on the under side. L. 0·125. B. 0·25.

Var. *fasciata*. Smaller, light-yellowish or creamcolour, with a spiral band of reddish-brown between the suture of the last whorl and the periphery.

HABITAT: Abundant on the fronds of *Laminaria saccharina*, and under loose stones, throughout the laminarian and lower part of the littoral zones, in Shetland, the Orkneys, both sides of Scotland, and the coasts of Berwickshire, Northumberland, Durham, and Yorkshire; Belfast (Hyndman); Dublin Bay (Warren and Kinahan); and Connemara (Farren). Brown says "also on the south coast of Devonshire," and Leach endorsed the statement; but this must have been a mistake. The variety was found by Mr. Bean at Scarborough, by Mr. Hyndman in the north of Ireland, and by myself in the west of Scotland. *T. helicinus* is fossil at Fort William (J. G. J.); Oban (Geikie); Clyde beds (Crosskey); and at Uddevalla. It inhabits the shores of Scandinavia, Iceland, Spitzbergen, the White Sea, Sea of Okhotsk, Greenland, Behring's Straits, Labrador, Canada, and the north-eastern coasts of the United States, at depths ranging from low-water mark to 40 f.

The animal is active and bold. It appears fond of crawling out of water. When floating with the shell downward, the tongue is seen to be continually protruded, as if in search of some microscopic food. The gill is visible through the opening on the left-hand side of the head, and resembles a miniature *Plumularia fal-*

cata. The spawn is deposited on sea-weed and the under side of stones; each egg is enclosed in a yellow membranous capsule, all of which are agglutinated together at their sides and form an irregular glairy mass. I counted above 100 eggs in one of these spawn-masses. The fry are clear white, and not unlike the young of *Cyclostrema serpuloides.* The shells of the two sexes are different, as will appear from my description. The globular form of the female, with the outer circumference of each whorl embellished not only by the invariable lustre, but occasionally by a variety of glowing tints, reminds us of the vision of Panthea in 'Prometheus Unbound,' in which were displayed

"Purple and azure, white, green and golden,
Sphere within sphere."

The shell is sometimes twisted or otherwise distorted. Zetlandic are much larger than English or Irish specimens; those from the Arctic Sea are comparatively giants. I dredged a specimen empty, but having the operculum in it, about 25 miles north of Unst in 80 fathoms; it was pierced, apparently, by some animal which had probably carried it off and dropped it in the far deep, after extracting the mollusk through the hole.

According to Fabricius this is the *Turbo neritoideus* of Olafsen. The *Trochus helicinus* of Gmelin (from Knorr and Chemnitz) is a large West-Indian shell, but still undetermined. Our species is the *Helix margarita* of Laskey, *Turbo inflatus* of Totten, *Trochus margaritus* of Gray, *Margarita vulgaris* of Leach (*fide* Sowerby) and certainly his *Margarites diaphana, Margarita helicoides* of Beck (*fide* Sowerby), and *M. arctica* of Gould. It is difficult to guess what was the *M. arctica* of Leach, described in the "Appendix No. II." to Sir John Ross's

Voyage. It may have been the present species; but the operculum is stated to be testaceous.

2. T. GRŒNLAN'DICUS*, Chemnitz.

T. grönlandicus, Chemn. Conch. Cab. v. p. 108, t. 171. f. 1671. *T. undulatus*, F. & H. ii. p. 528, pl. lxviii. f. 1, 2, and pl. lxxiii. f. 5, 6.

BODY creamcolour with a few light-purplish-brown streaks along the back and sides: *pallial lappets* small and thin: *head* broad, notched or divided into lobes at the front edge (as in *T. helicinus*), and furnished with a thin veil or hood in front: *tentacles* extremely slender, and continually in motion; tips blunt: *eyes* on short but prominent stalks: *foot* large, broad and somewhat truncated in front, bluntly pointed behind; tail keeled and having an eye-like tubercle at its extremity: *appendages* from 5 to 7 on each side, with an equal number of ocelli, one at the base of each filamental appendage. Every part of the body, except the snout, is ciliated in the most exquisite manner.

SHELL having a rounded contour, rather solid, opaque, somewhat glossy: *sculpture*, several narrow thread-like but not much raised spiral ribs, or occasionally a few impressed striæ on the upper side, and more numerous and fine striæ on the under side; the surface is also covered with microscopical and close-set transverse striæ, and below the suture of each (especially the last) whorl it is puckered or marked with short and curved folds in the same direction: *colour* yellowish-red or fleshcolour: *spire* moderately raised: *whorls* 6, rather tumid, gradually increasing in size: *suture* rather deep: *mouth* slightly angular above: *outer lip* thin and flexuous: *inner lip* thickened and angulated below, folded back over the pillar and umbilical cavity above: *inside* purplish and iridescent: *umbilicus* narrow, deep and obliquely angulated outside, exposing all the spire: *operculum* having from 10 to 12 volutions, which are separated from each other by a slight ridge. L. 0·2. B. 0·25.

Var. 1. *albida*. Shell of a whitish colour.

Var. 2. *dilatata*. More depressed and expanded at the sides, encircled on the upper part by only a few spiral striæ or impressed lines.

* Inhabiting the seas of Greenland.

Var. 3. *lævior*. Smaller, more conical, solid and glossy, quite smooth with the exception of one or two slight spiral ribs on the uppermost whorls, fleshcolour.

HABITAT: At the roots of *Laminariæ* and on stones, from low-water mark to 40 f., in the west of Scotland, the Orkneys, and Shetland; local but not uncommon. The Rev. Mr. Whyte, according to Dr. Gordon, found it in Dunnet bay, Caithness, and Mr. Hyndman has dredged dead specimens in Belfast Bay; but the latter are suspiciously like fossils from a submarine posttertiary deposit in that locality. Var. 1 is occasionally met with. Var. 2 was taken by Mr. Barlee at Skye, and by myself at Loch Carron. For the other variety I am also indebted to the same friend. *T. Grœnlandicus* occurs in the Clyde beds (Smith and others), Fort William (J. G. J.), Norwich Crag (Woodward), and at Uddevalla. It lives in every part of the Arctic Ocean, and on the coasts of the White Sea, Scandinavia, Iceland, Canada, and the States of Maine and Massachusetts.

Its habits are much the same as those of the last species. Their shells may be distinguished by this having a more conical form and greater solidity, by the spiral ribs and striæ on the upper surface, the deeper suture, and also by the deeper and angulated umbilicus. The size of some specimens considerably exceeds the average dimensions which I have given. The largest I have seen were obtained by Dr. Otto Torell in Iceland. The fry are white, and striated like the adult.

It is perhaps the *Turbo fuscus* of Müller's 'Prodromus' ("testa fulva striis elevatis transversis"), and *Trochus cinerarius* of Fabricius but not of Linné. The Rev. R. T. Lowe described it as *Turbo carneus*, G. B.

Sowerby as *Margarita undulata,* Couthouy as *Turbo incarnatus,* and Brown as *Trochus inflatus.*

3. T. AMA'BILIS *, Jeffreys.

BODY of a creamy-white hue, faintly speckled or tinged with yellowish-brown : *pallial lappets* small : *head* prominent and wedge-shaped at its extremity, which is finely and deeply fringed by about 20 digitations or points of different lengths and sizes, those in front being the largest ; it is semicircular in front, and expansile like the foot of *Nucula* or *Leda* : *mouth* lobed : *tentacles* filiform, remarkably long, and tapering to a fine point ; they are flexible and exquisitely setose: *eyes* conspicuous, set on short offsets: *foot* lanceolate, squarish in front, on each side of which it is furnished with two long conical processes, which project at a right angle to the tentacles ; it is sharp-pointed behind, and has a prominent triangular ridge, extending from the posterior edge of the opercular lobe to the tail : *appendages* 3 on each side, issuing from beneath the opercular lobe, and between these are a few small papillæ ; the two lateral filaments in front are ciliated, and resemble a second pair of shorter tentacles ; the foot is capable of being expanded to a size double that of the shell, so as to form a broad and solid fulcrum.

SHELL pyramidal, moderately solid, semitransparent, of a pearly and partially iridescent lustre: *sculpture,* two spiral ridges or keels on the upper part of each of the last three or four whorls, and one on the upper part of the next or smaller whorl, besides several finer but irregular ridges on the base of the last or largest whorl, and numerous minute spiral striæ between all the ridges ; the principal ridges are placed near the suture of each whorl, both above and below, leaving a broad flattened space in the middle and a narrow excavated space below the suture, thus imparting a tower-like appearance to the shell ; the upper whorls are also marked with numerous short and fine longitudinal ribs, which cross the ridges and make them crenellated: *colour* pure pearl-white : *spire* elevated ; apex semiglobose, prominent and slightly twisted: *whorls* 7, gradually increasing in size : *suture* very distinct : *mouth* nearly circular, but angulated or somewhat notched

* Lovely.

below by the umbilical ridge: *outer lip* thin and slightly expanded: *inner lip* folded a little back on the umbilicus, and adhering to the pillar: *inside* more or less iridescent: *umbilicus* large but not wide, funnel-shaped, and completely exposing the whole of the inner spire; it is encircled outside by a strong spiral ridge, which is often beaded, and winds like a staircase into the interior: *operculum* forming a spire of about a dozen whorls, the edges of which are imbricated and overlap one another in succession. L. 0·333. B. 0·275.

HABITAT: Fine sand, mixed with gravel, in 85–95 f., about 25 miles N.N.W. of Burra Firth lighthouse, Unst. The area in which it occurs appears to be limited to a few square miles. I discovered this new and beautiful species in 1861, while in company with my friend Mr. Waller; and we obtained specimens again in 1864 by dredging on the same ground. Living together with it were *Limopsis aurita, Cylichna alba, Buccinopsis Dalei* var. *eburnea*, and other treasures. I do not know any other place, at home or abroad, where it has been found.

The animal is active and crawls rapidly; if laid on its back, it twists its foot from side to side, until part of the sole touches the bottom of the vessel, when it regains its usual position. Mr. Alder has examined the tongue, and observes that it shows rather a departure from the generic character in the want of the numerous slender uncini which other species possess. When I mentioned the unique habitat of this species, it would probably not convey to the minds of my readers in general what is meant by dredging in Shetland, nor how many difficulties and disappointments beset the naturalist who ventures thus to explore that remote and wild tract of the North Sea. The weather is so uncertain, and the winds often so boisterous, even in the summer and autumn months, that, although provided with every appliance, and having plenty of time at his disposal, he will frequently be unable to leave harbour for many days together, or to remain any time out at sea. Hence arise continual disappointments, rarely alleviated by such a discovery as I have just described. In one of these periods of despondence there was a lull between a past and coming storm, when this loveable pearly shell made its appearance and gladdened our longing eyes : we realized the thought in 'Endymion'—

> . . . . " in spite of all,
> Some shape of beauty moves away the pall
> From our dark spirits."

We were the first of human race that beheld it ; although, for ages uncountable, generation after generation of it must have lived and died,

> " Full many a fathom deep,
> On thy wild and stormy steep,"
>   Hialtland !

Perhaps with our joy was not unmingled a secret feeling of pride in the discovery, against which, as little short of a sin, Professor Kingsley cautions us in his pleasant little book 'Glaucus.' Our "pearl of the deep" might have served to bedeck the mermaid in the lay of the 'Queen's Wake'; Burns would certainly have called it "a bonie gem." The eastern seas do not surpass our own in furnishing such a marvel of Nature's workmanship, although the oriental pearl and the northern shell are alike perfect in opaline lustre and purity. Their production, however, is a plain sphere. Ours is a pyramidal cone, encircled by a winding gallery, and more elegantly sculptured than the finest rood-screen; its base is hollow and exhibits a spiral staircase. The door or operculum is circular and transparent; it may be

compared to a rose-window in its exquisite tracery.
But the shell has also an inner life of beauty. The
builder is not less graceful than the edifice. A feathery
hood surmounts its arched head; two tapering horns,
clothed with most delicate hairs, project in front, and
three similar but shorter ones on each side of the body,
all of which wave and curl independently of each other,
and are apparently endued with the most exquisite sen-
sibility; the whole is supported by a slender foot, whose
softly gliding motion effects an almost imperceptible
progress. The sentient will is evidently not wanting
in our living pearl. Before I part with the subject, let
me have full vent for my enthusiastic admiration by
scattering a very few more flowers of poetry by way of
illustration:—

" Framed in the prodigality of Nature."—*Richard* III.

" Crown'd the nonpareil of beauty."—*Twelfth Night*.

. . . . " Like a pearl
Dropt from the opening eyelids of the morn
Upon the bashful rose."—*Middleton's* ' *Game at Chesse.*'

" These were tears by Naiads wept
For the loss of Marinel."—*Bridal of Triermain*.

When I first saw this shell, its sculpture appeared so
like that of *Margarita* (?) *maculata*, S. Wood, that I
considered them to be the same species. I have since
had reason to alter my opinion. A careful comparison
of the recent species with that of our Coralline Crag,
and with typical specimens of *Turbo moniliferus* or
Solarium turbinoides of Nyst (which Mr. Wood considered, and, as I believe, rightly, identical with his
species), has convinced me that, according to the modern
acceptation of the term species, the living and fossil
forms are distinct. The one is pyramidal and angulated,

with a rather narrow umbilicus, and is pure nacre; the other has a somewhat depressed spire and rounded periphery, with a very wide and open umbilicus, and is creamcolour with occasionally dark blotches. Possibly these markings were caused by fossilization or mineral action, and the prototype may have been as stainless as its modern representative :—

> "But no perfection is so absolute,
> That some impurity doth not pollute."

I am by no means prepared to assert that *T. amabilis* is or is not a descendant of the fossil and so-called extinct species, changed in the course of ages to a greater extent than *Terebratula caput-serpentis* and other persistent species; our knowledge of such infinitesimally small or differential gradations is at present too imperfect to justify an assumption that "descent by modification" has been the invariable or even the ordinary law of nature. It would be inconvenient to retain the name (*elegantulus*) which I once proposed for the present species, because there is already a *Trochus elegantulus*, belonging to the section *Ziziphinus*, as well as *T. elegantissimus* of the section *Margarita*. A figure of the shell will be given in the supplementary volume of plates.

T. cinereus, Couthouy (*Margarita striata*, Broderip and Sowerby, but not *T. striatus* of Linné) has been dredged by Mr. Waller on the Antrim coast, by Mr. Barlee in Shetland, by Mr. Dawson in the Moray Firth, and by Mr. Mennell in Berwick Bay; but it is a submarine fossil. It also occurs in the Clyde beds and at Uddevalla, and inhabits the Norwegian and North American coasts. This species differs from *T. amabilis* in its larger size, greater solidity, dull grey colour, coarser and cancellated sculpture, close-set and fine longitudinal striæ, flattened apex, and much smaller umbilicus. .

Another Clyde fossil, the *Margarita olivacea* of Brown, appears to be the *M. glauca* of Möller's Catalogue of Greenland Mollusca.

Margarita elegantissima of Bean, from the glacial deposit at Bridlington, also lives in the Arctic Ocean; it is the *M. plicata* of Sars, and *M. polaris* of Danielssen.

The *M. aurea* of Brown (described as " destitute of an umbilicus ") has been identified by Forbes and Hanley with *Turbo sanguineus* of Linné, a Mediterranean shell.

B. Low-spired and umbilicate. *Gibbula*, Leach.

4. T. MAGUS *, Linné.

T. magus. Linn. S. N. p. 1228; F. & H. p. 522, pl. lxv. f. 6, 7, and (animal) pl. D D. f. 3.

BODY yellowish, mottled with purple and brown, or speckled with reddish-brown and white, and closely covered with short papillæ: *mantle* sometimes forming an incomplete branchial fold on the right side; pallial lappets large and broad, sometimes orange bordered with yellow, left fringed, right plain: *head* broad, but not prominent, ornamented in front with a veil or hood, the centre of which is brown and its ends yellow; this veil is divided into two lappets with white fringed edges, which often hang over the head; the extremity of the snout is also fringed or setose: *tentacles* very long and slender, more or less annulated with black: *eyes* very large, turquoise or black in the centre, encircled with a bluish line; stalks short and somewhat angular: *foot* broad in front and bluntly pointed behind: *appendages* 3 on each side, springing from short sheaths, of a lighter colour than the tentacles, and with a white tubercle at the base of each.

SHELL forming a depressed cone, somewhat scalariform, solid, opaque, of a rough and rather dull aspect: *sculpture*, numerous but irregular spiral ridges crossed obliquely by minute and close-set striæ, which are laminar or imbricated

* From its supposed resemblance to the turban of a magician.

in the interstices of the ridges; the base of the shell is encircled by a much stronger and more prominent ridge, giving that part a keeled or angulated appearance, and the upper part of each whorl is frequently puckered lengthwise: *colour* pale yellowish-white, beautifully variegated or painted by short longitudinal streaks of pinkish-red or (rarely) purple: *spire* not much raised; apex small and pointed: *whorls* 8, regularly enlarging: *suture* deep and channelled: *mouth* very oblique, in consequence of the upper lip being placed far in advance of the lower: *outer lip* often broken and jagged: *inner lip* very thick, folded above over that part of the umbilical cavity, and furnished in the middle with a slight tooth-like projection: *inside* nacreous: *umbilicus* rather wide and bordered by a smooth broad ridge; it is very deep and shows all the inner spire: *operculum* having from 12 to 15 volutions, becoming somewhat concave towards the centre, the under side of which has a minute boss or point; each volution is microscopically striated in an oblique and somewhat curved direction. L. 0·85. B. 1·15.

Var. *alba*. Shell of a uniform white.

HABITAT: Rather common, from low-water mark to 40 f., in the southern and western counties of England, the Channel Isles, Bristol Channel, Ireland, west of Scotland, and the Orkneys and Shetland; Anglesea (Pennant); Isle of Man (Forbes). It does not appear to be a native of our eastern or north-eastern coasts, although Mr. Bean found a dead specimen at Scarborough. Sir Cuthbert Sharpe included it in his list of Hartlepool shells; and Miss Backhouse is said to have met with it at Seaton Carew, Durham. I agree with Mr. Alder in believing that these specimens may have been introduced in ballast. The variety occurs at Oban, Skye, Ullapool, and Lerwick. *T. magus* is fossil in the "post-pleistocene beds" at Belfast (Grainger), Clyde beds and Ireland (Smith), Strethill (Maw); higher and older deposits, 400–440 feet, in the Christiania district (Sars); Antibes (Macé); Subapennine tertiaries

(Brocchi); Sicily (Philippi). Lovén discovered it living in the south-west of Sweden after the publication of his 'Index': else all the foreign localities are southern, and comprise the coasts of France, Spain, Portugal, Italy, Greece, North Africa, Madeira, the Canaries, and Azores, at depths of from 4 to 40 f., besides the Red Sea (Forskål).

The animal is beautifully and variously coloured, and is tolerably active. Its prettily painted shell was the "Sorcière" of D'Argenville. Under a rude and dull exterior it has a thick layer of bright pearl, which is brought out by the process called "cleaning." Such improvements of Nature's work were placed by Scopoli foremost in the Catalogue of his "calamitates nobilis scientiæ." Mr. Barlee used to be proud of showing his fine collection of British shells, especially to young ladies, until one of them innocently asked him if he picked them up in the summer and polished them in the winter! Very young shells are equally convex on each side of the peripheral keel, and the umbilicus is then very small. They exhibit numerous fine longitudinal striæ, which are curved and not less conspicuous than the few spiral ribs formed at that period of growth. A cancellated appearance is the result; and the sculpture is not unlike that of *Margarita cinerea*, Couthouy.

This is the *T. tuberculatus* of Da Costa.

5. T. TU'MIDUS*, Montagu.

T. tumidus, Mont. Test. Brit. p. 280, tab. 10. f. 4. 4; F. & H. ii. p. 513, pl. lxv. f. 8, 9, and (animal) pl. D D. f. 2.

BODY pale yellowish-white, transversely streaked with brown or fine dark-purplish lines, which are sometimes

* Swollen.

arranged diagonally (so as to give the upper surface a partially granulated appearance), and minutely but irregularly speckled with flake-white: *pallial lappets* large and unequal in size, the left one being the smaller and slightly scalloped, the other plain-edged: *head* semicircular, lineated or wrinkled transversely, closely scalloped at its edge; front lobes small and white: *tentacles* white, filiform, very long, slender, flexible, somewhat contractile, and finely setose; tips blunt: *eyes* proportionally large, seated on angular bulbs or short tubercles ("capable of twisting about in various directions," Montagu): *foot* lanceolate, thick, rounded at each end, with small angular points at the corners; edges delicately scalloped; top fringe or ridge on each side thin and wavy; sole flake-white: *appendages* 3 on each side, white, issuing from beneath the dorsal ridge; they resemble the tentacles, and are nearly as long and more pointed; each of the filamental appendages or vibracula has at its base a small cup-shaped tubercle. The animal is exquisitely ciliated all over.

SHELL turreted but not much elevated, solid, opaque, of rather a dull hue: *sculpture*, numerous fine spiral ribs, which are often alternately larger and smaller, and vary in size and their relative proximity; the surface is crossed by minute and close-set oblique striæ; the base of the shell, and usually the upper part of each whorl, are encircled by a more or less distinct keel, giving an angulated appearance: *colour* varying from white to citron, closely spotted or speckled with reddish-brown (the spots being arranged in spiral lines), and often marked with more or less irregular dark longitudinal blotches or streaks: *spire* moderately raised: *whorls* 6 or 7, their convexity being in an inverse ratio to the height of the spire; they gradually increase in size: *suture* frequently slight, deeper in more turreted specimens: *mouth* oblique, in consequence of the upper lip advancing considerably beyond the lower; it is notched in the middle of the outer lip, and channelled below the pillar: *outer lip* thin and plain: *inner lip* thick, folded back on the umbilicus, and furnished in the middle with a slight tooth-like tubercle: *inside* beautifully iridescent: *umbilicus* large but not wide, obliquely excavated, and exposing a considerable part of the inner spire: *operculum* having from 10 to 12 whorls, and mostly becoming concave towards the centre. L. 0·333. B. 0·333.

HABITAT: Oozy ground in the laminarian zone, and

on a stony or shelly bottom in deeper water, in every part of our seas from 4 to 95 f.; off the Mull of Galloway in 50–145 f. (Beechey). It occurs in all our upper tertiary strata, from Fort William (J. G. J.) to the Red Crag (S. Wood); Christiania district, in the higher and older deposits, at a height of 400–440 feet above the sea-level (Sars). Its distribution in a recent or living state extends from Iceland (Steenstrup and Torell) to the Ægean (Forbes), at depths varying from 4 to 60 f. M'Andrew and Barrett found it living on the shore in Upper Norway.

The animal of this rather common species is active and restless. Northern greatly exceed southern specimens in size; but those from deep water in every locality are invariably dwarfed. Some have no umbilicus; in others the spire is either pyramidal or depressed. The fry are often marked with spiral pink lines.

It is the *T. Racketti* of Payraudeau, and probably the *T. Nassaviensis* of Chemnitz and *T. patholatus* of Gmelin. The fry was figured by Walker as *T. fuscus*, and described by Macgillivray as *Skenea Serpuloides*.

6. T. CINERA'RIUS*, Linné.

T. cinerarius, Linn. S. N. p. 1229; F. & H. ii. p. 516, pl. lxv. f. 1–3, and (animal) pl. D D. f. 1 & 1*a*.

BODY purplish-grey minutely speckled with yellow, or yellowish speckled with flake-white, and marked with purplish-brown lines or streaks in front and blotches of the same hue at the sides (in southern examples barred with violet and white): *mantle* rather thick, yellowish; lappets thin, leaf-like and folded, that on the left being split into branched pectinations, the other plain: *head* semicircular, finely scalloped at the edges; veil forming two fringed lobes above the tentacles,

---

* For *cinereus*, ash-coloured.

one on the inner side of each; the veil is a continuation of the foot-crest and " when erected has the appearance of an awning or semipavilion hanging over the disk of the muzzle" (Clark): *tentacles* filiform, long and tapering, marked across with purplish-brown (in southern examples alternately violet and white) rings, and sometimes down the middle with a dark line; they are covered with whitish cilia, and contractile: *eyes* placed on short angular stalks which are white in southern examples: *foot* thick, broader and rounded in front, and bluntly pointed behind, with finely and closely ciliated edges; sole yellow; the ridge or crest on the upper part of each side is irregularly fringed; lateral appendages 3 on each side, with frequently several shorter intermediate ones; the principal filaments resemble the tentacles, but are usually shorter and slighter (white in southern examples); each is sheathed, and has sometimes at its base a small tubercle on each side, which are occasionally of a darker colour and might be taken for ocelli or eye-specks.

SHELL varying in height, according to the nature of habitat (being more depressed when living among *Laminariæ* than among stones between tide-marks or in the coralline zone), solid, opaque, and of a rather dull hue: *sculpture*, 7 or 8 thread-like spiral ridges on the upper part of the body whorl, with often one or two finer striæ between each ridge, and about a dozen fine ridge-like striæ on the under side; the intermediate surface is covered with numerous very minute longitudinal hair-like striæ, which are set obliquely; the basal keel is blunt, but distinct: *colour* light grey or pale yellowish, variegated by close-set narrow and oblique streaks of dark purplish-brown, the continuity of which is mostly interrupted by the spiral ridges, so as to give a somewhat speckled appearance: *spire* more or less raised, with a blunt apex: *whorls* 6 or 7; the lower ones are flattened and expanded, and the top ones rounded and moderately convex: *suture* narrow, although rather deep and sometimes channelled: *mouth* large, squarish and oblique, as in other species: *outer lip* bevelled to a thin edge: *inner lip* thick, somewhat reflected, especially over the upper part of the umbilicus, and usually furnished in the middle with a slight tubercular projection: *inside* highly nacreous except near the edge of the mouth, which is white and dull: *umbilicus* rather small and narrow, obliquely funnel-shaped and colourless, not exposing the spire of the penultimate whorl: *operculum* having from 10 to 12 volutions, which appear a little imbricated, and each is marked by a raised line or ridge; they are microscopically striated across in a radiating direction. L. 0·5. B. 0·55.

Var. 1. *electissima*. Smaller and more regularly conical. *T. electissimus* (Bean), Thorpe, Brit. Mar. Conch. p. 264.

Var. 2. *variegata*. Also smaller, and ornamented by a few short and broad dark reddish-brown rays on the upper part of each whorl, besides the ordinary coloured streaks.

HABITAT: Abundant everywhere, on stones and seaweed at low-water mark and in the laminarian zone. Var. 1 inhabits deep water; the other variety is found in the Channel Isles, as well as in the Mediterranean. This species frequently occurs in our latest tertiary strata, including the Clyde, Belfast, and Sussex beds, and the Red Crag; Christiania, in lower and younger deposits, 100–150 feet (Sars); Piedmont (Brocchi). Living in Iceland (Mohr); Scandinavia (Linné and others); Heligoland (Frey and Leuckart); North of France (De Gerville and others); Vigo and the North Spanish coast (M'Andrew); Mediterranean (Linné and others); Adriatic (Chiereghini); Mogador (M'Andrew); Black Sea (Krynicki and others). The bathymetrical range given in these foreign localities extends from low-water mark to 60 f.

When crawling it moves each side of its foot by turns. The left-hand pallial lappet serves for aërating the gill, like the semitubular fold in the *Muricidæ* and other Siphonobranchiata. According to Lovén the eggs are yellowish and numerous, not enclosed in capsules, but laid indiscriminately. M. Lespés detected one of the Trematode parasites (*Cercaria brachiura*) in the animal of this species at Arcachon. Its strong shell does not protect it from also becoming the prey of creatures larger than itself. Fishes devour it wholesale; and Macgillivray tells us that on the shores of the

Hebrides the throstle feeds on this kind of *Trochus*, as well as on the common periwinkle, holding one in its beak and breaking the shell by sharp and repeated strokes against a stone. A small living specimen which I dredged in Loch Alsh was thin, pearly, and lustrous, owing to the greater part of the outer layer having been removed by some natural cause. Some have no umbilicus or perforation. Those on *Laminaria saccharina* in Shetland are remarkably large, nearly an inch in breadth. The fry are not angulated at the base.

T. cinerarius of Born is an Indian shell, and that of Olivi appears to be a variety of *T. varius*. The present species is the *Trochus* (not *Turbo*) *lineatus* of Da Costa, and *Gibbula striata* of Leach. *T. littoralis* of Brown is scarcely a variety; and his *T. perforatus* was probably a specimen encrusted with a zoophytic growth, which he mistook for an epidermis. The variety *variegata* corresponds with the description and figure of Payraudeau's *T. ægyptiaca*; but it is not Lamarck's species of that name. This variety was described by Récluz as *T. Philberti*.

T. cinereus of Da Costa has the inner or pillar lip plaited, and is a species of *Clanculus*. It is "said to be from the South Seas" (Donovan) and "a native of the West Indies" (Forbes and Hanley); but assuredly it is not British. Mr. Dillwyn possessed and gave me one of the original specimens.

7. T. umbilica′tus*, Montagu.

N. umbilicatus, Mont. Test. Brit. p. 286; F. & H. ii. p. 519, pl. lxvi. f. 1–4 (as *T. umbilicalis*).

Body light yellowish-brown, marked transversely with purplish lines, and tinged in front with fleshcolour: *mantle* thin, edged with short purplish streaks; lappets leaf-like, the

* Umbilicate.

inner one on the left irregularly pectinated or fringed, and the other plain but folded; each of these lappets is continued along the upper part of the foot, where it forms a jagged crest: *head* semicylindrical and short, wedge-like at the extremity, streaked across, and notched at the front edge; veil composed of two membranous and fringed lobes or expansions above the tentacles, one on each side of the intermediate space: *tentacles* slender and bluntly pointed; they are thickly covered with short cilia, and marked with purplish rings, which are alternately large and small, and often interrupted or broken as well as scalloped; these rings somewhat resemble the joints of an *Equisetum*: *eyes* rather large; stalks angular and yellow: *foot* rather oval than oblong, sparingly granulated on the upper part and sides; edges fringed with minute cirri; sole slightly furrowed down the middle: *appendages* 3 on each side, the hinder two being rather close together, and the other in the middle of the lateral space; they are indistinctly annulated and slightly setose; each is encircled at its base by a jagged sheath, and provided with a small whitish and raised tubercle on each side, which issues out of the foot-crest.

SHELL more depressed than *T. cinerarius*, and (although the base is flatter) never inclined to a pyramidal form; the spiral ridges are sharper and fewer, especially in the young: the colouring is different; both have a similar kind of marking, but in the present species the longitudinal rays or streaks are red, besides being broader and not half so many as in the other species; and they are sometimes zigzag, instead of being broken into spots or interrupted by the sculpture; this is striped and the other lineated; just within the outer lip are two borders, one of yellow and the other of green, variegated by red spots; this edging is minutely tubercled, like shagreen. L. 0·55. B. 0·7.

Var. 1. *atro-purpurea*. Always depressed and of a dark-purplish hue.

Var. 2. *decorata*. More conical, and speckled like the variety *variegata* of the last species.

Var. 3. *Agathensis*. Smaller, with the spire more raised, less angular, and somewhat glossy on the underside; colouring purple instead of red; base usually not umbilicate (except in the young), but occasionally perforated. *T. Agathensis*, Récluz, in Rev. de Zool. for 1843.

HABITAT: Gregarious among stones, and on *Fucus serratus*, just below the brink of high-water mark at neap tides, on our southern coasts, in the Bristol Channel, Isle of Man, all around Ireland, and west of Scotland as far north as Loch Alsh. The following localities are doubtful: — " North Britain " (Laskey); Aberdeenshire, Banff, and Kincardine (Macgillivray). The variety *atro-purpurea* was found by Mr. Clark at Exmouth; *decorata* by myself at Weymouth; and *Ayathensis* is not uncommon in the Channel Isles, and remarkably plentiful in Fermain bay, Guernsey. This last variety frequents a lower part of the littoral zone than the typical form; the young are distinctly umbilicate, and resemble in shape and sculpture those of *T. cinerarius*. It is the variety *læta* of the Rev. R. T. Lowe. The fossil localities for the present species are questionable. Mr. J. Smith enumerates Ireland, and Mr. Maw Strethill; but possibly the latter geologist was deceived by the " navvies " who brought him specimens. The case of the Macclesfield deposit has served as a useful warning not to place too much reliance on the discoveries of those ingenious workmen. *T. umbilicatus* inhabits the north and north-west of France; Vigo, and Faro in Algarve, on *Zostera* (M'Andrew); Gulf of Lyons (Martin); Toulon (Gay); south coast of the Crimea, in the Black Sea (Middendorff). I found the variety *læta* at Rochelle, Mr. M'Andrew at Corunna, and the Rev. R. T. Lowe at Mogador.

This littoral species lives in company with *T. cinerarius*, but always retains its distinctive character: their mode of locomotion is the same. If either is taken from the shore, and immersed in sea-water, it will expel bubbles of air through the right-hand lappet or fold of the mantle. The fry of *T. umbilicatus* is white, nearly flat, and has only two or three prominent ribs.

It is the *T. obliquatus* of Gmelin, *T. umbilicaris* of Pennant, *T. cinerarius* of Pulteney and Lamarck (though neither of the two latter are Linné's species so named), and *Gibbula lineata* of Leach.

C. Very small, circular, nearly flat-spired, with an exceedingly wide and open umbilicus. *Circulus.*

8. T. Dumi'nyi *, Requien.

Delphinula Duminyi, Req. Cat. Cors. p. 64.

Animal not known.

SHELL orbicular, rather solid, but semitransparent and somewhat glossy: *sculpture*, 8-10 sharp and narrow spiral ridges on the upper part of the last whorl, half that number on the penultimate whorl, and two or three on the next, the upper two whorls being smooth; the lowest ridge is placed just under the periphery, and is usually stronger and more prominent than any of the rest (from which it is frequently separated), and it encloses the umbilical area; sometimes this part is also ridged; the furrow between each ridge is crossed by curved microscopical striæ: *colour* white: *spire* scarcely raised, but the apex is well defined: *whorls* 5, cylindrical and gradually enlarging: *suture* distinct, although not deep: *mouth* squarish, obliquely truncated as in other species of *Trochus* belonging to the last section: *outer lip* flexuous, with a sharp edge, strengthened a short distance within by a slight rib: *inner lip* somewhat thickened and reflected towards the umbilicus, and adhering to a considerable part of the periphery of the penultimate whorl: *inside* porcellanous and polished (not nacreous), exhibiting the outside ridges as dark lines: *umbilicus* extending more or less over the base of the shell; it shows nearly as much of the internal spire as is seen of the spire outside; in some specimens the inner whorls are concentrically striated: *operculum* circular, with about a dozen volutions, which wind spirally and gradually, and converge to the centre. L. 0·035. B. 0·1.

* Named in honour of Professor Duminy of Ajaccio.

HABITAT: Bundoran, in Donegal Bay, where it was first found by Mr. Waller. As yet only about a dozen specimens have been met with. Searles Wood discovered this characteristic and interesting species in the Coralline Crag at Gedgrave and Sutton; and Philippi recorded a single specimen from clay at Cefali near Catania. Requien briefly described it as recent from Ajaccio, on the authority of M. Brice and Professor Duminy; Weinkauff has enumerated it as an Algerian species; and M. Honoré Martin procured a few specimens from the Gulf of Lyons. The kindness of this last-named excellent conchologist has enabled me to describe the operculum.

It differs from the fry of *T. umbilicatus* (which also inhabits Donegal Bay) in being equally convex on both sides, the whorls being cylindrical and never angulated as in that species, having twice as many and much finer spiral ridges, the periphery being rounded and not keeled, the suture not so deeply channelled, and in its remarkably wide and open umbilicus. The two species cannot well be confounded. Being anxious to confirm and extend the discovery of my friend, Mr. Waller, I made a purpose-journey to Bundoran, a few years ago when I was last in Ireland, in the hope of procuring more specimens of this rare shell. I had but a single day, which turned out to be about the worst ever known in that rainy climate; but by leaving Enniskillen at four in the morning, I got two or three hours at Bundoran, and attained my object. Should you see any one acting in a manner apparently so eccentric, do not straightway set him down as out of his senses, but suppose that he may be devoted to an uncommon pursuit. Perhaps your ideas with regard to his conduct may even be more charitable if you consider that such pursuits advance knowledge of some kind; you might then do more than excuse him, and, with no feeling of disparagement,

"You would say, it hath been all in all his study."

Philippi described this species in 1836 as *Valvata? striata*, in consequence of its occurring in the same deposit with *Corbicula fluminalis*. He afterwards, however, suspected its being a *Delphinula*. Wood placed it in his genus *Adeorbis*; but the typical species (*A. subcarinatus*) has a paucispiral and horny operculum, with a lateral nucleus, and is probably allied to *Solarium*. The specific name *striatus* is preoccupied by a well-known Linnean species. *A. supranitida* and *A. tricarinata* of Wood appear to be fossil varieties of the present species. Requien's Catalogue and Wood's Monograph were published in the same year, 1848.

D. Spire moderately raised; base slightly umbilicate in the adult, and perforated in the young: pillar-lip furnished with a strong tubercular tooth. *Trochocochlea*, Klein.

9. T. LINEA'TUS *, Da Costa.

Turbo lineatus, Da Costa, Brit. Conch. p. 100, t. vi. f. 7. *Trochus lineatus* F. & H. ii. p. 525, pl. lxv. f. 4, 5 (as *T. crassus*.)

BODY dark-ashcolour, with a greenish tint: *mantle* thin, yellowish-brown; lappets leaf-like, the left unequally pectinated, and the right plain and usually folded: *head* semicylindrical, rather long, transversely streaked, notched at the front edge; veil above the tentacles membranous, and irregularly digitated or fringed: *tentacles* slender, bluntly pointed; they are annulated with purple lines variable in the intensity of their colour, and alternately large and small, sometimes interrupted or partly zigzag; they are clothed with fine short cilia: *eyes* large, placed on angular stalks or processes, which are more or less tinged with orange: *foot* oval, with a bluntly pointed tail, closely and finely granulated at the sides; margin purplish and thickly fringed with short cilia; dorsal crest jagged; sole divided down the middle by a whitish line,

* Decked out.

and when at rest showing on each side similar lines of different lengths, which are rather less numerous and more irregularly disposed towards the tail; these lateral lines represent folds or creases that disappear when the foot is in action: *appendages* 3 on each side (sometimes 4 on one side and 3 on the other, Clark), tapering, ringed and setose, like the tentacles; there is frequently at the base of each appendage a white or yellow tubercle on either side of it.

SHELL regularly conical, very thick, opaque, and of a dull hue: *sculpture*, none in the adult; but the young have spiral ridges and minute cross striæ, as in *T. cinerarius* and other species of the same section: *colour* yellowish or light-grey, with a greenish tinge, variegated by numerous and close-set zigzag purplish markings, arranged in longitudinal rows or streaks, giving the surface an obscurely tessellated appearance; apex (which is always eroded) of a yellowish hue, and sometimes partly exposing the inner layer of nacre: *spire* more or less raised, and bluntly pointed: *whorls* 6, rather quickly enlarging, and convex; the upper part of the last whorl is compressed or somewhat flattened: *suture* slight: *mouth* large, obliquely oval: *outer lip* rounded, and sharp-edged, with a slight notch or angular point at the upper corner: *inner lip* extremely thick and broad, reflected a little over the umbilicus; it is furnished below the middle with a remarkably strong tubercular prominence, which is nacreous and apparent in all states of growth: *inside* beautifully iridescent, except at the margin, the outer zone of which is mottled with black and green, and is microscopically pustulated, and the inner is white and almost pearly: *umbilicus* rather large but shallow, partly covered by the inner lip; the base of the shell is more or less worn away by the continual friction of the upper part of the foot: *operculum* yellowish-horncolour, with about 15 volutions, each of which is obliquely and minutely striated in the line of growth. L. nearly 1. B. 1.

Var. *minor*. Smaller, and eroded.

HABITAT: Local, but not uncommon, on rocks and stones just below high-water mark at neap tides in the counties of Dorset, Devon, and Cornwall; Channel Isles (Hanley); bays near Swansea (J. G. J.); Pwllheli, Carnarvonshire (Da Costa); Anglesea (Donovan); Ireland, as far north as Donegal Bay (Waller and

J. G. J.); Dunbar, where "one specimen of this shell was taken by the dredge from deep water" (Laskey); Peterhead (Macgillivray); Cumbrae, Clyde district (J. Smith). Da Costa also gives Hampshire and Norfolk; but these and the Scotch localities want confirmation. With similar hesitation I must cite my friend Mr. Smith as the authority for considering this species fossil in the Paisley beds. The variety is from Instow, North Devon (J. G. J.), and Arran Isles, co. Galway (Barlee). The typical form inhabits the north of France (De Gerville and others); Rochelle (J. G. J.); Vigo, and Faro in Algarve (M'Andrew); Santander, in the north of Spain (E. J. Lowe); Hyères (Sir W. C. Trevelyan, Bart.); Mogador (M'Andrew and R. T. Lowe).

The motion of the foot is wave-like, each side alternately. On leaving the water this *Trochus* takes in a supply of air, which (if the animal be again immersed) is expelled or escapes in bubbles by the right-hand lappet of the mouth. The erosion of the shell, which is not unfrequent, seems to be caused, and is certainly increased, by the perforations of a minute kind of seaweed or its spores; water enters the orifices thus formed, and gradually effects a disintegration of the outer layers, one after another. The whole fabric not being of a homogeneous nature, or equally compact, some parts are more easily acted on than others. Mr. Clark found that every specimen in a particular spot near Exmouth had a distorted operculum; this was irregularly paucispiral, and sometimes ear-shaped, but always had a central nucleus. He accounts for it in the following way. "The animal either sheds the operculum, or is deprived of it by the attacks of enemies, perhaps from its own pulli, white masses of which, in the genial season, I have seen deposited on the foot, and they may possibly feed on and destroy it." I should be disposed to attribute the malformation to an epidemic disease of the operculigerous lobe. It also occurs in *Buccinum undatum*, the fry of which are separately enclosed in capsules, and are therefore incapable of feeding on the maternal operculum. Besides, all the specimens, male and female, are affected in the same manner. *T. lineatus* may be known from its congeners by its size, colour, tooth, and peculiar umbilicus; and the periphery is never angulated.

Pulteney described it as *T. crassus*, and the young as *T. lineatus*. *Monodonta articulata* of Lamarck and *M. Draparnaudi* of Payraudeau are closely allied to the present species, if all of them are not the same. *T. lineatus* of Lamarck is a New Holland shell. According to Bouchard-Chantereaux, ours is the *T. punctulatus* of De Blainville. *T. (Monodonta) sitis* of Récluz appears to be the young of the European shell.

E. Spire pyramidal; base imperforate; pillar-lip notched or angulated at the lower part. *Ziziphinus*, Leach.

10. T. Montacu'ti*, (*Montagui*) W. Wood.

T. Montagui, Wood, Ind. Test. Suppl. pl. 6. f. 43; F. & H. ii. p. 511, pl. lxv. f. 10, 11.

Body yellowish-white, speckled with purplish-brown and milk-white flakes: *mantle* thin and semitransparent, marked with greenish spots; lappets large in proportion, forming two saucer-shaped lobes, one on each side of the tentacles; both these lobes appear plain, although of different sizes: *head* semicylindrical, with the front edge minutely notched; veil bilobed, scarcely perceptible: *tentacles* slender, sometimes finely pointed, in other examples having club-shaped tips: *eyes* rather large, on short hairy tubercles: *foot* lanceolate and thick, rounded in front, with somewhat angular corners, and

---

* Another tribute to the memory of Col. Montagu.

bluntly pointed behind; sides granulated; margin fringed; dorsal ridge serrated; tail keeled: *appendages* 3 on each side, filiform, with an eye-like tubercle at the hinder base of every filament, besides an extra or supernumerary eye-spot in front between the outer base of each tentacle and the filament next to it.

SHELL narrow at the base in proportion to the height, somewhat convex on the underside, with a bluntly angulated periphery, moderately solid, opaque, and slightly glossy: *sculpture*, fine spiral ridges, of which there are six or seven on each whorl except those forming the apex, and about the same number encircle the base; the space between each ridge (and sometimes the ridges also) is crossed by minute close-set imbricated striæ, which are curved or lie obliquely in the line of growth, and are occasionally finer and less distinct on the last whorl: *colour* yellowish-white, with a row of small dark reddish-brown spots on each ridge, or with longitudinal streaks of that colour on the last whorl and rarely on the others; now and then may be seen a greenish or partially iridescent hue: *spire* bluntly pointed: *whorls* 7, gradually enlarging, compressed but not flattened; those forming the apex of the spire are rather convex: *suture* slight but distinct: *mouth* obliquely squarish: *outer lip* rather thin: *inner lip* thick, reflected on the pillar, which is angulated below, and furnished with a scarcely prominent tubercle that seems to form a slight notch at the base: *inside* silvery and iridescent, except towards the margin, where it is either whitish or coloured like the outside: *umbilicus* none in the adult, but deep in the young, and obliquely margined by a whitish ridge: *operculum* having from twelve to fifteen volutions, which are defined by raised lines, and indistinctly striated across. L. 0·3. B. 0·25.

Monstr. Scalariform; whorls somewhat angular, and separated by a deep suture.

HABITAT: All our coasts, chiefly in the coralline zone, from 7 to 95 f.; local, but tolerably common in Guernsey and the west of Scotland. A specimen of the monstrous variety (which is very elegant) was dredged by Mr. Waller and myself at Larne, co. Antrim; and another, less symmetrical in its irregularity, was taken by Dr. Lukis in deep water at Guernsey. *T. Montacuti* occurs, according to S. Wood, in the Red and Coralline Crag. It inhabits the north of France (Macé, Cailliaud, and J. G. J.); Portugal and Spain (M'Andrew); Gulf of Lyons (Martin); Ajaccio (Requien); Malta and Sicily (M'Andrew); Algiers (Weinkauff); and M'Andrew obtained a dwarf variety at Tunis. Its range of depth in the Mediterranean is from 12 to 50 f.

When placed on its back, with the shell underneath, it twists about actively, in order to regain a footing. The edges of the foot in this and other species of *Trochus* are occasionally folded inwards and brought together, so as entirely to conceal the disk or sole. I put a live specimen of *T. Montacuti* into fresh water for three minutes; it withdrew into the shell, and by keeping its door shut suffered no inconvenience, as soon appeared upon its being restored to its native element. The spiral ridges in the fry are frequently marked with reddish-brown lines.

This species is the *T. Cyrnæus* of Requien, and *Montagua Danmoniensis* of Leach.

11. T. STRIA'TUS*, Linné.

T. striatus, Linn. S. N. p. 1230?; F. & H. ii. p. 508, pl. lxvi. f. 5, 6.

SHELL proportionally narrow at the base, more or less flattened on the underside, with a rather sharply keeled periphery, solid, opaque, and somewhat glossy : *sculpture*, fine spiral ridges, of which there are eight or nine on the last and next two whorls, the number decreasing towards the apex; the lowest ridge is the largest, and forms the basal keel; there are also from ten to twelve similar ridges on the base; the whole surface is covered with delicate and numerous imbricated striæ, which obliquely cross the ridges, but are stronger in their interstices; sometimes the ridges are partly nodulous in consequence of this decussation : *colour* pale yellowish, or white with oblique streaks of dull red or very dark brown (nearly

* Striated or grooved.

black) in the line of growth; in some specimens the streaks are interrupted and give a speckled appearance, or there is a greenish tint, and in others the apex is reddish: *spire* bluntly pointed: *whorls* 7, gradually enlarging, flattened, all but the two apical ones, which are rounded: *suture* very slight and inconspicuous: *mouth* obliquely squarish: *outer lip* rather thin: *inner lip* short, broad and thick, undistinguishable from the pillar; it is slightly reflected above, and notched below by a small blunt tubercular tooth, as in the last species: *inside* silvery and iridescent, except towards the margin, where it is frosty-white and thickened by an indistinct angulated rib; the young are slightly umbilicate: *operculum* as in *T. Montacuti*. L. 0·35. B. 0·3.

Monstr. Scalariform; whorls convex, each having a keel-like ridge in the middle, and separated from the one next to it by a deep suture; base rounded.

HABITAT: Laminarian zone (especially on *Zostera marina*), from low-water mark to 15 f., in the Channel Isles, Dorset, Devon, Cornwall, Cork, Baltimore, and Bantry. Dublin Bay (Turton). The monstrosity was found by Mr. Hockin at Falmouth, and by him kindly presented to me; it is similar to that of *T. Montacuti*. The present species has only been noticed as fossil in the Sicilian tertiaries (Philippi). Recent on the coasts of France, Portugal, Spain, Italy, Algeria, the Adriatic, Madeira, and the Canaries, from the shore to 20 f.

The animal of this rather common species does not appear to be known. The shell differs from *T. Montacuti* in its larger size, remarkably flattened whorls and base, and in having a greater number of spiral ridges.

In all probability the *T. striatus* of Linné was intended for the next species—if indeed that is not a variety of the one which I have now described. Gmelin and his followers named the present species *T. erythroleucos*, Da Costa *T. parvus*, Donovan *T. conicus*, and Deshayes *T. depictus*.

12. T. exaspera´tus*, Pennant.

T. exasperatus, Penn. Brit. Zool. iv. p. 126. *T. exiguus*, F. & H. ii. p. 505, pl. lxvi. f. 11, 12.

"The animal has the sides of the foot, the tentacles, and lateral cirrhi tinged with madder red. The eye-peduncles are white, as is also the disk of the foot" (Forbes and Hanley).

SHELL of the same size and general shape as *T. striatus*. It is rather more pyramidal, and decidedly more solid; the sculpture is much coarser, and the basal ridge is longer and stronger, and encircles each whorl; it has only half as many ridges and cross striæ, and the former are frequently nodulous; the colour is different, having usually a good deal of red or pink in it, and is sometimes prettily decorated by occasional concentric rows of rose-red and white spots, or it is now and then of an ashy or olive hue; the apex is mostly, but not always, red or pink.

HABITAT: Channel Isles, among loose stones at low-water mark (Lister and others); Lulworth, 7–12 f. (J. G. J.); Weymouth (Pulteney and others); Land's End (Maton and others). The following localities are doubtful, or some of them belong to *T. striatus*:—Margate (Hanley); Hants (Forbes); Sussex and Devonshire (Da Costa); Torquay (Hanley); Bantry Bay (Dillwyn); Cork (Humphreys); Dublin Bay (Turton, Warren, and Walpole); north of Ireland (Thompson); Ayr and Firth of Clyde (J. Smith). Further information is also desirable as to the only British locality where the present species has been recorded as fossil, viz. Wexford (Col. Sir H. James, *fide* Forbes). Brocchi noticed it from the tertiary strata in the Isle of Ischia, and Philippi from those of Sicily. It inhabits the coasts of France, Portugal and Spain, every part of the Mediterranean, the Ægean, Madeira, and the Canaries, at

* Roughened.

depths ranging from 3 to 105 f.; Black Sea (Kutorga, *fide* Middendorff); Azores (Drouet).

At Lulworth this little *Trochus* enters the lobster-pots, along with *T. cinerarius* var. *conica, Buccinum undatum, Nassa reticulata, N. incrassata,* and *Murex erinaceus*—all of them apparently being attracted by the bait, which consists of soft crabs or pieces of fish. It is therefore highly probable that the *Trochi* are sarcophagous. It may turn out that this so-called species is only a variety of *T. striatus*, owing to a difference of habitat—although the young and fry of each are distinguishable, and exhibit the same relative characters as the adult.

The present species is the *T. conulus* of Da Costa (but not of Linné), *T. exiguus* of Pulteney, *T. crenulatus* of Brocchi (not of Lamarck), *T. pyramidatus* of the last named author, and *T. Matonii* of Payraudeau.

13. T. MILLEGRA'NUS*, Philippi.

T. millegranus, Phil. Moll. Sic. i. p. 183, t. x. f. 25; F. & H. p. 502, pl. lxvi. f. 9, 10.

BODY yellowish-white, streaked or spotted with purplish-brown, and sometimes faintly tinged with green, covered all over with short prickly points, so as to appear pustulated: *mantle-lappets* large and expanded: *head* wrinkled, finely scalloped at its edges; veil small, bilobed, and serrated: *tentacles* filiform, long and slender, with blunt tips, marked lengthwise with three purplish-brown lines, one in front and another on each side: *eyes* large, on the underside of whitish tubercles at the external bases of the tentacles: *foot* thick, oblong, truncated, slightly angulated at the corners in front, and rounded behind; the upper part is flat and edged with a serrated ridge, the operculum resting on the posterior extremity of this level space; sole pale lemoncolour: *appendages* 3 on each side of the foot, issuing from beneath the top

* Covered with numerous granules.

ridge; they closely resemble the tentacles in every particular, except in being more slender; each has a brownish eye-speck at its hinder base, and there is an extra pair of such ocelli between the tentacles and front pair of appendages. Every part of the animal is exquisitely and closely ciliated.

SHELL broad and flattened at its base, and regularly tapering to a rather fine point, solid, opaque, not glossy: *sculpture*, six or seven concentric rows of granules and as many intermediate rows of a smaller size on the upper part of the last whorl, nearly as many on the next four whorls, and fewer on the apical or top whorls, the first two of which are ridged instead of granulated; the lowest row in each whorl is much the largest and most prominent, and it forms a conspicuous keel on the basal circumference of the body-whorl, and at the suture of the next two whorls; the granulation arises from the intersection of spiral ridges by fine and obliquely longitudinal striæ; the base of the last whorl is encircled by about a dozen ridges, which are imbricated, and alternately large and small, as well as decussated by the oblique striæ; these basal ridges are seldom, or but slightly, granulated: *colour* whitish, with a very faint tinge of yellow, usually more or less spotted or speckled with reddish-brown or light purple: *spire* considerably raised, but not elevated, except in the variety; apex somewhat pointed: *whorls* 8, flat, and gradually enlarging: *suture* very slight, only marked by the ridge at the base of each whorl: *mouth* obliquely squarish: *outer lip* thin and mostly broken: *inner lip* white, and folded over the pillar, which is extremely thick and short, with an obscure tubercular excrescence near the base: *inside* nacreous, except towards the margin: *operculum* rather concave, having from twelve to fifteen volutions; it is membranous, and microscopically reticulated, like the scales of certain fishes. L. 0·6. B. 0·6.

Var. *pyramidata*. Smaller, and narrower at the base.

HABITAT: Hard ground, and among Tunicata, from 2 to 70 f., on the coasts of Northumberland and Durham, Aberdeen, Orkneys, Shetland, west coast of Scotland, Mull of Galloway (50–145 f., Beechey), and all Ireland. The variety occurs in Hants (Forbes); Shetland, Fishguard, and Guernsey (J. G. J.); Plymouth (Barlee); west bay of Portland (Forbes and M'Andrew); Cornwall (Hockin); Scilly Isles (Lord Vernon). This species has been found by me fossil at Fort William, and by S. Wood in the Coralline Crag; Antibes (Macé); Sicily (Philippi). It is Swedish and Norwegian (with a range of from 15 to 50 f.); but the extent of its distribution south of Great Britain is not well ascertained. M'Andrew has taken it off Lisbon in 7-12 f., and between Cadiz and Cape Trafalgar in 30 f.; Gay obtained it at Toulon; and Forbes dredged it in the Ægean, from 41 to 110 f. All the southern specimens that I have seen belong to the variety.

It is rather plentiful in the west of Scotland, but apparently not so much at home elsewhere. If it had not been for the far and wide researches of my friend Mr. M'Andrew, our knowledge of the geographical distribution of this species would be very scanty. His experience, as a dredger, surpasses that of the Shipman (in the 'Canterbury Tales') as a mariner, who had explored what was then reckoned the greater part of the European seas—

"Fro Scotland to the Cape of Fynystere,
And every creek in Brittain and in Spain."

The fry of *T. millegranus* has an umbilical perforation. This is probably the *T. miliaris* of Brocchi, and certainly *T. Clelandi* of W. Wood, *T. Martini* of Smith, my *T. elegans*, and *T. Clelandiana* of Leach.

14. T. GRANULA′TUS*, Born.

T. granulatus, Born, Ind. Mus. Cæs. Vind. p. 343; F. & H. ii. p. 499, pl. lxvii. f. 7, pl. lxviii. f. 3, and (animal) pl. D D. f. 4.

BODY pale yellowish or whitish, speckled with reddish-brown: *mantle-lappets* very large, white, pendent, and slightly scalloped: *head* strong and thick, finely fringed at the ex-

* Granulated.

tremity; veil slight, and bilobed: *tentacles* marked with a broad red-brown line down the middle: *eyes* dark-blue with black pupils, placed on short, but stout, white stalks: *foot* very large, truncated in front, and lobed or angulated at each corner, granulated at the sides, and pointed behind; sole fringed; crest white and puckered: *appendages* 3 on each side, white, shorter than the tentacles, but equally flexible.

SHELL exceedingly dilated and rounded at the base, with a slight incurvation towards the apex, moderately solid, opaque, and scarcely glossy: *sculpture*, from six to eight concentric ridges, and about as many smaller intermediate ones, on the upper part of the last whorl, besides an equal number on the lower part; the next whorl has nearly the same number and alternate disposition of ridges as are visible on the upper half of the last whorl, the ridges on the succeeding whorls becoming gradually fewer; the larger ridges, or some of them, are usually granulated, and invariably those at the apex; the whole surface is covered with very minute and close-set oblique longitudinal striæ; in younger specimens the periphery is encircled by a stronger ridge, which gives a keeled or angulated appearance to that part, and forms a kind of crest on the top of each of the upper whorls: *colour* yellowish-white, with a faint tinge of reddish-brown, and speckled with the latter colour on all or most of the principal ridges, or else irregularly marked lengthwise by blotches of the same hue; the larger ridges on the base are always prettily spotted: *spire* moderately raised, and tapering somewhat abruptly to a fine point: *whorls* 10, rather flattened; the last considerably exceeds all the others put together: *suture* slight, defined by a shallow furrow or level space between the uppermost ridge of each whorl and the lowest ridge of the preceding one: *mouth* obliquely truncated, slightly effuse or spread out below, rounded without, and angulated within: *outer lip* thin: *inner lip* white, and reflected on the pillar, which is extremely thick and somewhat curved, with occasionally an obscure tubercle near the base; behind the pillar is a slight depression, like a rudimentary umbilicus: *inside* highly nacreous: *operculum* rather concave, with a small cup-shaped pit in the centre, and having from fifteen to eighteen volutions; it is microscopically, but indistinctly, striated in a radiating direction. L. 1·5. B. 1·5.

Var. 1. *lactea*. Milk-white and spotless.

Var. 2. *conoidea*. More regularly conical and solid, with

the last whorl not so broad or large in proportion to the rest.

Monstr. Outer lip irregularly notched at its junction with the penultimate whorl, like a *Pleurotoma*.

HABITAT : Coralline zone, in Cornwall, Devon, Dorset, and the Channel Isles ; Isle of Man (Forbes and Walpole); south and east of Ireland (Turton and others); Belfast Bay, " two broken specimens, but probably introduced accidentally " (Hyndman); 50 f. off the Mull of Galloway, and living at a depth of 145 f. in Beaufort's Dyke, the species having been determined by the late Mr. William Thompson of Belfast (Beechey). The varieties and monstrosity are from Exmouth. Lamarck says it is found fossil in England; but he was probably misinformed. I do not agree with Mr. Wood in considering the Red Crag shell, which he named in his catalogue *T. granosus*, to be a variety of our recent species. Whether it was the progenitor of *T. granulatus* is another question. The fossil species is much smaller and more solid; it never has an incurved outline towards the apex, or a prominent tuberculated ridge on each whorl; and the spire is more depressed. Mr. James Smith has enumerated the present species as an Irish fossil, Mr. Woodward as occurring at Bramerton and Thorpe in the Norwich Crag, and Brocchi from Piacentino. It is not uncommon on the coasts of France, Spain, Italy, Algeria, Madeira, and the Canaries, at various depths ranging from 4 to 60 f.

This handsome shell is frequently procured by trawling. The fry has the first whorl smooth, and the second regularly and strongly cancellated; and it exhibits a conspicuous and rather deep umbilicus. The lingual ribbon is comparatively short; its outer extremity is covered by two large, oval, horny jaws.

It is the *T. papillosus* of Da Costa, *T. fragilis* of Pulteney (but not of Gmelin), and *T. tenuis* of Montagu. Born's publication bears the same date (1778) as that of Da Costa; the name given by the former is generally preferred or best known.

15. T. ZIZY'PHINUS *, Linné.

T. zizyphinus, Linn. S. N. p. 1231; F. & H. ii. p. 491, pl. lxvii. f. 1–6.

BODY yellowish, tinged with purple or crimson, and streaked or mottled with reddish-brown: *mantle* plain-edged; lappets as in *T. granulatus*, but not scalloped: *head* large, prominent, and flexible, wrinkled transversely; veil bilobed, but so small as to be almost rudimentary: *tentacles* sometimes pinkish, more or less distinctly streaked with a brown line down the middle: *eyes* rather large and prominent, with black pupils; stalks short, stout, and often white: *foot* thick and rather broad, slightly cloven in front and angulated at the corners, pointed behind; sole fleshcolour; crest fringed: *appendages* 4 on each side, and in some specimens several short intermediate ones; they are mostly white.

SHELL regularly pyramidal, with a level outline and a somewhat flattened or compressed base, solid, opaque, slightly glossy: *sculpture*, from six to eight concentric and imbricated ridges, besides as many smaller intermediate ones, on the upper part of the last whorl, and about a dozen grooves or impressed lines on the base; the preceding whorls have similar ridges, which gradually decrease in number towards the apex; the ridge which girds the base of each whorl is larger and broader than the rest, and gives the periphery an angulated appearance; the ridges on the upper whorls are granulated; the entire surface is covered with very minute and close-set, but obscure, oblique longitudinal striæ: *colour* pale yellow with a reddish tint, or fleshcolour (sometimes purple, flecked with white), with longitudinal streaks of reddish-brown, which are mostly interrupted or zigzag, and frequently mark each of the basal ridges with a line of spots; the underside of the shell is not thus decorated, except at the periphery; the apex or point is usually purplish: *spire* more or less raised, and tapering

---

* From the resemblance of its colour to that of the jujube.

to a rather sharp point: *whorls* 10-12, flattened, gradually diminishing in size towards the apex: *suture* slight, defined by the basal ridge of each whorl: *mouth* rhomboidal, spread out a little at the base of the pillar: *outer lip* thin: *inner lip* pearly, and reflected on the pillar, which is extremely thick, curved, and now and then furnished with a blunt tubercle; behind the pillar is an oblique and shallow excavation: *inside* nacreous: *operculum* slightly concave, with a small central pit, having from fifteen to eighteen turns, and microscopically striated in the line of growth. L. 1. B. 1.

Var. 1. *Lyonsii.* White, with occasionally a purplish tip. "*T. Lyonsii*" (Leach), Fleming, Brit. An. p. 323.

Var. 2. *humilior.* Spire depressed.

Var. 3. *lævigata.* Smooth and polished, with strong sutural ridges, considerably expanded towards the base, and having a depressed spire. *T. lævigata*, J. Sowerby, Min. Conch. t. 181. f. 1.

Var. 4. *granulifera.* White, with the ridges granulated.

Var. 5. *elata.* Dwarf, having the spire elevated, a narrow base, and the longitudinal striæ flexuous.

Monstr. Scalariform, with a rounded periphery and convex base. *T. discrepans*, Brown, in Mem. Wern. Soc. ii. p. 519, pl. xxiv. f. 4.

HABITAT: Rocks and stony ground, from low-water mark to 85 f.; common everywhere, especially in the laminarian zone. The 1st variety is equally diffused, although not so generally abundant; the other varieties are also occasionally white. Var. 2. Exmouth (Clark); Bantry Bay (Humphreys); Oban (Barlee). Var. 3. Anglesea (M'Andrew and Mrs. Hanmer Griffith); Loch Carron (Barlee and J. G. J.). Var. 4. West coast of Scotland; a single specimen (Barlee). Var. 5. Deep water on the coasts of Antrim and Shetland (J. G. J.). The monstrosity occurs with the ordinary form, but is rare. Fossil in the Caithness boulder-clay (Peach); Ireland (J. Smith); Norwich Crag (Woodward); Red and Coralline Crag (S. Wood); Antwerp Crag (Nyst). Its foreign distribution in a living state comprises all the North Atlantic from Finmark and the Faroe Isles to the Canaries, the Mediterranean, Adriatic, and Ægean, at depths ranging between the shore and 60 f.

The shell is subject to much variation in the height of the cone, as well as in the number and size of the ridges. Specimens procured by trawling on the Devon coast are more than an inch and a half in length and breadth; the smallest are from Guernsey. The fry are slightly umbilicate, and the topmost whorl is reticulated.

The spelling of the specific name has partaken of the variability of the object designated. *Zezyphinus, Zyziphinus, Ziziphinus,* and *Sisyphinus* are the readings proposed by Chemnitz, Born, Montagu, and Macgillivray. The last of these writers imagined that the name was derived from the rolling stone of Sisyphus, and not from *Zizyphum,* the fruit of the jujube-tree.

This species is the *T. conuloides* of Lamarck, *T. Cranchianus* and *T. irregularis* of Leach, and *Ziziphinus vulgaris* of Gray. Risso seems to have manufactured half a dozen species out of it or of *T. conulus*. Cantraine comprehended both, with a number of allied species, under the name of *T. polymorphus*. The fry is probably the *T. parvus* of Adams.

Whether Philippi was right or wrong in uniting *T. zizyphinus* with *T. conulus* is a moot question; but there is not, in my opinion, sufficient evidence of the latter species or form being British. Mr. Bean says that many years ago his son took a living specimen of it, attached to the sounding-lead, off the Lincolnshire coast, during his voyage in a collier from Newcastle to London. It appears that the discoverer had not long previously been in the Mediterranean, where *T. conulus* is common on the shore at low water.

It resembles the variety *elata* of *T. zizyphinus* in size, shape, and every other particular, except in having a bright polish, and darker or more vivid hues. Linné noticed that *T. conulus* was so very much like *T. zizyphinus* as to be almost a dwarf variety of it, but that the former had a prominent ridge or line between each whorl. Pennant figured a small-sized *T. zizyphinus* as *T. conulus*. The one may be the northern, and the other the southern form of the same species; and some of my references in respect of the geographical distribution of *T. zizyphinus* may be applicable to *T. conulus* only.

16. T. OC'CIDENTA'LIS*, Mighels.

T. occidentalis, Migh. in Proc. Boston Soc. Nat. Hist. i. p. 49. *T. alabastrum*, F. & H. ii. p. 497, pl. lxvi. f. 7, 8 (as *T. formosus*).

BODY creamcolour or white, with irregular streaks and specks of purplish-brown, or tinged with yellowish-brown: *mantle* slightly projecting beyond the shell, and finely ciliated at its edge; lappets roundish-oval and thin, one between each eye and the foot: *head* conical, thick, flexible (like the trunk of an elephant), closely fringed or scalloped in front; no veil was perceptible in any of the specimens examined by me, although Forbes says that "the capital lobes are minute and imperfectly developed:" *tentacles* filiform and slender, finely setose, with often a brown line down the front, and another on each side: *eyes* large, on short stalks: *foot* thick and broad, truncated in front, with a triangular expansion or lobe (like an auricle) on each side of this part, and bluntly pointed behind; the posterior half is raised into a long triangular ridge, whence there is a gradual slope to each side, with a depression in the middle; the upper edges are irregularly fringed or studded with short papillæ, between which issue the lateral filaments or appendages; sole exquisitely fringed: *appendages* 3 (sometimes 4) on each side, resembling the tentacles in every respect except in being smaller. The whole body is covered with cilia.

SHELL pyramidal, with a somewhat turreted outline and a

* Belonging to the west.

rounded base, rather thin, semitransparent and glossy: *sculpture*, four or five concentric, prominent, and sharp ridges on the upper part of the last whorl; on the base are three ridges immediately below the periphery, and three or four more (separate from the last) on the umbilical area; the penultimate and preceding whorls have similar ridges, which gradually decrease in number upwards; those on the three or four smaller whorls, and occasionally some of the other ridges, are granulated or beaded; the apex is rounded, and pitted like the top of a thimble; the furrow or space between each ridge on the larger whorls is flat, three or four times as broad as the ridge, and indistinctly lineated in a spiral or concentric direction: the whole surface is covered with very minute close-set and oblique longitudinal striæ: *colour* opaline, with the ridges of a pale golden or light yellowish-brown hue: *spire* gradually raised, and terminating in a sharp point: *whorls* 7–8, somewhat convex; the last is proportionally much larger than the next, and the same as to each of the rest in succession: *suture* slight, but well defined in consequence of the convexity of the whorls: *mouth* roundish, angulated above, and spread out below: *outer lip* thin, indented or scalloped by the ridges: *inner lip* nacreous and reflected on the pillar, which is thick and curved, with an oblique but slight excavation behind it: *inside* iridescent: *operculum* very thin, slightly concave, with a small central pit, having from 15 to 18 turns, and microscopically striated in the line of growth. L. 0·5. B. 0·4.

Var. *pura.* Altogether pearl-white.

HABITAT: Stony or "hard" ground on the fishing-banks of Shetland, in 40–90 f.; not uncommon in some places. Also from 40 to 80 f. on both sides of the Orkneys, and in 60 f. off Troup Head, Aberdeenshire (Thomas); among the refuse of a long-line fishing-boat at Peterhead (Peach). Two other British localities have been published; but the first has since been admitted to be erroneous, and the second is very questionable. These are Moray Firth (Gordon), and Lamlash Bay in the Clyde district (Eyton). The variety is Zetlandic, and occurs with specimens of the usual colour. Red and Coralline Crag (S. Wood); Lillo, on the banks of the Scheld near Antwerp (De Wael). It has been taken on many parts of the Scandinavian coast, as far north as Havösund, at depths varying from 25 to 150 f. (Lovén and others); and off Grand Manan and Casco Bay, in Maine, at 30 f. and more (Mighels and Stimpson).

Although an inhabitant of the deep-sea zone, its first impulse, when taken from it and placed in a vessel of water, is to crawl out into the open air, or to float with the sole of the foot uppermost and the shell downwards. The eagerness thus shown to get to the surface, apparently for the purpose of respiration, does not accord with the general notion that the water at the bottom of the sea is less aërated or oxygenated than that on the shore. However, exactly the reverse has been ascertained by means of some experiments conducted on board the French surveying-ship 'Bonite'; and it is now a recognized fact that the quantity of atmospheric air increases with the depth of water. According to Dr. Wallich ('North Atlantic Sea-bed,' p. 120), "hydrogen and oxygen, both of which gases in their separate state resist all pressure that has been applied to them, when combined to form water continue liquid under a pressure considerably below that of a single atmosphere." We do not yet exactly understand the mode in which the solution of atmospheric air in sea-water is brought about; but the tendency of fluids to absorb gaseous matter is constant under all circumstances, and their capability of appropriating it is facilitated by the pressure of the overlying stratum. This may account for deep-sea mollusks not finding in water drawn from the surface of the ocean a supply of oxygen equal to that which they had been accustomed to enjoy, and for their escaping into the open air to avoid a sensation which we should call stifling or suffocating. Another peculiar habit of such mollusks is worthy of notice, and is one which I cannot pretend to explain. It is the faculty of floating. Now it is very certain that in their native habitat, at a depth of from 150 to 540 feet, these shell-fish, being ground-dwellers and having no organ or means by which they can rise to the surface, could never exercise this faculty. Is it instinct that teaches them to float after having been forcibly dragged from the bottom of the sea and put into a shallow vessel of water? and if so, when was it implanted? Two living specimens, which I took in the same spot, differed in the colour of the animal, although the shells were undistinguishable. One was of a uniform yellowish-white, while the other was milk-white and had the sides of the foot streaked with brown. Mr. Alder says that the tongue is very beautiful and of a complicated structure, and that the uncini on each side are extremely numerous. It agrees in general character with that of *T. zizyphinus*: indeed the animals of both are much alike. The first whorl of the fry is exquisitely reticulated, like *Lagena squamosa*.

The present species is the *T. alabastrum* of Beck (according to Lovén), *T. quadricinctus* of S. Wood, and *Ziziphinus alabastrites* of Gray. No wonder that Forbes, who described this shell as a new species, gave it the name of *formosus*. It is truly beautiful; and we offer but faint praise in saying of such splendid prizes of the dredger—

. . . . . ." There's not a gem,
Wrought by man's hand to be compared to them."

Family VIII. TURBI'NIDÆ, (*Turbonidæ*) Fleming.

BODY resembling that of the *Trochidæ*.

SHELL conical or oval, and spiral: *operculum* calcareous and solid, convex on the outer side, flat or concave and paucispiral on the inner side.

For the mere purpose of classification, it is immaterial whether the characters which serve to distinguish one family or group from another allied to it are many or (as in the present instance) consist of a single feature. In the *Trochidæ* the operculum is horny, thin, and multispiral. The *Turbinidæ* have their home in southern climes; a single straggler, and that a very small one, inhabits the British seas.

Although the founder of the family was a good naturalist, the breed was at first decidedly mongrel, and included *Turritella, Odostomia, Scalaria, Skenea,* and *Paludina,* with other equally incongruous genera, which agreed only in being holostomatous univalves. The family circle is now more restricted and select.

Genus PHASIANELLA*, Lamarck. Pl. VIII. f. 1.

BODY elongated.

SHELL oval or oblong, rather solid, polished, and beautifully variegated in colour, imperforate at the base : *mouth* having its lips or edges disunited : *operculum* ear-shaped, concave on the inner side, with a short excentric spire.

It appears from Woodward's excellent 'Manual of the Mollusca' that the number of recent species belonging to this genus is 25, and of fossil species 70.

George Humphreys gave it the name of *Eutropia* and Risso described it as *Tricolia*.

* Speckled like a hen-pheasant.

Phasianella pulla*, Linné.

Turbo pullus, Linn. S. N. p. 1233. *P. pullus*, F. & H. ii. p. 538, pl. lxix. f. 1-3, and (animal) pl. D D. f. 5.

Body yellowish-white, marked transversely with pink or purplish-brown lines, and tinged with green: *mantle* thick, emerald-green; margin plain; lappets placed between the eyes and front pair of the pedal filaments, fan-shaped and digitated or frilled; the pectinations are delicately ciliated, those of the right-hand lappet being deeply divided, and those of the other lappet slighter or less distinct: *head* reddish-brown, terminating in a semicylindrical snout, which is short and does not project beyond the foot; it is sometimes lineated lengthwise; front edge scalloped: *tentacles* rather flattened, long, slender, and tapering to a blunt point, frequently edged with a brown line, thickly clothed with fine and short cilia: *eyes* raised on short, yellow, white, or green tubercles or stalks, one at the outer base of each tentacle: *foot* oblong, folding inwards towards the front, tapering at each end, and divided down the middle by a narrow groove; margin double-edged: *appendages* 3 on each side (the middle one being usually very short and sometimes inconspicuous), equidistant from each other, about half the length and size of the tentacles, and likewise setose.

Shell conic-oval and somewhat pointed at each end, semitransparent, and glossy: *sculpture* none when examined with an ordinary lens of a one-inch focus; but under a stronger microscopical power the surface appears covered with close-set but irregular longitudinal striæ and with a few very slight and indistinct spiral lines: *colour* various, usually yellowish with reddish or purple flame-like and obliquely longitudinal streaks of different widths, which are frequently broken or zigzag, interspersed with spots, sometimes altogether spotted with red; the ground-colour is occasionally white; and rarely the colour is uniform chocolate: *spire* short and rather abrupt: *whorls* 5-6, convex, but slightly compressed towards the suture; the last exceeds in size all the others put together: *suture* well defined: *mouth* roundish-oval, spread out at the base: *outer lip* thin, incurved above: *inner lip* flat and white, reflected on the pillar, which is thick and curved: *inside* partially nacreous but not iridescent: *operculum* porcelain-white, gibbous outside, somewhat flexuous on the other side, and having on the lower side of the mouth a small spire of a few rapidly

* Dark-coloured.

increasing turns, the outer edges of which are raised and keel-like. L. 0·35. B. 0·25.
Var. *oblonga*. Narrower, with the spire more protruded.

HABITAT: Common in the lower part of the littoral and upper part of the laminarian zones, in the Channel Isles, south and west of England, Bristol Channel, St. George's Channel, and on the coasts of Ireland; Oban and Mull (J. G. J. and Bedford); Stonehaven, Aberdeen, and Cruden (Macgillivray); Dunnet bay, Caithness (Gordon). I found the variety at Lulworth; it may be the male. This species has been noticed by Mr. James Smith as fossil in Ireland, and by Philippi as occurring in the Sicilian tertiaries. It has essentially a southern range, extending on the east to the Ægean and on the west to the Canaries; Black Sea (Middendorff). Forbes records it as living in the Archipelago from 2 to no less than 80 f., and M'Andrew has enumerated different depths from 3 to 60 f.

P. pulla is usually found on *Chondrus crispus*; Mr. Templer says that it feeds on *C. mammillosus*. Mr. Clark, however, found in the stomachs of all the individuals examined by him a number of minute Foraminifera, including *Truncatulina lobatula* and *Textularia variabilis*, which were entire, and did not appear to have been acted on by the tongue of the *Phasianella*. He has observed that the animal " is sometimes infested with a longish, strong, cylindrical, dark-brown parasite with a clavate termination, which hangs to the side of the opercular lobe, and may be mistaken for a vibraculum." Its mode of locomotion is like the amble of a horse. The foot being divided in the middle, each side advances in its turn, the stationary half serving as a point d'appui. This shows its affinity to the *Littorina* family, many of which have the same peculiarity of gait. The fry of the present species is globular and distinctly umbilicate; it might almost be mistaken for that of a *Lacuna*. The shell and animal are equally pretty. Now and then the former is pearl-white; and both may have sat for their portraits when Tennyson sketched the 23rd Canto of 'Maud.'

"Stanza 1.

"See what a lovely shell,
  Small and pure as a pearl,
  Lying close to my foot,
Frail, but a work divine,
  Made so fairily well
With delicate spire and whorl,
  How exquisitely minute,
  A miracle of design!

2.

"What is it? a learned man
  Could give it a clumsy name.
Let him name it who can,
  The beauty would be the same.

3.

"The tiny cell is forlorn,
  Void of the little living will
  That made it stir on the shore.
Did he stand at the diamond door
  Of his house in a rainbow frill?
Did he push, when he was uncurled,
  A golden foot or a fairy horn
  Through his dim water-world?

4.

"Slight, to be crush'd with a tap
  Of my finger-nail on the sand,
Small, but a work divine,
Frail, but of force to withstand,
  Year upon year, the shock
Of cataract seas that snap
  The three-decker's oaken spine
  Athwart the ledges of rock,
  Here on the Breton strand!"

The presumed subject of these exquisite lines is the *Turbo pictus* of Da Costa, *P. pulchella* of Récluz, and *Eudora varians* of Leach. Lamarck placed it in the old genus *Turbo*, and not in *Phasianella*.

Turbo rugosus of Linné (*T. calcar*, Montagu) was said to have been taken by Captain Laskey in Iona, one of the Western Islands. It is a rather common Mediterranean shell, but not British. *Turbo castanea* of Gmelin (*T. mammillatus*, Donovan) is West-Indian, and supposed to have been picked up by a Mr. Platt on the Scilly rocks.

Family IX. LITTORI'NIDÆ, Gray.

BODY spiral: *mantle* plain: *head* snout-shaped; lingual ribbon armed with numerous hook-like teeth, as in the preceding families of the same order: *tentacles* long, one on each side of the head: *eyes* placed on very short stalks or tubercles at the outer bases of the tentacles: *gills* forming a single plume, which is composed of several flat laminar plates: *foot* having the usual operculigerous lobe, from the hinder part of which in certain genera issue one or two tentacular processes or filaments.

SHELL conical, never nacreous: *mouth* obliquely squarish or oval: *operculum* horny, thin, ear-shaped, and few-whorled, with a lateral nucleus.

This family, as their name imports, are for the most part littoral :—

> "Huge ocean shows, within his yellow strand,
> A habitation marvellously planned
> For life to occupy."

The *Littorinæ*, which live on the beach, exposed to frost and cold, snow and rain, do not hibernate, but appear to pass the dreary season of winter without discomfort. The equal temperature of the sea and the thickness of their shells protect them from the vicissitudes of climate; and (what is of more consequence to them) they are supplied all the year round with an abundance of food. It is otherwise with some of their small cousins, the *Rissoæ*, which depend for their subsistence on the *Zostera marina* or sea-grass. These must either perish, like the greater number of the insect tribe, or remain in a torpid state until

> " To mute and to material things
> New life revolving summer brings."

The former supposition is more probable. Homer, with his tendency to view all nature in relation to ourselves, illustrated the idea of such annual reappearance of life by some well known lines, which I will venture to paraphrase.

> Men are like the race of falling leaves,
> That winds in autumn whirl and sweep away:
> Yet spring, with joy and freshness ever rife,
> Nature will soon restore to former life.
> Each year the same unvaried tissue weaves
> Of birth and death, of verdure and decay.

Several species of *Littorina* abound on every stony part of our coast; and the seaweeds swarm with different kinds of *Lacuna* and *Rissoa*. All live together in perfect harmony; there is here no " struggle for existence," nor intermixture of races. Similar conditions may reasonably be presumed to have continued ever since the formation of the Crag—a period of incalculable antiquity—because we find associated in this formation certain species of *Littorina* and *Rissoa* unquestionably identical with those which still inhabit the same area, and even exhibiting a variability of form precisely analogous to what is observable at the present time. The prevalent hue of the animals in the present family (which indeed may be said of the Gasteropoda in general) is yellowish, with frequently a tinge of purplish brown. The sexes are separate. The males are distinguishable from the females by being of a smaller size. This is notoriously also the case with the Crustacea and most of the insect tribe, as well as with many other animals, including our own race. The food of the *Littorinidæ* consists of vegetable matter, either fresh or in various states of putridity. They crawl in a peculiar fashion, moving first one and then the other side of the foot by turns; the line of such division is marked in the middle of the sole.

Genus I. LACU′NA*, Turton. Pl. VIII. f. 2.

BODY stout: *head* short: *tentacles* flattened and smooth: *eyes* nearly sessile, owing to the smallness of the stalks: *foot* oval and rounded at each end, with a sharp pointed tail: *opercular appendages* two, one at each side or corner of the tail.

SHELL more or less channelled or grooved at the base, and slantingly umbilicate: *mouth* obliquely squarish: *pillar* rather broad and flattened, so as to receive the channel or groove above mentioned: *operculum* furnished on the under side with a cartilaginous rib which nearly follows the direction of the spire.

Da Costa was the first to notice the peculiar character of the channelled pillar in the shell of *Lacuna*, finding it difficult to assign his *Cochlea parva* (*Lacuna puteolus* of Turton) to any Linnean genus. The only species known to us (four in number) were placed by their respective discoverers in as many different genera, viz. *Turbo*, *Trochus*, *Cochlea*, and *Nerita*. They are phytophagous According to Lovén, those which live on brown seaweeds have green bodies, while others found on red seaweeds are rosecolour. They occasionally secrete slimy threads (like the *Limax arborum*), by which they suspend themselves from the frond or stalk

* From the excavation of the pillar.

of a seaweed; and they may sometimes be observed floating in a reversed position, the sole of the foot being on a level with the surface of the water. The spawn forms a gelatinous but firm cylindrical mass, and is curved in a semicircle. As soon as the fry emerge from their receptacle they swim about freely by means of a ciliated and vibratile bilobed veil, which occupies the front of the body. The otolites are circular and simple. Clark proposed to merge this apparently natural genus in *Littorina*; Leach, on the other hand, divided it into *Temina*, *Epheria*, and *Medoria*. The principle of classification advocated by the one was synthetical; he reduced genera to species. The other pushed the analytical system to an opposite extreme; consequently in his hands species became raised to genera.

1. Lacuna cras'sior*, Montagu.

Turbo crassior, Mont. Test. Brit. p. 309, t. 20. f. 1. *L. crassior*, F. & H. iii. p. 67, pl. lxxii. f. 5, 6.

Body yellowish-white, or pale yellow, with an orange tint on the upper part; there is sometimes a dark brown triangular spot a little behind the point of the muzzle: *mantle* thick: *head* produced into a rather long, narrow, and compressed square-pointed muzzle, having an oval disk in front, which contains the mouth: *tentacles* slender, tapering gradually to a rather sharp point: *eyes* black, seated on short tubercles, one at the outer base of each tentacle: *foot* broader and slightly curved in front, with small lobe-like corners, occasionally sinuated in the middle on one side or the other, and rounded behind; sole double-edged, apparently slit in three or four places behind: *appendages* curved and white, very much shorter than the tentacles, which they resemble in shape.

Shell turreted, bluntly angulated at the base, solid, opaque, lustreless when covered with the epidermis, under which it is somewhat glossy: *sculpture*, numerous slight and sinuous spiral impressed lines or wrinkles, which are to a great extent con-

---

* More solid than other species.

cealed by the epidermis: *colour* yellowish with a faint tinge of brown; the apex is sometimes of a darker hue: *epidermis* membranous; it is usually puckered lengthwise into irregular folds, and it is most commonly rubbed off or absent on the top of the shell: *spire* more or less raised, terminating in a blunt point: *whorls* 6–7, rather convex but compressed, somewhat angular above, and gradually increasing in size: *suture* deeply excavated: *mouth* rather large, considerably expanded below and angulated at the base: *outer lip* very thin and fringed by the epidermis, incurved above on the pillar side: *inner lip* filmy, spread over the lower part of the body-whorl, and partly covering the canal when present; it is not united with the outer lip: *pillar* white, sometimes faintly tinged with pink; canal or groove more frequently wanting, but when existing it is rather wide, oblique, and ends in a small but deep perforation: *inside* porcelain-white and polished: *operculum* having 5 or 6 whorls, the outermost of which occupies nearly the whole area, the others being disproportionately small; it is marked across with curved lines of growth, and lengthwise with microscopical and close-set striæ, which last radiate from the nucleus. L 0·5. B. 0·3.

HABITAT: Among stones and old shells in sandy ground mixed with mud, from low-water mark to considerable depths, on all our coasts, including the Channel Isles and Shetland; it is rather local. Mr. Grainger has recorded it as fossil from a deposit at Belfast. I found it at Etretat, in Normandy; and what I consider a variety of *L. crassior*, connecting it with *L. divaricata*, has been described by Möller as a Greenland shell under the name of *L. glacialis*. Middendorff gives the White Sea, coasts of Russian Lapland, Sea of Okhotsk, and Sitka Island as habitats of the present species and of the variety *glacialis*.

The animal is active, hardy, and seemingly fond of getting out of the water. Mr. Dawson has observed that it moves at the rate of about two inches per minute; as it progresses the shell is carried along at a slow swinging pace. This arises from the peculiar action of the foot, which jerks forwards, first on one side and then on the other.

It is in all probability the *Turbo pallidus* of Donovan; and if his description (part of which is, "whorls very slightly bicarinated") were recognizable with sufficient certainty, that name ought to have precedence of the one proposed by Montagu. I willingly avail myself of the doubt, in order not to alter the name by which this shell is now generally known. Leach called it *Medoria Walkeri* and *M. Danmoniensis*.

2. L. DIVARICA′TA*, Fabricius.

Trochus divaricatus, Fabr. Fn. Grœnl. p. 392. *L. vincta*, F. & H. iii. p. 62, pl. lxxii. f. 10–12, lxxiv. f. 7, 8, lxxxvi. f. 6–8, and (animal) pl. G G. f. 4.

BODY yellowish-brown faintly streaked with purple or tinged with pink: *head* fleshcolour, large, broad, prominent, and becoming wedge-shaped towards the extremity: *tentacles* tapering, with blunt tips; owing to their contractility they are sometimes finely, but irregularly, scalloped at the edges: *eyes* raised on short stalks: *foot* angulated at each of the front corners, behind which it is contracted; sole edged with a broad white border: *appendages* short and ribbon-like.

SHELL obliquely conical, expanded and more or less bluntly angulated at the base, usually thin, semitransparent, and somewhat glossy: *sculpture*, numerous slight and sinuous spiral impressed lines or striæ, as in *L. crassior*, but always perceptible and more regular: *colour* varying from white to yellowish-brown, and often diversified by reddish-brown spiral bands of different widths; there are generally four of these bands on the largest whorl (viz. two above the peripheral keel and two below it), two on the penultimate, and one on the antepenultimate whorl; the bands are sometimes confluent, and so disposed as to exhibit a white or yellowish-white zone just below the suture of the last three whorls; the apex is often reddish-brown or horncolour: *epidermis* membranous and thin: *spire* considerably raised and terminating in a blunt point: *whorls* 6, compressed, the last occupying about two-thirds of the spire: *suture* distinct, but not excavated: *mouth*

---

* Spread out.

wide, expanded outwards and below, slightly angular at the base: *outer lip* very thin, occasionally strengthened a little way inside by a slight white rib or callus: *inner lip* also thin, united with the outer lip, and partly covering the canal: *pillar* white ; canal wide, oblique, funnel-shaped, and exposing a considerable part of the spire: *inside* polished, of the same colour as the outside: *operculum* as in *L. crassior* and similarly sculptured. L. 0·45. B. 0·3.

Var. 1. *canalis*. Without coloured bands, and usually of a thinner texture. *Turbo canalis*, Mont. Test. Brit. p. 309, t. xii. f. 11.

Var. 2. *quadrifasciata*. Smaller, more conical and solid, with a keeled periphery; outer lip thickened within its edge by an inside rib. *Turbo quadrifasciatus*, Mont. l. c. p. 328, t. xx. f. 7.

Var. 3. *gracilior*, Metcalfe. Smaller and much elongated.

HABITAT: Seaweeds and *Zostera*, at low-water mark and in the laminarian zone throughout the British seas; abundant. The first two varieties are also everywhere common; the third was found by Mr. Metcalfe in Guernsey, and by me in Langland bay near Swansea. Fossil in the Clyde beds (Smith and others); Fort William (J. G. J.); Aberdeenshire (Jamieson); Moel Tryfaen (Darbishire) ; Norwich or Mammalian Crag (S. Wood); Uddevalla (J. G. J.), and 40 feet above the sea (Malm); Christiania district, newer deposits, 100 feet (Sars). Its distribution in a living state is mainly northern, and comprises Greenland, the White Sea, Russian Lapland, Iceland, the Faroe Isles, Scandinavia, Heligoland, Normandy, Brittany, and Gulf of Gascony, besides the eastern and western coasts of North America.

It appears to be the favourite food of many sea-birds. Dr. Saxby took specimens from the stomach of a Black Guillemot at Unst, each of which had the operculum in its usual position, although nearly all the soft parts had disappeared. This is a shy but restless mollusk; and it has a very shambling and awkward gait. In some specimens the canal does not exist, and in others it is very slight and scarcely perceptible. Those from Shetland are considerably larger than the average dimensions I have given. The fry are nearly globular and widely umbilicate. The shell differs from *L. crassior* in neither being turriculate nor having a thick epidermis; and the last whorl in the present species is always very much larger in proportion to the rest. Its texture also is usually thinner; but this last character varies, and cannot be depended on as a ground of distinction.

Fabricius was right in suspecting that his *Trochus divaricatus* was not that of Linné. This great clerical zoologist described the present species with such accurate minuteness as fully to justify my following Lovén and other northern writers in preferring that name to the subsequent one (*vincta*) given by Montagu. According to Gould it is the *L. pertusa* of Conrad. Brown described some of the variously coloured specimens as *Phasianella fasciata*, *P. bifasciata*, *P. cornea*, and *P. striata*; and I cannot distinguish, in a specific sense, the *L. solidula*, *L. labiosa*, or *L. frigida* of Lovén. The *L. albella* of the last named author is intermediate between the present species and *L. puteolus*; it is different from the thickened slender specimen of *L. divaricata* found by Mr. Alder at Cullercoats and doubtfully referred by him to Lovén's species. Leach called the variety *canalis Epheria Bulweriana*; the variety *quadrifasciata* is his *E. Goodallii*.

3. L. pute′olus*, Turton.

Turbo puteolus, Turt. Conch. Dict. p. 193, f. 90, 91. *L. puteolus*, F. & H. iii. p. 58, pl. lxxii. f. 7-9, and lxxiv. f. 9.

Body yellowish-white faintly tinged with pink (sometimes

* A little pit.

with purplish-brown), or uniform pale yellow: *mantle* thick, fleshcolour: *head* broad, projecting beyond the foot, pale red or edged with greenish-brown: *tentacles* white, ribbon-like, with blunt tips and jagged edges: *eyes* rather large: *foot* thickened, opaque, of a dusky hue towards the sides, double-edged in front, narrower or contracted in the middle, and ending in a minute bluntly pointed tail; sole irregularly bordered with white, and divided down the middle by a slight groove: *appendages* small and flattened, like miniature tentacles.

SHELL globular, slightly expanded at the base, with an angulated periphery, rather solid, opaque and glossy: *sculpture* similar to that of the two foregoing species, usually not so conspicuous or regular as in *L. divaricata*; the present species has frequently also numerous slight striæ in the line of growth: *colour* yellowish-white with the upper whorl sometimes purplish, dull reddish-brown, or whitish with three rufous bands on the body-whorl, the middle one of which is much broader than either of those which encircle the upper and lower part of that whorl; the colour when uniform, and the bands when present, are of various shades and degrees of intensity; occasionally the uppermost band is continued on the penultimate whorl: *epidermis* membranous and thin: *spire* scarcely raised, but prominent, terminating in a blunt point: *whorls* 3-4, convex, the last occupying about four-fifths of the spire: *suture* rather deep: *mouth* slightly expanded outwards and below, and more or less angular at the base: *outer lip* very thin, incurved towards the pillar: *inner lip* slight, not united with the outer lip, but spread over the base above the canal, which it partly covers: *pillar* white; canal generally wide and forming a deep excavation in the base of the shell, so as to expose nearly all the interior of the spire: *inside* polished, of the same colour as the outside: *operculum* resembling in every respect those of *L. crassior* and *L. divaricata*. L. 0·2. B. 1·5.

Var. 1. *conica*. Banded, rather thin, and having the spire longer than usual.

Var. 2. *auricularis*. Light horncolour or dirty white, thin and transparent. *Turbo auricularis*, Mont. Test. Brit. p. 308.

Var. 3. *lactea*. Milk-white and solid.

Var. 4. *clausa*. Base of the shell pointed; pillar not exhibiting any canal or excavation.

Var. 5. *expansa*. Of various colours; last whorl extended and partly separated from the rest.

HABITAT: Small seaweeds (chiefly *Chondrus crispus* and *Nitophyllum laciniatum*) at low-water mark, in the Channel Isles, south and west of England, and Bristol Channel; local and gregarious. Isle of Man (Forbes); Filey (J. G. J.); Northumberland and Durham (Alder); north, east, and west of Ireland (Turton and others); west coast of Scotland (Barlee and others); Dunbar (Laskey); Dunnet bay, Pentland Firth (Gordon and Peach); Nordwick bay, Unst (Dawson). Var. 1. Exmouth (Clark); Manorbeer, Pembrokeshire (J. G. J.); Scarborough (Bean); co. Antrim (Hyndman); Skye (Barlee). Var. 2. Southampton (Montagu and J. G. J.). Var. 3. Guernsey and the Hebrides (Barlee). Var. 4. Sark (Barlee). Var. 5. Exmouth (Clark); Torbay (Mrs. Wyatt). This species occurs in the newer pliocene, "Ireland" (Forbes); Clyde beds (Crosskey and Robertson); Fort William (J. G. J.); Norwich Crag (Woodward). Its foreign distribution comprises the Scandinavian coasts, from Bohuslän to Finmark (Lovén, as *L. Montagui*, and Malm); Normandy (J. G. J.); Brittany (Macé, Delaunay *fide* Taslé, and Cailliaud); Rochelle (D'Orbigny père); Corunna and Vigo (M'Andrew). The variety *auricularis* has been dredged in Kiel Bay by Meyer and Möbius.

Mr. Spence Bate watched some spawn which he procured on the 24th of January. He could distinguish the eyes on the 10th of February; and ten days afterwards the fry were fully developed, and crawled out of their gelatinous covering. His note refers to *L. pallidula*; but the fry evidently belong to the present species.

It is the *Cochlea parva* of Da Costa, *Helix fasciata* of

Adams, *Helix lacuna* and *Nerita rufa* (young) of Montagu, *L. Montacuti* (without bands) of Turton, *L. Montagui* of Brown, *Temina Turtoniana*, *T. rufa*, and *T. variabilis* of Leach; *L. sulcata* of Macgillivray is the young shell. Neither of the two earliest specific names (*parva* and *fasciata*) appears to have been used by any writer except those who respectively proposed them; and they may therefore be regarded as obsolete. These being disposed of, the name ought in strictness to be *Montacuti*, which was given by Turton in the ' Zoological Journal ' (vol. iii. p. 191) to the typical form; his *Turbo* or *L. puteolus* is the variety which I have noticed as *expansa*. But it does not seem necessary, or desirable, to change the name adopted by the authors of the ' British Mollusca.'

4. L. PALLI'DULA*, Da Costa.

Nerita pallidulus, Da Costa, Brit. Conch. p. 51, t. iv. f. 4, 5. *L. pallidula*, F. & H. iii. p. 56, pl. lxii. f. 1, 2, and (as *L. patula*) f. 3, 4.

BODY whitish: *mantle* tumid at the margin: *head* nearly cylindrical, projecting a little beyond the foot ["the upper part of the neck has two short flake-white diverging lines imbedded in the ground-colour."—Clark]: *tentacles* resembling in miniature the leaves of the water-flag: *eyes* rather small: *foot* double-edged in front, behind which it is somewhat contracted, thickened and opaque towards the edges, and ending in an extremely short pointed tail; sole grooved lengthwise: *appendages* nearly of the same shape as the tentacles, but smaller and very much shorter, although extending beyond the foot.

SHELL somewhat triangular, largely and obliquely expanded in front, rather thin, opaque, and glossy: *sculpture*, fine but irregular striæ in the line of growth, which are for the most part concealed by the epidermis, and are more conspicuous just below the suture; there are also a few remote and scratch-like lines in a spiral direction: *colour* pale yellowish-green: *epidermis* not very thin, resembling oilskin: *spire* very small

* Palish.

and depressed, sunk within the upper margin, the base of the shell or lower part of the mouth being placed in front of the observer: *whorls* 3–4, convex, the last disproportionately large and occupying nearly the whole of the spire: *suture* well defined: *mouth* exceedingly large and capacious, equal in size to the closed part of the shell; the base is somewhat angular, especially in immature specimens: *outer lip* thin, incurved towards the pillar: *inner lip* thickened or callous above the canal, over the upper part of which it is folded, not united with the outer lip: *pillar* white; canal very wide and extending funnel-wise into the interior of the spire, so as to expose the greater part of it: *inside* polished, and coloured like the outside: *operculum* having from three to four whorls, wrinkled across, and indistinctly marked with very minute and close-set spiral lines. L. 0·45. B. 0·35.

Var. 1. *neritoidea*. Grass-green, much smaller, and less expanded, resembling in shape *Neritina fluviatilis*, and having a rather prominent but short and eccentric spire. *L. neritoidea*, Gould, Inv. Mass. p. 263, f. 170.

Var. 2. *patula*. Olive-green, rather more solid, with a flat spire and the expansion outwards being not so much in front as above and below, making the outline that of an equal-sided triangle; canal nearly closed in the adult. " Variety ? *Patula*," Thorpe, Brit. Mar. Conch. p. 37, f. 83.

Var. 3. *albescens*. Of a paler hue or white, smaller but shaped like the last variety.

HABITAT: On *Laminariæ* and other sea-weeds having flat and smooth fronds, at low-water mark and in a few fathoms seawards, chiefly on our southern and western coasts, but also in St. George's Channel, all round Ireland, the Clyde district, and Frith of Forth; Kent (Da Costa and Boys); Northumberland and Durham (Alder); Aberdeenshire (Macgillivray); Orkneys (Forbes). Var. 1. West of Scotland and Shetland. Var. 2. Torquay and Sunderland (Hanley); Guernsey, Langland bay near Swansea, and Barmouth (J. G. J.); Bantry bay (Barlee). This variety seems to connect the present species with *Littorina obtusata*. Var. 3. Skye and the

Hebrides (Barlee). In a fossil state *L. pallidula*, var. *neritoidea*, was found by me at Fort William; and the ordinary form is enumerated in Godwin-Austen's list of shells from an upper tertiary deposit in Sussex. In a recent state the latter has been recorded from the Boulonnais (Bouchard-Chantereaux), Quibéron and Belle-île in Brittany (Taslé), and Loire-Inférieure (Cailliaud); and the former ranges from Heligoland northward to Iceland, Greenland, Spitzbergen, New England, and Massachusetts.

Mr. Clark says that the opercular lobe has occasionally four caudal filaments. I never saw more than two in the numerous specimens which I have examined of the typical form and principal varieties. He also describes the tentacles as "setose." This character I have likewise failed to detect, although I used the same optical aids for observation that he did. The edges of the tentacles are more or less uneven and sometimes serrated, arising (as I believe) from the contractility of these organs; possibly such appearances may have misled Mr. Clark, and induced him to consider them as indicating hairs or setæ. A specimen in my cabinet of the variety *neritoidea* is distorted by having a rather deep and irregular indentation down the front. The fry of this variety are of a light horncolour; the colour of the animal in the adult state is greyish, with a faint tinge of purple. Perhaps this may be a distinct species; but as I am not satisfied on this point, I prefer leaving it to the judgment of my brother conchologists. To add another species to the list of any local fauna or flora, unless on conclusive grounds, would indeed be unworthy of a naturalist.

L. retusa of Brown appears to have been described and figured from a half-grown shell of the present species.

Genus II. LITTORI′NA*, Ferussac.   Pl. VIII. f. 3.

BODY stout, twisted into a short cone: *head* strong: *tentacles* conico-cylindrical and smooth: *eyes* placed on globular expansions of the tentacles at their outer bases, or sessile: *foot* oval, rounded at each end, plain-edged: *opercular lobe* smaller than the operculum, and destitute of appendages.

SHELL rather solid, not umbilicate: *spire* short: *mouth* oval, with the lips usually disunited: *pillar* even, never channelled or grooved: *operculum* having underneath a process of attachment on or near the nucleus of the spire.

The presence of these shells in a fossil state affords a useful criterion to the geologist, and invariably indicates littoral conditions. They inhabit only

"The beachy girdle of the ocean,"

and are seldom found at a greater depth than low-water mark of spring-tides. *L. neritoides* and some of the varieties of *L. rudis* take up their abodes above highwater mark, where they probably subsist on *Lichina pygmæa* and other minute sea-weeds, which cover the rocks in such situations. They have never been observed to go down to the sea when the tide comes in. This peculiar habit of truly marine mollusks frequenting places beyond the reach of the tide induced Dr. Johnston to make the following quaint remarks on a subject which has of late much engaged the attention of naturalists. After mentioning the case of certain Gasteropods, furnished with gills, that pass so large a portion of their term of life completely out of the water as almost to be amphibious, he says, "The *Patellæ* and *Littorinæ* are also good examples. Our common species of the latter genus seem, indeed, to prefer

* From *littus*, the sea-shore.

spots where they can be covered only at high water, and I have seen myriads of them, when young, clustered in hollows of rocks that were many feet above the highest tides. Still, their respiratory organs are, as they have ever been, branchial; nor does it seem easy, on the Lamarckian hypothesis, to account for their non-improvability; why these shell-fish, so fond of air, have not acquired, by their residence in it, the lungs of the snail, and betaken themselves to the land; why their shells have not become lighter to enable them to move with more alacrity; and why their eyes have not risen to a higher elevation than the base of the tentacula, that they might scan the landscape and avoid its perils." The gill-plume is composed of from 45 to 60 strands or pectinations, which are very long, slender, and close-set. Adanson appears to have considered the *Littorinæ* hermaphrodite; but, on his return home from Senegal, he was undeceived in this respect by the great botanist, Jussieu, who showed him that the sexes were certainly distinct in the common European periwinkle. Most of the species are oviparous, and deposit their spawn on seaweeds, rocks, or stones; the eggs are enveloped in a glairy mass, which is just firm enough to retain its shape in the water, and adheres to the nidus with considerable tenacity. Each egg has its own globule of jelly, and is contained within an extremely thin and transparent membrane, so as to be separated from the rest. They are hatched after a short exposure to the water, air, and sun, and soon exhibit the shells completely formed and occupied by the ciliated fry. Some species are ovoviviparous or viviparous, and develope their spawn in the branchial cavity. We find, therefore, in this genus, examples of both kinds of propagation. The same fact has been observed with respect to species of *Helix* and *Pupa* among the Pulmonobranchiata*. The *Littorinæ* are extremely prolific, and found in all parts of the world. According to Nyst, out of the 59 species known in 1843 (when he published his excellent catalogue of the fossil shells and polypes of Belgium) 37 were recent, 8 from tertiary, 10 from cretaceous, and 1 from carboniferous strata. The Messrs. Adams have lately enumerated 56 recent species; but some of these are only recognized by other conchologists as varieties.

Menke changed the spelling of the generic name to *Litorina*—a pedantic and unnecessary innovation. *Littus* and *Litus* were used indifferently by the best Latin writers. Cicero seems to have preferred the former mode of spelling; Ovid has both.

1. LITTORINA OBTU'SATA †, Linné.

Turbo obtusatus, Linn. S. N. p. 1232. *L. litoralis*, F. & H. iii. p. 45, pl. lxxiv. f. 3–7, and p. 49, pl. lxxxvi. f. 2, 3.

BODY yellowish-white, lemon or orange-yellow, often tinged with purple or violet, rarely sootcolour, and marked across by lines of a paler hue: *mantle* sometimes edged with orange or black: *head* narrow, occasionally reddish or fleshcolour on the upper part or neck: *tentacles* tapering, with blunt whitish tips; their sides are in some specimens bordered by fine lead-coloured lines: *eyes* small, with pearl-white irides and black pupils: *foot* broader in front, and bluntly pointed behind, somewhat contracted at about one-third of the way down; sole pale yellow, yellowish-white, or whitish, divided lengthwise in the middle by a slight line, which resembles a crack in the glaze of an earthenware dish: *opercular lobe* now and then sinuated or finely cloven.

SHELL nut-shaped, thick, opaque and lustreless: *sculpture*, numerous minute fine, but irregular, spiral wavy striæ, which are mostly observable on young and immature specimens; the crossing of these striæ by the lines of growth causes a slight

---

* See vol. i. pp. 222 and 248. † Blunted.

decussation: *colour* most variable, yellow, brown, red, green, and purple of all shades, diversified by bands, streaks, tessellated or reticulated and zigzag markings of every conceivable kind: the predominant hues are yellow and brown; it is rarely milk-white: *epidermis* membranous, yellowish or horn-colour, usually thin: *spire* very blunt and sometimes flattened: *whorls* 5–6, convex, but somewhat compressed or squeezed together; the last embraces nearly the whole spire: *suture* narrow although distinct: *mouth* large, occupying nearly half the lower portion of the shell, sharply angulated below in young specimens: *outer lip* thick, a little incurved above, and forming with the inner lip an acute angle in that part: *inner lip* thin, spread like glaze over that side of the mouth, and indented in the middle: *pillar* curved, sloping outwards, white and thick: *inside* polished, coloured like the outside; edges often stained with purple: *operculum* having 4 or 5 whorls, the outermost of which occupies nearly the entire area; it is marked across with microscopical and close-set curved striæ or wrinkles, which are not quite regular, but frequently anastomose or interlace. L. 0·65. B. 0·5.

Var. 1. *neritiformis*. Shell squeezed together at the sides, so as to make it longer and the periphery angulated. *L. neritiforma*, Brown, Ill. Conch. G. B. & I. p. 17, pl. x. f. 24.

Var. 2. *ornata*. Smaller and rather more convex, having the spire somewhat more produced, and ornamented with broad reddish-brown bands on a white or yellowish-white ground. *L. palliata*, F. & H. iii. p. 51, pl. lxxxiv. f. 8–10.

Var. 3. *fabalis*. Dwarfed or young, inclined to a globular shape. *Turbo fabalis*, Turton, in Zool. Journ. ii. p. 366, tab. xiii. f. 10.

Var. 4. *compacta*. Smaller, thick set, and also subglobular.

Monstr. Scalariform, with a very broad base and a keel encircling the upper part of each whorl, or having the suture deeply and widely excavated.

HABITAT: Among stones and *Fuci* on all beaches below high-water mark of neap tides. The 1st variety is not uncommon in the west and north of Scotland, and in Shetland; and Captain Brown has given Downpatrick as an Irish locality. The 2nd abounds in the Isle of Wight and at Southampton; it appears to have been mistaken by the authors of the 'British Mollusca' for the *L. palliata* of Say, which I shall presently have occasion to notice. The 3rd was discovered by Mr. Bean at Filey; and I also found it not only there, but plentifully at Larne in the north of Ireland, and in Shetland. Lilljeborg has taken the last in Norway. I believe it represents the young males of the ordinary form. The body of this variety is dark grey and lineated, with a tinge of purple on the upper part, and whitish underneath; the head is thick, edged with yellow above; tentacles marked across with dark rings; eyes proportionally large, each surrounded by a pale yellow circle; foot oval, with a creamcolour sole; verge falciform. The fanciful name *fabalis* (derived from that of the well-known conchologist at Scarborough) may be matched with the punning mottoes in heraldry. Geologists have also their little weaknesses of this kind,—for example the "Genista" cave at Gibraltar, which was so designated, not from its mouth being concealed by the shrub of that name, but from its discoverer or explorer, Captain Broome. Macgillivray with greater sobriety, but less attention to the rules of nomenclature, changed the name of this variety to *Beanii*. The 4th variety inhabits Loch Torridon and other parts of the Ross-shire coast; Meyer and Möbius found it in Kiel Bay. Examples of the monstrosity were in Mr. Clark's collection of Exmouth shells, and occurred to me on the coast of Antrim. Another malformation, from Unst, has the outer lip remarkably flexuous, and the upper angle of the mouth converted into a long and narrow notch. In a fossil state this species has been enumerated by Mr. J. Smith from the Clyde beds, by Mr. Rose from the brick-earth of the Nar in Norfolk, by Mr. Grainger from Belfast, by Mr. Darbishire from Macclesfield, by Captain Drury Lowe from Moel Tryfaen, by Sars from older and younger glacial deposits in the Christiania district (at heights varying from 100 to 440 feet above the present level of the sea), and by myself from Lilleherstehagen, near Uddevalla. I have particularized these localities, in order that the range of *L. litoralis* (or *L. palliata*), which is a peculiarly arctic fossil, may be ascertained. In consequence of the doubt which I entertain with regard both to the identity of that with the present species, and to the correct assignment of each of these so-called species to the recorded localities, I give the range of northern distribution provisionally and subject to future correction. Iceland (Mohr and Steenstrup); Faroe Isles (Landt); White Sea (Middendorff); Scandinavia (Müller and others); Heligoland (Frey and Leuckart); Holland (Waardenburg); North of France (Lamarck and others); Rochelle (D'Orbigny père, Aucapitaine, and J. G. J.); Santander (E. J. Lowe); Vigo (M'Andrew); ? Toulon (Gay); ? Adriatic (Olivi); ? Sicily (Philippi, *fide* Bivona, Gemellari, and others). The habitat of *L. palliata* is the North-American sea-board from the St. Lawrence and Cape Cod.

Lister noticed the habit of this species (as well as of *L. litorea*) of copulating on the dry part of the shore. Individuals of *L. obtusata* were found by Mr. William Thompson at Weymouth in union with others of *L. rudis*; and Dr. Battersby tells me that he has seen the same in Ireland. It does not appear that any hybrid form resulted from the coition in any of these cases. The 'Opuscula subseciva' of Baster (1769) contain excellent figures of the spawn and fry of the present species. Newly born shells have a small umbilicus, which is closed in the course of growth, and concealed by the broad pillar-lip. The males are invariably smaller than the females, and have the spire more produced. Clark described the tentacles as "setose." May not this have been a *lapsus typographicus* for "slender"? Our remote ancestors appear to have used the shells as personal ornaments. They made necklaces of them, probably by rubbing the points on a stone, and stringing them together, when thus perforated, with a fibre or sinew. An account is given in Wilson's 'Prehistoric Annals of Scotland' of the remains of such necklaces having been found underneath a Cromlech, which was discovered on levelling a tumulus in the Phœnix Park at Dublin in 1837; this disclosed two male skeletons, and beside the skull of each lay perforated shells of *L. obtusata* in such a position that they must have been placed around the necks of the buried chieftains. A portion of the vegetable fibre with which they had been strung together remained through some of the shells. The only other relics found in the sepulchre were a small fibula of bone and a knife or lance-head of flint. Our patriotic poet, old Michael Drayton, in the 20th song of his 'Polyolbion,' gave these shell-ornaments a mythological air, when he described the fair Norfolcean

> "Nymphs trick'd up in tyers, the sea gods to delight."
> . . . .
> "With many sundry shells, the scallop large and fair,
> The cockle small and round, the periwinkle spare,
> The oyster wherein oft the pearl is found to breed,
> The mussel which retains that dainty orient seed:
> In chains and bracelets made, with links of sundry twists,
> Some worn about their waists, their necks, some on the wrists."

I believe that the *Nerita littoralis* of Linné and Fabricius is a Scandinavian, Arctic, and North-American species, known as the *Turbo palliatus* of Say, *T. expansus* of Brown, *L. arctica* of Möller, and *L. limata* of Lovén. Both Linné and Fabricius say that the animal has cirrous excrescences from the foot; and their descriptions of the shell accord much better with those given by Say and the other writers, and with typical specimens of *L. palliata*, than with the present species. The other species is common in the Clyde beds, and I found it fossil also at Fort William; it does not now inhabit our seas. Middendorff considered it a variety of the *Turbo tenebrosus* of Montagu. I am inclined to regard it as intermediate between that variety (or rather the variety *patula*) of *L. rudis* and *L. obtusata*.

For the reasons above stated, and following Deshayes, Menke, Lovén, Philippi, and Middendorff in their adoption of the name *obtusata*, we avoid the confusion necessarily incident to so many declensions of the word "*littus*" or "*litus*" in this genus and its species. Pulteney, Lamarck, and other authors called this species *Turbo neritoides*; but it is not Linné's species of that name. Lamarck described it as *T. retusus*.

2. L. NERITOI'DES*, Linné.

Turbo neritoides, Linn. S. N. p. 1232. *L. neritoides*, F. & H. iii. p. 26, pl. lxxxiv. f. 1, 2.

BODY dark-grey above with a tinge of purplish-brown [dusky marked with white, especially about the eyes, Philippi]: *head* extensile and projecting beyond the foot: *tentacles* awl-shaped and slender, very broad and bulbous at the base, light-grey and lineated above with two dusky streaks: *eyes* rather large, sessile, one on the middle of the thickened base of each tentacle: *foot* broad, with the front corners very slightly auricled; sole whitish and partly furrowed in the middle.

SHELL forming a pointed cone, rather solid, opaque, glossy in the young and half-grown state, but of a dull hue when

* Having the aspect of a *Nerita*.

adult: *sculpture*, only the usual lines of growth, when viewed by the naked eye or an ordinary lens, but if examined with a high microscopical power the surface is seen to be indistinctly and slightly striated in a spiral direction; these striæ are wanting in full-grown specimens, which are always more or less eroded in consequence of their exposure to the atmosphere and sea-spray: *colour* chocolate or dark reddish-brown, usually paler or variegated by a yellowish zone at the base, sometimes of a greyish or lighter hue at the top of each whorl or in other parts of the shell: *epidermis* very slight, horncolour: *spire* rather short, sharp-pointed: *whorls* 5–6, somewhat convex, but compressed towards the suture, so as to make that part of each whorl considerably overlap the one next above it; the last occupies about two-thirds of the spire: *suture* narrow and slight: *mouth* equal to nearly two-fifths of the lower portion of the shell; it is acute-angled above, somewhat expanded outwardly, and strengthened inside by a rim or ledge; the base is more or less angulated, and in young specimens sharply peaked: *outer lip* thin: *inner lip* forming a glazed coating over that side of the mouth: *pillar* thick, reddish-brown or dirty white, sloping downwards in a direct line for nearly its whole length, and bevelled outwards from the above described rim or ledge: *inside* glossy, chocolate-coloured or dark-brown: *operculum* having three or four whorls, proportionally more solid than in other species of *Littorina*, horncolour, rather strongly but irregularly striated in the line of growth; the inside edge is surmounted by a rim which is partly continued round the spire. L. 0·275. B. 0·225.

HABITAT: Rocks above high-water mark, on all our coasts from Jersey to Shetland; local but abundant. Godwin-Austen included it in his list of newer pliocene shells from Sussex. Geikie has lately quoted it as fossil in the undermentioned places—" Paisley; Kyles of Bute; Lochgilphead (common)." I suspect that there has been some error here with regard either to the determination of the species, or to this being a glacial fossil; it inhabits at present the Clyde district. According to Lovén, it occurs in a living state on the Scandinavian coast from Kullen to Norway; and various writers have described or enumerated it as ranging from Heligoland to the Ægean, along the sea-board of the Atlantic, Mediterranean (including Algeria), Adriatic, and Black Sea, westward to Madeira and the Canary Isles.

This is probably the only kind of *Littorina* common to the north and extreme south of Europe. It congregates in families or clusters, and in dry weather adheres to the rock by means of a membranous film or epiphragm in front of the operculum, of the same nature as that which is secreted by some of the herbivorous *Helices* and *Bulimi*. This state of æstivation sometimes lasts many days, during which the little periwinkle appears to fast. The foot is all this time kept withdrawn, in order to prevent any evaporation of the water by which the gill-plume is kept moist and fit for action. The smaller varieties and young of *L. rudis* are frequently attached in the same manner to rocks beyond the reach of the tide. Bouchard-Chantereaux noticed this singular habit about thirty years ago. Some individuals, which I immersed in fresh water for eighteen hours, crawled about vigorously after being restored to the open air. My largest specimens were collected by Mr. Barlee in Arran Isle on the coast of Galway; they are four lines long. The shell is frequently eroded or fretted, like the limestone on which it is commonly found; for this reason it often appears distorted. The outermost layer of the shell (owing probably to its constant exposure) occasionally exhibits in certain parts a ramified or efflorescent appearance, as if it were permeated by an extraneous tubular organism. I submitted specimens to the examination of Mr. Berkeley and Dr. Bowerbank. The former thought this appearance might be a condition of some parasitic sponge; but the latter considered it " a nacreous deposit of carbonate of lime natural to the shell." The tongue is remarkably long and slender in proportion to the size of the body—more than three times its length. The operculum, resting on a ledge, is never sunk within the shell.

It is the *Turbo petræus* of Montagu, *T. cærulescens* of Lamarck, *L. Basterotii* of Payraudeau, *Helix neritoidea* and *T. Lemani* of Delle Chiaje, *T. petreus* of Fleming, *L. melanostoma* of Krynicki, and *T. petricola* of Leach.

3. L. RU'DIS*, Maton.

Turbo rudis, Maton, Nat. Hist. and Antiq. West. Count. i. p. 277. *L. rudis*, F. & H. iii. p. 32, pl. lxxxiii. f. 1-7, & lxxxvi. f. 1.

Body of various hues, white, yellow, brown, or fleshcolour, usually more or less clouded or streaked across with dark purple: *head* thick, wrinkled transversely, often tinged with violet on the upper part and neck: *tentacles* rather slender, with blunt tips, frequently marked with a pale-yellowish or black stripe in front down the middle, and with another of a similar colour on the under side: *eyes* globular and prominent, on short and thick stalks, which are amalgamated with the tentacles at their outer bases; pupils black, within gelatinous and transparent irides: *foot* double-edged in front; sole light-yellow or whitish, bordered by a clear hem at the sides and behind, and divided down the middle by a slight fold.

SHELL forming a short cone, solid, opaque, and lustreless: *sculpture*, several flattened spiral ribs, crossed obliquely by slight, irregular and laminar marks of growth; the surface is covered with close-set minute spiral wavy striæ or wrinkles, which are always discernible in every form of this extremely variable species: *colour* most diversified, consisting chiefly of yellow, brown, red, orange, and purple, sometimes jet-black or pure white, and usually variegated by zones or spiral bands of different hues and widths: *epidermis* not observable, and (if formed) probably a mere film and caducous: *spire* moderately pointed: *whorls* 6-9, convex, somewhat flattened or compressed just below the suture; the last whorl occupies in the female

* Rough.

at least two-thirds, and in the male not much more than one-half of the spire: *suture* more or less deep, and always distinct: *mouth* equal to about one-third of the lower portion of the shell in females, but proportionally much smaller in males, angulated and slightly channelled above, and considerably expanded as well as angulated below: *outer lip* thin, a little reflected, incurved towards the pillar: *inner lip* united with the outer lip, and forming a thin glaze on the upper part of the mouth: *pillar* short, but thick and very broad, especially at the base; it shelves inwards, and is white or light-coloured: *inside* of a darker hue in coloured specimens: *operculum* horn-colour, having four or five volutions, which are crossed by curved and rather numerous striæ in the line of growth: the under side has an irregular boss in the centre of the spire, but no rim as in the last species. L. 0·65. B. 0·5.

Var. 1. *saxatilis.* Stunted, nearly globular, usually smooth or finely ribbed; colour greyish with a white base. *L. saxatilis*, Johnston, in Proc. Berw. Nat. Club, i. p. 268; F. & H. iii. p. 43, pl. lxxxvi. f. 4, 5.

Var. 2. *sulcata.* Ribs flattened; colour yellow, with purplish-brown furrows. *Turbo sulcatus*, Leach, Syn. Moll. G. B. p. 187, tab. ix. f. 6.

Var. 3. *jugosa.* Smaller than usual, having strong and sharp spiral ridges, which are variable in number, and sometimes alternately larger and smaller. *T. jugosus*, Mont. Test. Brit. pt. ii. p. 586, tab. xx. f. 2.

Var. 4. *patula.* Ear-shaped and expanded, thinner; spire not prominent, placed somewhat obliquely; mouth wide. *L. patula*, Jeffreys (erroneously), Thorpe's Brit. Mar. Conch. p. 259; F. & H. iii. p. 36, pl. lxxxv. f. 6-10, and (animal) pl. G G. f. 2.

Var. 5. *globosa.* Larger, globular, thick, and nearly smooth.

Var. 6. *tenebrosa.* Smaller, thinner, smoother, more turreted, and having a deeper suture, dusky and often tessellated or chequered. *T. tenebrosus*, Mont. Test. Brit. pt. ii. p. 303, tab. xx. f. 4. *L. tenebrosa*, F. & H. iii. p. 39, pl. lxxxiv. f. 11, 12, and lxxxv. f. 1-5.

Var. 7. *similis.* Resembling the last variety in size and shape, but more distinctly ribbed.

Var. 8. *lævis.* Oval, solid, and smooth.

Var. 9. *compressa.* Oval, compressed or squeezed together; ribs flattened, defined by impressed lines instead of furrows; last whorl extended lengthwise and disproportionately large, with the base consequently more angular than in the ordinary form.

Monstr. Keeled on the upper part of each whorl (especially the last), or else in the middle or lower part.

HABITAT: Stony beaches everywhere; plentiful. Var. 1. Nestling in the crevices of rocks above high-water mark.=*L. sexatilis,* Brown, and *L. neglecta,* Bean. The *Turbo saxatilis* of Olivi is *L. neritoides.* Var. 2. Land's End (Turton); Channel and Scilly Isles (Barlee, and Cranch *fide* Leach); St. David's (J. G. J.) Another prettily marked variety from the Scilly Isles is grey with white ridges and black furrows. Var. 3. Exposed and high rocks; my largest specimens are from Shetland. Var. 4. Eddystone lighthouse (Mrs. Barbor); Penzance (Bingham, *fide* Brown); Unst, three times the usual size of this variety (J. G. J.). I never called or considered it a distinct species. This appears to be the *Turbo labiatus* of Brown and *L. Sitchana* of Philippi. Var. 5. Dublin Bay (Branscombe, *fide* Clark); Oban (J. G. J.). I have specimens nearly an inch long. Var. 6. Mud-banks and salt-marshes in estuaries, with *Hydrobia ulvæ.* It is the *T. ventricosus* of Brown, *T. obligatus* and *T. vestitus* of Say, and *L. marmorata* of Pfeiffer. Var. 7. Occasionally on rocks in Cornwall, South Wales, Aberdeenshire, and Shetland. Var. 8. Sark and Shetland, on sheltered rocks. Var. 9. Not uncommon on various parts of our shores. I have now and then met with the monstrosity. This very common species or some of the varieties have been found in most of the English, Irish, and Scotch quaternary and newer pliocene strata, from Moel Tryfaen to the Norwich Crag;

Uddevalla (J. G. J.). They inhabit both sides of the Atlantic, from Spitzbergen (Torell) to Lisbon (M'Andrew) in the eastern hemisphere, and from Hamilton's Inlet (Wallich) to Massachusetts (Say) in the western hemisphere. Kutorga has enumerated *L. rudis* as a South-Crimean species, and Mr. Lord has brought home specimens from Vancouver's Isle.

Lister distinguished this from the common eatable periwinkle by the name of *Nerita reticulatus*, &c.; it was figured by Chemnitz as a variety of the former species. Schröter seems to have mistaken it for a freshwater shell. I have taken it in places overflowed by streams during the recess of the tide, together with the common mussel and limpet. There are three distinct forms, resulting from a difference of habitat. One of them lives among loose stones and pebbles on the beach; another on mud; and the third on rocks,

" And all along the indented coast
Bespattered with the salt sea foam."

These forms have given birth to a multiplication of species, the details of which fill, but do not improve every book and treatise on our native mollusca. " 'Tis sixty years since" the viviparous habit of *L. rudis* was noticed by Boys*. It seems to breed throughout the whole of the summer. Mr. Bate observed couples engaged in procreation while the females contained not only eggs in every stage of development, but perfectly formed young, which were about to enter on their own separate errand of life. According to Dr. Johnston this function is continued far on in November, both in the present species and *L. obtusata*. The male is, as usual, smaller, and has a longer spire. It may be presumed

---

* Mr. Rich has enabled me to add *Clausilia biplicata*, and probably *C. rugosa* and *Balia perversa*, to the list of viviparous mollusca.

that the reason for the female having a larger body is that she requires more space to develope the fry within it than if she had merely to produce eggs. The shells of the fry are not umbilicate. A section of the spire in the adult shows that the apex is solidified, in consequence of the first two whorls (which had become too small to contain the upper fold of the liver, and were therefore useless) being filled up with new shelly matter. The shell, when injured, can be repaired to a great extent. A specimen which I picked up in Shetland had been cracked and broken in two, probably by some bird, which may have been interrupted in its meal: the fracture appeared to be too extensive to admit of a complete renewal of the severed portion, but it was patched up, so that the remnant of the shell served the purpose of the surviving and lucky periwinkle.

I consider the present species to have been the *Nerita littorea* of Fabricius, *L. grœnlandica* of Menke and others, and *L. sulcata* also of the last-named author. *L. zonaria* and *L. rudissima* of Bean can hardly be called varieties (much less distinct forms) of this proteiform species.

4. L. lito′rea*, Linné.

Turbo littoreus, Linn. S. N. p. 1232. *L. littorea*, F. & H. iii. p. 29, pl. lxxxiii. f. 7, 8, and (animal) pl. G G. f. 3.

BODY sootcolour, or pale-yellowish, marked with close-set transverse stripes of purplish-black, and irregularly cross barred with lines of the latter colour: *mantle* thick, yellowish-white, lining the inside of the mouth or opening of the shell: *head* semicircular and projecting: *tentacles* annulated or streaked across with black; they are contractile or compressible to such an extent as to be sometimes flattened; tips blunt: *eyes* pro-

* Living on the shore.

minent, on short, thick, and somewhat angular stalks: *pupils* black, within yellowish or dull pearly irides: *foot* large, double-edged in front, striped like the rest of the body: sole light yellowish-brown or pale fleshcolour, divided lengthwise in the middle by a transparent line.

SHELL forming a cone of moderate height, thick, opaque, and mostly of a dull hue: *sculpture*, numerous fine spiral flattened ridges, crossed obliquely by slight irregular striæ or lines of growth; the surface is also covered with close-set minute spiral wrinkles, as in *L. rudis*, but these are in the present species more strongly marked, and are slightly decussated or even punctured by the intersection of the longitudinal striæ: *colour* not so various as in the last species, commonly bistre, yellowish with dark-brown zones or rings, or greyish-yellow, occasionally reddish-orange, and very rarely white: *epidermis* light yellowish-brown, usually obscure or not visible, sometimes thick and velvety: *spire* sharp-pointed: *whorls* 7-8, more convex in female than in male individuals, compressed upwards towards the suture, so that the top of each lower whorl overlaps the periphery of the one above it; the proportional difference between the size of the last whorl in the two sexes is not so great as in *L. rudis*, although in the present species each sex is also distinguishable by its shape: *suture* slight and indistinct: *mouth* equal to nearly one-third of the lower portion of the shell in females, but rather smaller in males, narrowly angulated above, and considerably expanded as well as bluntly angulated below: *outer lip* rather thin, somewhat reflected in full-grown males, flexuous, but not incurved, towards the pillar: *inner lip* forming a white glaze on the upper part of the mouth: *pillar* short but thick; it is always white, and shelves inwards: *inside* or throat usually chocolate-colour, now and then of a pale hue or whitish; margin exhibiting the coloured bands when present; intermediate space white: *operculum* dark-horncolour, having the same number of volutions and lineated in the same manner as *L. rudis*; the under side is likewise similar. L. 1·25. B. 1.

Var. 1. *paupercula*. Somewhat dwarfed, with the whorls more convex, of a dusky hue.

Var. 2. *brevicula*. Smaller and ventricose, with a short spire.

Var. 3. *turrita*. Spire turreted, the whorls being divided by a deep and channelled suture.

Var. 4. *sinistrorsa.* Spire of the shell turned to the left; that of the operculum dextrorsal or regular.

Monstr. Keeled as in *L. rudis*—the body-whorl furrowed, or irregularly puckered lengthwise below the suture—the spire much elongated—a new mouth thrown out or formed at the side, and twisted backwards—or distorted in other ways.

HABITAT: Among stones and *Fuci*, and on rocks, below high-water mark of neap tides; extremely common. The 1st variety frequents mud-flats in estuaries and tidal inlets of the sea; the 2nd was found by me on Llanrhidian salt-marsh near Swansea, at Southend, and in Christiania fiord; the 3rd occurred rather plentifully to Mr. Barlee and myself in Loch Carron, and I have solitary examples from other places; of the 4th I procured two specimens at Billingsgate, and Mr. Rich obtained a third which is now in Mr. Leckenby's collection. It is rather surprising that, considering the enormous number of periwinkles brought every year to this market, the reversed kind should be so excessively rare. I was assured by all the dealers in shell-fish that only these three specimens had ever been heard of. 3 and 4 are perhaps monstrous rather than varietal forms. The distortions above noticed are found now and then with the ordinary sort. Mr. S. Wood has figured many of these monsters in his 'Monograph of the Crag Mollusca.' *L. litorea* finds a place in almost every list of our upper tertiary fossils, from Moel Tryfaen to the Red Crag; Sars has recorded it from the Christiania district, in both older and newer deposits, at heights varying from 100 to 460 feet; and I observed it at Uddevalla. The limits of its extra-British distribution are comprised within Greenland (Mörch), White Sea (Baer and Middendorff), and Lisbon (M'Andrew); and Stimpson gives it as a New England shell. The undermentioned localities are suspicious :—Nice (Risso, and "subfossile"); Palermo (Philippi, who however doubted this species being indigenous to Sicily); and Algiers (Weinkauff).

The old English name of "periwincle" is supposed to have been a corruption of petty winkle or wilk. Lister says that the Scarborough fishermen called them "couvins"; and he adds that they were much sought after by the Flemings. According to Dale, they were called in Suffolk "pinpatches." The ancient vernacular names for them were in Swedish "kupunge," in French "bigourneau," "vignot," or "vignette," and in the Breton dialect "vrélin" or "brélin." Throughout Shetland they are known as "wilks." In Ström's time th Scandinavian peasants used to believe that, whenever these shell-fish crept far up the rocks, it indicated a storm from the south. The habits and anatomy of the common periwinkle, and of some other marine testaceous mollusca, were carefully described by the late Mr. Osler in the 'Philosophical Transactions' for 1832. With respect to the phytophagous kinds, he states that they have three distinct modes of feeding. "They browse with opposite horizontal jaws—they rasp their food with an armed tongue, stretched over an elastic and moveable support—or they gorge it entire. *Trochus crassus* [*T. lineatus*] is a convenient example of the first, *Turbo littoreus* [*L. litorea*] of the second, and *Patella vulgata* of the third." With respect to the tongue of *L. litorea* ("a flat strap-shaped organ and more than two inches long") he observes, " It presents three longitudinal ranges of teeth, which recline backwards, and are set like scales, with very little elevation of their edges. In the two outer rows the teeth are single, irregularly crescentic in shape, and set by their convexity; in the middle one each transverse range contains several, which are small and nearly square. All are too minute to be distinguished, except under a high magnifying power. The magnified lingual membrane appears beautifully reticulated." And he further remarks that the periwinkle " feeds upon the softest algæ. I have observed it devouring a minute filament, which entered the mouth by a succession of jerks, repeated at very short intervals. In this case it is probable that the filament passes undivided into the stomach. When browsing upon larger fragments, the portions cut away are so very small that the impressions left can be seen only by a close inspection." M. Beudant's celebrated experiments show that the present species has a greater capability than *L. obtusata* of living in fresh water. There was probably some mistake in the assertion of Bouchard-Chantereaux that the present species is viviparous, like *L. rudis*. Although this peculiarity may have been wrongly attributed by him to *L. litorea*, instead of to a variety of the last-named species, the particulars which he gives are sufficiently interesting to justify their being transferred to these pages, and they are as follows. The female produces about 600 young ones, which are clustered in a vascular ovary, situate on the upper part of the body, and extending from the liver to the right tentacle where the orifice or duct lies; the fry are expelled one by one during a period of many hours in succession, so that about six or seven months elapse before the entire birth is completed; the growth of the year's brood is therefore very unequal, the first born being eight or ten times the size of the last. This statement that the common periwinkle is viviparous seems to be disproved by the fact that it is eatable at all seasons of the year and is never gritty, which last would certainly be the case if it contained testaceous fry. It is sometimes striped like the zebra. In one individual which I examined the right-hand tentacle was branched like a stag's antler; and Dr. Johnston mentions a specimen " in which the tentacula were divided into two branches." In another individual the left-hand tentacle had been mutilated, and appeared not to be of more use to the periwinkle than the stump of an arm would be to a crippled soldier, who had lost that limb on the field of battle. Besides the monstrosities or malformations above specified, and which appear to have resulted from some injury sustained by the mantle, the shell is liable to be affected by chemical action and other causes. On one part of the shore of the Thames at Southend I found almost every specimen of *L. litorea* more or less eroded, some of them to so great an extent as to be distorted. This could not have been owing to the admixture of fresh and salt water, because on another part of the same shore, where a stream flowed into the sea, none of the specimens which I found were eroded. In many places on the open coast, where there is no fresh water, all the shells, as well as the limestone rocks, are fretted. An explanation of this curious phenomenon was offered in the Introduction (pp. l–liv) to the first volume of this work. Shells thicker than usual are often attacked and penetrated, sometimes by minute *Algæ*, and at other times by a species of *Cliona*, or by a small cylindrical annelid; the latter frequently destroys the upper whorls. One specimen in my collection is so encrusted with bleached nullipore as to be easily mistaken for a small lump of chalk. I have a pearl which was extracted from the common periwinkle; it is round and white, the tenth of an inch in diameter. Petiver noticed the large size of the "periwincles" on our northern coasts; he figured a specimen in his Natural history patchwork the "Gazophylacium," with a note that it came from the Orkneys, and "resembles our Scarborow covins, but four times bigger." Many from Shetland are an inch and three-quarters long. Males are narrower and smoother than the females, and have a contracted mouth. The operculum is often irregularly laminated. Mr. Rich found one that was double, the original operculum only being spiral. The animal is sometimes infested by Trematode parasites. M. Lespés detected *Cercaria proxima* in the liver, and *C. linearis* in the kidney of *L. litorea* at Arcachon. Man has utilized periwinkles as well as everything else in creation. They are employed by some of the Essex oyster merchants to keep the grounds clear of seaweeds; Mr. Smith of Burnham informs me that he lays down every year scores of bushels for that purpose. They are also very serviceable in the same way for cleaning an aquarium. The periwinkle is a favourite delicacy of the poor. Drayton, of course, did not omit it in his catalogue of our edible mollusca. According to Swammerdam, it was eaten in Holland during the months of April and May only; it was said to excite thirst. In Mr. Hyndman's Report to the British Association on the operations of the Dredging Committee at Belfast (1857) we find that at that place " the periwinkles are gathered and exported in large quantities to London. Mr. Getty, Secretary to the Harbour Commissioners, informs me that this trade has been carried on for the last twenty-five years by one person, who employs three horses and a mule to draw them, besides employing boats, &c., paying about £60 weekly in wages during the season. The periwinkles are assorted and put into sacks, of which one hundred are often shipped by one steamer weekly. The quantity exported in 1854 amounted to 400 tons, and in 1855 to 459 tons. During this long period there appears to have been no diminution in the supply until this last season [1856], when it is stated that they are not so plentiful as formerly." I was lately told at Kirkwall and Stromness that more than 1000 bushels are exported weekly, every spring and autumn, from those ports to London. At Lerwick, also, vast quantities are shipped by the steamer, and sent to Leith. The bags are occasionally soused with sea-water during the passage, in order to keep the stock alive and fresh. Messrs. Baxter & Son of Billingsgate have kindly furnished me with particulars of the home periwinkle trade. The supply is about 2000 bushels per week for six months, from March until August inclusive, and about 500 bushels per week for the remaining six months. The number of persons employed in gathering is at least 1000 (chiefly women and children), and quite as many more in selling. The best gathering-grounds are the coasts of Scotland, Orkneys, Shetland, and Ireland. The trade-price varies from two to eight shillings per bushel of eight gallons heaped measure; the larger the "winkles" are, the higher the price. Those gathered from rocks keep a fortnight in summer and a month in winter; mud-winkles will not live much more than half that time. When the supply is greater than the demand, Messrs. Baxter now and then send their surplus stock to Southend, and have it laid on some ground of theirs between tide-marks; but the cost of carriage, and of gathering the stock and bringing it again to market, is considerable, and it is often cheaper to throw away what is unsaleable. My informants send large quantities to about thirty provincial towns, and give credit to retail dealers to the amount of from £50 to £60 a week during the season. *L. litorea* may be always known from *L. rudis* or any of its varieties in every state of growth by being at least twice the size, having flatter whorls, a much slighter suture, a more elongated and sharply pointed spire, and a straight outer lip. The two species are frequently found together.

It is the *Turbo littoralis* of Baster, *Castanea tosta* or "marron roti" of D'Argenville, *L. vulgaris* of Sowerby and Reeve, and *L. communis* of Thompson.

The following two species of *Littorina* have been erroneously introduced into the list of British mollusca: both are West-Indian.

1. *L. ziczak* (*Trochus ziczak*, Chemnitz), said to have been found by Miss Hutchings in Bantry Bay. I agree with Mr. Alder in assigning the supposed small variety of this species, without the dark zigzag lines, which was found by Lady Wilson near Sunderland, and mentioned by Maton and Rackett, to *L. neritoides.*

2. *L. dispar* (*Turbo dispar*, Montagu). "Poole" (Rev. Mr. Bingley); "Portmarnock and Teignmouth" (Turton).

Inest in explicatione Naturæ insatiabilis quædam e cognoscendis rebus voluptas, in qua una, confectis rebus necessariis, vacui negotiis, honeste ac liberaliter possumus vivere.—CICERO *de Finibus*, Lib. IV. c. 5.

ERRATA.

Page 254, line 6 from top, for "ANCYLOIDE," read "ANCYLOIDES."
„ 258, line 9 from top, for "*T. Noachina,*" read "*P. Noachina.*"
„ 312, line 16 from bottom, for "*T. ægyptiaca,*" read "*Monodonta ægyptiaca.*"

TABLE of geographical and geological distribution.
(See Vol. I. pp. 314–316, and Vol. II. p. 448.)

Species.	Northern.	Southern.	Upper Tertiary.	Extra-European localities.
Conchifera (*continued from vol. ii. p. 451*).				
Solecurtus candidus	—	—		North Africa, Canary Isles, and Madeira.
antiquatus	—	—	—	North Africa and Canaries.
Ceratisolen legumen......	—	—	—	North Africa.
Solen pellucidus	—	—	—	North Africa.
ensis....	—	—	—	North Africa and North-east America.
siliqua	—	—	—	North Africa, Behring's Straits, and North-east America.
vagina	—	—	—	North Africa, Red Sea, and Azores.
Pandora inaequivalvis	—	—	—	North Africa and Canaries.
Lyonsia Norvegica	—	—		Sea of Okhotsk, North Africa, and Madeira.
Thracia praetenuis	—	—	—	
papyracea	—	—	—	Canaries.
pubescens	—	—		
convexa	—	—		
distorta	—	—	—	North Africa.
Poromya granulata	—	—	—	North Africa and Madeira.
Neaera abbreviata	—	—		
costellata	—	—		North Africa, Madeira, and Azores.
rostrata	—	—		
cuspidata	—	—	—	North Africa, Madeira, Azores, and Greenland.
Corbula gibba	—	—	—	Canaries.
Mya arenaria	—	—	—	China, Greenland, and North-east America.
truncata............	—	—	—	Kamtschatka, Greenland, and both sides of North America.
Binghami	—	—	—	North Africa.
Panopea plicata	—	—		
Saxicava Norvegica	—	—	—	Sea of Okhotsk, Newfoundland, and North-east America.
rugosa	—	—	—	Asia, Africa, America, and Australia.

378 TABLE OF DISTRIBUTION.

Species.	Northern.	Southern.	Upper Tertiary.	Extra-European localities.
Conchifera (*continued*).				
Venerupis Irus	—	—	—	North Africa and Canaries.
Gastrochæna dubia	—	—	—	North Africa, Madeira, and Canaries.
Pholas dactylus	—	—	—	North Africa.
candida	—	—	—	North Africa.
parva	—	—		North Africa.
crispata	—	—	—	Both sides of North America.
Pholadidea papyracea		—		
Xylophaga dorsalis	—	—		
Teredo Norvegica	—	—	—	North Africa.
navalis	—	—		North Africa and North-east America.
pedicellata	—		—	North Africa.
megotara	—		—	Greenland and North-east America.
	38	36	35	29	
Solenoconchia.				
Dentalium entalis	—	—	—	Both sides of North America.
Tarentinum	—	—		North Africa.
	2	2	2	1	
Gasteropoda.				
Chiton fascicularis	—	—	—	North Africa and Canaries.
discrepans	—	—		North Africa.
Hanleyi	—	—?	—	West Indies?
cancellatus	—	—		
cinereus	—	—	—	North Africa and Greenland.
albus	—	—		Greenland and North-east America.
marginatus	—	—		North Africa and both sides of North America.
ruber	—	—?	—	Greenland and North-east America.
lævis	—	—		North Africa.
marmoreus	—	—		North-east America.
Patella vulgata	—	—	—	North Africa.
Helcion pellucidum	—	—	—	North Africa.
Tectura testudinalis	—			Nova Zembla, Greenland, and North-east America.
virginea	—	—		North Africa, Canaries, and Azores. Sitka I.?

TABLE OF DISTRIBUTION.   379

Species.	Northern.	Southern.	Upper Tertiary.	Extra-European localities.
Gasteropoda (*continued*).				
Tectura fulva	—			
Lepeta cæca	—			Sea of Okhotsk and both sides of North America.
Propilidium ancyloides	—			
Puncturella Noachina	—			Greenland and North-east America.
Emarginula fissura	—	—	—	Canaries.
rosea	—	—	—	North Africa.
crassa	—			
Fissurella Græca	—	—	—	Canaries.
Capulus Hungaricus	—	—	—	North Africa.
Calyptræa Chinensis	—	—	—	North Africa, Madeira, and Canaries.
Haliotis tuberculata	—	—		North Africa, Canaries, and Azores.
Scissurella crispata	—	—		Greenland.
Cyclostrema Cutlerianum	—	—		
nitens	—			
serpuloides	—	—		North-east America.
Trochus helicinus	—			Sea of Okhotsk, Behring's Straits, Greenland, and North-east America.
Grœnlandicus	—		—	North-east America.
amabilis				
magus	—	—		Red Sea, North Africa, Madeira, Canaries, and Azores.
tumidus	—	—	—	
cinerarius	—	—	—	North Africa.
umbilicatus	—	—	—?	North Africa.
Duminyi		—		
lineatus	—	—	—?	North Africa.
Montacuti	—	—	—	North Africa.
striatus	—	—		North Africa, Madeira, and Canaries.
exasperatus	—	—	—?	North Africa, Madeira, Canaries, and Azores.
millegranus	—	—		
granulatus	—	—	—?	North Africa, Madeira, and Canaries.
zizyphinus	—	—		Canaries.
occidentalis	—	—		North-east America.
Phasianella pulla	—	—	—?	North Africa and Canaries.
Lacuna crassior		—		Sea of Okhotsk and Sitka I.
divaricata	—	—		Greenland and both sides of North America.

Species.	Northern.	Southern.	Upper Tertiary.	Extra-European localities.
Gasteropoda(*continued*).				
Lacuna puteolus	—	—	—	
pallidula	—		—	Greenland and North-east America.
Littorina obtusata	—	—	—	
neritoides	—	—	—?	North Africa, Madeira, and Canaries.
rudis	—	—	—	Both sides of North America,
litorea	—	—?	—	Greenland and North-east America.
54	51	37	34	
Total 94	89	74	64	

Of the above species (not taking into account doubtful cases of distribution) 71 may be considered northern as well as southern, 19 peculiarly northern, and 3 peculiarly southern; this distribution is, of course, irrespective of their British habitat. One species (*Pholadidea papyracea*) has not yet been noticed on any foreign coast. Eight other species (viz. *Thracia truncata, Dentalium abyssorum, Piliscus commodus, Cyclostrema costulatum, Trochus cinereus, T. olivaceus, T. elegantissimus,* and *Littorina litoralis* or *L. palliata*) have been noticed in the present volume as occurring only in our newer tertiaries; all these still exist in high northern latitudes. Such recent species as are also enumerated as fossil in the list now given, comprise 15 peculiarly northern, and but one peculiarly southern; the rest are common to both divisions.

INDEX TO VOL. III.

The synonyms, as well as the names of spurious species, and of species, genera, and other groups which are not described in this volume, are in italics.—The figures in smaller type refer to the page in which the description of species, genera, and higher groups will be found.

Acanthochætes vulgaris, Leach, 213.
Acanthochites, Leach, 210.
  æneus, Risso, 213.
  carinatus, Risso, 215.
  communis, Risso, 215.
Acanthopleura, Guild., 205, 210.
Acephala, 201, 202.
Acmæa, Esch., 246.
  testudinalis, F. & H., 246.
  virginea, F. & H., 248.
Adasius Loscombeus, Leach, 5.
Adeorbis, Wood, 317.
  subcarinatus, 317.
  supranitida, Wood, 317.
  tricarinata, Wood, 317.
Adesmacea, De Bl., 101.
Agina, Turt., 77.
Aloides, Mühlf., 55.
Amphidesma corbuloides, Lam., 31.
  phaseolina. Lam., 38.
  truncata, Brown, 43.
Amphisphyra, 204, 246.
Anatina, Lam., 29, 32, 47.
Anatina, Sch., 32.
  brevirostris, Brown, 55.
  intermedia, Clark, 38.
  longirostris, Lam., 52.
  myalis, Lam., 39.
  oblonga, Ph., 36.
  rupicola, Lam., 43.
  truncata, Lam., 36.
  truncata, Macg., 38.
  villosiuscula, Macg., 37.
Anatina? pusilla, Ph., 43.
ANATINIDÆ, D'Orb., 28, 29, 32, 55.
Anatomus, De Montf., 283.
Anchomasa Pennantiana, Leach, 112.
Ancylus, 254.
  fluviatilis, 235.

Ancylus (continued).
  Gussonii, Costa, 249.
  lacustris, 235.
Anodonta, 152.
Anomia, 98.
  tabacca, Meusch., 28.
Aporrhais, 201.
Arca, 98.
Arcinella, Ok., 76.
Arcinella, Ph., 76.
Arcinella, Sch., 76.
Aspergillum, 90.
Astarte sulcata, var. *elliptica*, 32.
Auris, Kl., 279.
  vulgaris, Kl., 281.
Avicula, 23.
Azor, Leach, 3.

Dalia perversa, 367.
Barnea, Leach, 107.
  spinosa, Risso, 109.
Biapholius, Leach, 77.
Bontæa, Leach, 36.
Bontia, Leach, 36.
Brachiopoda, 184.
Bruma dell' oceano, Vall., 180.
Buccinopsis Dalei, var. *eburnea*, 301.
Buccinum undatum, 320, 325.
Buccinum undatum, var. *Zetlandica*, 27.
Bulimi, 363.
Bullidæ, 200.
Byssomya, Cuv., 77.
Byssonia, De Bl., 77.
Bythinia tentaculata, 66.

Cadmusia Solanderia, Leach, 118.
Cæcum, 201.
Calopodium, Bolt., 24.

Calyptra, Kl., 242, 273.
   canaria, Bon., 275.
CALYPTRÆA, Lam., 201, 254, 256, 269, 273, 276.
   Chinensis, 269, 273, 275, 379.
   lævigata, Lam., 275.
   mamma, Kr., 276.
   Polii, Sc., 276.
   Sinensis, F. & H., 273.
   succinea, Risso, 276.
   vulgaris, Ph., 276.
CALYPTRÆIDÆ, Brod., 272.
CAPULIDÆ Fl., 268, 272.
CAPULUS, De Montf., 201, 265, 268, 269.
   fallax, S. Wood, 272.
   Hungaricus, 269, 270, 379.
   militaris, 271, 272.
   militaris, Macg., 271.
   obliquus, S. Wood, 272.
Cardita, 77.
Cardium, 98, 216.
   edule, 66.
   striatum, &c., Walk., 58.
Carinaria, 200.
Castanea tosta, D'Arg., 376.
Cemoria, Leach., 256.
Cemoria, Risso, 256.
   Flemingiana, Leach, 258.
   Montaguana, Leach., 267.
   princeps, Migh. & Ad., 257.
CERATISOLEN, Forb., 2, 8, 9.
   legumen, 10, 376.
Chæna, Retz, 90.
Chama, 77.
   parva, Da Costa, 93.
   prætenuis, Petiver, 36.
CHITON, L., 100, 203, 204, 205, 208, 209, 213, 219, 222, 226, 229, 235.
   abyssorum, Sars, 216, 285.
   achatinus, Brown, 227.
   albus, L., 210, 220, 378.
   albus, Pult., 218.
   alveolus, Sars, 218.
   aselloides, Lowe, 221.
   asellus, Sp., 218, 220.
   cancellatus, Leach ?, 210, 217, 218, 219, 378.
   cimex, Ch., 224.
   cimicinus, Landt., 224.
   cinerea, L., 218.
   cinereus, 210, 218, 219, 221, 224, 378.

CHITON (*continued*).
   cinereus, Laskey, 221.
   corallinus, Risso, 227.
   Cranchianus, Leach, 227.
   crinitus, Penn., 213, 214, 215.
   dentiens, Gould, 222.
   discors, Mat. & Rack., 227.
   discrepans, Brown, 210, 212, 213, 214, 215, 378.
   Doriæ, Cap., 227.
   fascicularis, L., 210, 211, 213, 214, 215, 378.
   Flemingius, Leach, 229.
   fulminatus, Couth., 229.
   fuscatus, Brown, 220, 224.
   fuscatus, Leach, 220.
   gracilis, Jeffr, 212.
   Hanleyi, Bean, 210, 215, 216, 285, 378.
   islandicus, Gm., 220.
   lævigatus, Fl., 229.
   lævis, Penn., 210, 225, 226, 227, 378.
   latus, Leach, 226.
   latus, Lowe, 226, 229.
   marginatus, Penn., 206, 207, 210, 221, 223, 224, 225, 378.
   marmoreus, Fabr., 209, 210, 225, 227, 229, 378.
   minimus, Sp., 226.
   Nagelfar, Lov., 216, 285.
   nucleus, Lam., 56, 58.
   onyx, Sp., 220.
   oryza, Sp., 221.
   pictus, Bean, 229.
   punctatus, L., 224.
   punctatus, Turt., 223.
   punctatus, Str., 229.
   quinquevalvis, Brown, 224.
   Rissoi, Payr., 218, 219.
   ruber, L., 210, 224, 225, 226, 378.
   ruber, Turt., 223.
   sagrinatus, Couth., 221.
   Scoticus, Leach, 220.
   septemvalvis, Mont., 227.
   strigillatus, S. Wood, 216.
   tuberculatus, Leach, 218.
   variegatus, Leach, 224.
   variegatus, Ph., 224.
CHITONIDÆ, Guild., 199, 203, 204.
CHITONS, 205, 210, 228.
Cladopoda, Gr., 101.
Clanculus, 312.
Clausilia, 367.

Clausilia (continued).
  biplicata, 367.
  rugosa, 367.
Clavagella, 90.
Clotho, Fauj., 77.
Cochlea, 343.
  parva, Da Costa, 343, 350.
Cochlodesma, Couth., 34, 43.
  prætenerum, S. Wood, 36.
  prætenue, F. & H. 34.
Cochlolepas antiquata, 272.
Conchifera, 1, 123, 184.
Conidæ, 201.
Conus, 201.
Coramya, Leach, 77.
Corbicula fluminalis, 317.
CORBULA, Brug., 44, 55, 59.
  costellata, Desh., 49, 51.
  cuspidata, Brown, 52.
  gibba, 56, 57, 58, 376.
  granulata, Nyst & West, 45.
  labiata, 59.
  mediterranea, Costa, 58, 59.
  olympica, Costa, 58.
  ovata, Forb., 58.
  physoides, Desh., 58.
  rosea, Brown, 57.
  vitrea, Desh., 47.
  Waelii, Nyst, 51.
Corbulacea, Hinds, 44.
Corbuladæ, Flem., 43.
Corbulæ, 66.
Corbulæa, Latr., 44.
Corbulées, Lam., 44.
CORBULIDÆ, 43, 44.
Crania anomala, 250.
Crepidula plana, Say, 276.
  sinuosa, Turt., 276.
  unguiformis, Lam., 276.
Crucibulum. Sch., 273.
  extinctorium, 273.
Ctenobranchiata, Gr., 201.
Cultellus, Sch., 14.
Cumingia parthenopæa, Tib.. 47.
Cuneus foliatus, Da Costa, 88.
Cuspidaria, Nardo, 48.
  typica, Nardo, 55.
CYCLOBRANCHIATA,Cuv.,199,203,235.
Cyclostoma, 292.
Cyclostrema, Fl., 287.
CYCLOSTREMA, Marr., 286.
  Cutlerianum, 287, 289, 379.
  nitens, Ph., 289, 379.
  serpuloides, 290, 297, 379.

CYCLOSTREMA (*continued*).
  costulatum. Möll., 291, 292, 380.
Cylichna, 204, 246.
  alba, 301.
  cylindracea, 204.
  truncata, 204.
Cypræa, 201, 271.
Cypræidæ, 200.
Cyrtosolen, Herrm., 3.

Dactylina, Gr., 104.
Delphinoidea, Brown, 287.
Delphinula, De Roissy, 287, 317.
  Duminyi, Req., 315.
  lævis, Ph., 291.
  nitens, Ph., 289.
Delphionoidea, Brown, 287.
DENTALIA, 193, 194.
DENTALIIDÆ, H. & A. Ad., 191.
DENTALIUM, L., 132, 166, 171, 172, 173, 186, 189, 191, 193, 198, 207, 268.
  abyssorum, Sars. 197, 379.
  album, Turt., 198.
  attenuatum, Say, 197.
  bifissum, S. Wood, 171.
  clausum, Turt., 198.
  dentalis, Forb., 197.
  dentalis, L., 196, 197.
  eburneum, Turt., 198.
  entale, S. Wood, 197.
  entalis, L., 191, 192, 195, 196, 197, 378.
  gadus, Mont., 198.
  Indianorum, P. Carp., 194.
  labiatum, Turt., 196.
  læve, Turt. 196.
  novemcostatum, Lam., 197.
  octangulatum, Don., 197.
  octogonum, Lam., 197.
  octohedra, Leach, 198.
  politum, Turt., 196.
  pretiosum, Nutt., 193.
  semipolitum, Brod. and Sow., 198.
  semistriatum, Turt., 198.
  semistriolatum, Guild., 198.
  striatulum, Turt., 197.
  striatum, Born, 196, 197.
  striatum, Mont., 196.
  striatum, J. Sm., 197.
  striolatum, St., 197.
  subulatum, Desh., 198.
  Tarentinum, Lam., 186, 192, 194, 195, 196, 197, 378.

DENTALIUM *(continued)*.
  variabile, Desh., 198.
  vulgare, Da Costa, 196.
  vulgare, H. and A. Ad., 197.
Didonta, Sch., 77.
Diodora, Gr., 257.
Donaces, 88.
Donax, 77.
  Irus, L., 86.
Dorididæ, 200.

EMARGINULA, Lam., 257, 259, 266, 282, 283.
  capuliformis, Ph., 263.
  conica, Sars, 261.
  conica, Sch., 261, 263.
  Costæ, Tib., 263.
  crassa, J. Sow., 263, 264, 265, 379.
  curvirostris, Desh., 263.
  decussata, Ph., 264.
  fissura, 259, 260, 261, 262, 263, 264. 379.
  fissurata, Récl., 261.
  lævis, Récl., 261.
  Mülleri, Forb., 261.
  Müllerii, F. & H., 259.
  pileolus, Mich., 263.
  reticulata, J. Sow., 259, 261, 262.
  rosea, Bell, 261, 262, 263, 379.
  rubra, Lam., 263.
  tenuis, Récl., 261.
Embla, Lov., 45.
  Korenii, Lov., 47.
Ensatella Europæa, Sow., 18.
Ensis, Sch., 16.
  magnus, Sch., 18.
Epheria, Leach, 343.
  Bulweriana, Leach, 348.
  Goodallii, Leach, 348.
Erycina anodon, Ph., 43.
Eucharis, Récl., 45.
Eudora varians, Leach, 340.
Eulima, 204, 246.
Eutropia, Humphr., 337.

FISSURELLA, Brug., 190, 265, 277, 282, 283.
  cancellata, Gr., 267.
  cancellata, G. B. Sow., 267.
  Europæa, Sow., 267.
  Græca, 256, 266, 268, 378.
  Listeri, Woodw., 267.
  nimbosa, Lam., 268.
  nimbosa, Ph., 268.

FISSURELLA *(continued)*.
  nubecula, L., 267, 268.
  occitanica, Récl., 267.
  Philippii, Req., 268.
  reticulata, F. & H., 266.
  rosea, Lam., 268.
  rosea, Ph., 268.
  striata, Récl., 267.
Fissurelladæ, Fl., 255.
FISSURELLIDÆ, Fl., 255, 256, 268, 270, 276, 285.
Fistulana, Brug., 90.
  corniformis, Lam., 171.
Fusus, 201.

Galaxura, Leach, 36.
Galeomma, 29.
Galerus, Humphr., 273.
GASTEROPODA, 199, 201, 202, 342.
GASTROCHÆNA, Sp., 90, 91, 92, 94, 138.
  cuneiformis, Lam., 93.
  cuneiformis, Phil., 93.
  dubia, 91, 92, 377.
  fulva, Leach. 93.
  modiolina, Lam., 91, 93.
  mumia, Sp., 90.
  pelagica, Risso, 93.
  Poliana, Lam., 93.
  Polii, Lam., 93.
  tarentina, Costa, 93.
GASTROCHÆNIDÆ, Gr., 89.
Gastrochina, Sw., 90.
Gibbula, Leach, 294, 305.
  lineata, Leach, 315.
  striata, Leach, 312.
Glicimeris, Kl., 75.
Glycimeris, 75.
  arctica, Lam., 81.
Goniodoris nodosa, 275.

HALIOTIDÆ, Fl., 276, 285.
Haliotides, 277.
HALIOTIS, L., 194, 265, 277, 278, 281, 282, 283, 293.
  tuberculata, L., 277, 279, 280, 379.
HELCION, De Montf., 230, 242, 245.
  pectinatum, 245.
  pellucidum, 242, 377.
Helices, 363.
Helicidæ, 200.
Helix, 355.
  depressa, Mont., 287.

Helix (continued).
 fasciata, Ad., 350.
 incarnata, 142.
 lacuna, Mont., 351.
 margarita, Lask., 297.
 nemoralis, 103.
 neritoidea, Delle Ch., 364.
 serpuloides, Mont., 287, 290.
 strigella, 142.
Heteropoda, 200.
Hiatella, Daud., 77.
Hipponice, Defr., 271.
Hipponyx, Defr., 271.
Hyalæa (Carolina) tridentata, 167.
Hydrobia ulvæ, 366.
Hypogæa, Poli, 24, 104.
 crinita, Poli, 20.
 falcata, Poli, 18.
 gibba, Poli, 28.
 hirudo, Poli, 11.
 tentaculata, Poli, 22.
 verrucosa, Poli, 107.

Ianthina, 167.
Iothia, Forb., 246.

Jouannetia, 115.

Kellia, 69.
Kuphus arenarius, 123, 156.

LACUNA, Turt., 340, 343.
 albel'a, Lov., 348.
 crassior, 344, 345, 346, 348, 349, 379.
 divaricata, 346, 348, 349, 379.
 frigida, Lov., 348.
 glacialis, Möll., 345.
 labiosa, Lov., 348.
 Montacuti, Turt., 351.
 Montagui, Brown, 351.
 neritoidea, Gould, 352.
 pallidula, 350, 351, 380.
 patula, F. & H., 351.
 pertusa, Conr., 348.
 puteolus, Turt., 343, 348, 351, 380.
 retusa, Brown, 353.
 solidula, Lov., 348.
 sulcata, Macg., 351.
 vincta, F. & H., 346.
Leda, 300.
LEPETA, Gr., 230, 251, 253.
 cæca, 251, 252, 379.
Lepidopleurus, Leach, 210.

VOL. III.

Lepidopleurus (continued).
 carinatus, Leach, 224.
 punctulatus, Leach, 227.
Ligula, Mont., 33.
 pubescens, Mont,, 38.
Lima, 98.
Limacidæ, 200.
Limapontia nigra, 66.
Limapontiidæ, 200.
Limax arborum, 343.
Limnæa auricularia, 66.
 peregra, 66.
 stagnalis, 66.
Limopsis aurita, 301.
Lingula, 235.
Lithophaga dactylus, 102, 114.
Litorina, Menke, 355.
LITTORINA, Fér., 293, 339, 341, 342, 344, 354, 356, 363, 376.
 arctica, Möll., 361.
 Beanii, Macg., 358.
 communis, Th., 376.
 dispar, 376.
 grœnlandica, Menke, 368.
 limata, Lov., 361.
 litoralis, F. & H., 356.
 litoralis, 359, 379.
 litorea, 359, 368, 370, 371, 372, 373, 376, 380.
 marmorata, Pf., 366.
 melanostoma, Kryn., 364.
 neg'ecta, Bean, 366.
 neritiforma. Brown, 357.
 neritoides, 354, 361, 366, 375, 380.
 obtusata, 66, 352, 356, 359, 361, 367, 372, 380.
 palliata, F. & H., 357.
 palliata, 358, 359, 361, 379.
 patula, Jeffr., 365.
 rudis, 354, 359, 361, 363, 364, 367, 372, 376, 380.
 rudissima, Bean, 368.
 saxatilis, Johnst., 365.
 sexatilis, Brown, 366.
 Sitchana, Phil., 366.
 sulcata, Menke, 368.
 tenebrosa, F. & H., 365.
 vulgaris, Sow., 376.
 ziczak, 376.
 zonaria, Bean, 368.
LITTORINÆ, 213, 341, 354, 355.
LITTORINIDÆ, Gr., 340, 343.
Lophyrus, Poli, 206.
Lottia, Gr., 246.

S

Lutraria, 1.
  elliptica, 90.
LYONSIA, Turt., 28, 29, 52.
  Norvegica, 29, 30, 35, 377.
Macha, Oken, 3.
Mactra solida, 27.
  solida, var. *elliptica*, 27.
Mactridæ, 32, 44.
Magdala, Leach, 29.
Mangelia, 204, 246.
Margarita, Leach, 292, 294, 295, 304.
  arctica, Gould, 297.
  arctica, Leach, 297.
  aurea, Brown, 305.
  cinerea, Couth., 307.
  elegantissima, Bean, 305.
  glauca, Möll., 305.
  helicoides, Beck, 297.
  olivacea, Brown, 305, 379.
  plicata, Sars, 305.
  polaris, Dan., 305.
  pusilla, Jeffr., 289.
  striata, Brod. & Sow., 304.
  undulata, Sow., 300.
  vulgaris, Leach, 297.
Margarita? costulata, Möll., 291.
Margarita? maculata, S.Wood, 303.
Margarites diaphana, Leach, 297.
Martesia, 115.
  striata, 111, 114.
Medoria, Leach, 344.
  Danmoniensis, Leach, 346.
  Walkeri, Leach, 346.
Melania, 59.
Meleagrinæ, 281.
Mitra Hungarica, Kl., 269.
Mitrularia, Sch., 273.
Mölleria, Jeffr., 292.
  ægyptiaca, Lam., 312.
  ægyptiaca, Payr., 312.
Monodonta articulata, Lam., 320.
  Draparnaudi, Payr., 320.
  sitis, Récl., 320.
Montacuta, 42.
Montagua Danmoniensis, Leach, 322.
Murex erinaceus, 325.
  trunculus, 61.
Muricidæ, 199, 311.
MYA, L., 1, 44, 60, 61, 63, 66, 69, 75, 77, 101, 114.
  acuta, Say, 66.
  arctica, L., 82.

MYA (*continued*).
  arenaria, L., 33, 61, 64, 65, 66, 67, 74, 377.
  Binghami, 70, 72, 75, 76, 377.
  convexa, W. Wood, 39.
  declivis, Penn., 39.
  distorta, Mont., 41.
  dubia, Penn., 91.
  inæquivalvis, Mont., 58.
  lata, J. Sow., 64.
  membranacea, Gm., 32.
  mercenaria, Say, 66.
  nitida, Fabr., 31.
  nitida, Müll., 31.
  Norvegica, Ch., 29.
  norvegica, Sp., 78.
  ovalis, Turt., 70.
  pellucida, Brown, 31.
  Pholadia, Mont., 93.
  prætenuis, Pult., 34.
  pubescens, Pult., 38.
  pullus, S. Wood, 70.
  punctulata, Ren., 38.
  rostrata, Sp., 51.
  striata, Mont., 31.
  truncata, Chier., 38, 68.
  truncata, L., 39, 66, 67, 68, 69, 71, 72, 81, 377.
Myadæ, Flem., 60.
Myatella Montagui, Brown, 31.
MYIDÆ, 60, 101.
Mytilus, 77, 91.
  Adriaticus, 255.
  ambiguus, Dillw., 93.
  carinatus, Brocchi, 76.
  edulis, 61, 66, 88, 103.
  modiolus, 80.
  pholadis, L., 82.
  plicatus, Mont., 75.
  præcisus, Mont., 82.
  rugosus, L., 81.

Nacella, Sch., 242.
Nassa incrassata, 325.
  reticulata, 325.
Nausitora Dunlopei, P. Wr., 147.
NEÆRA, 44, 47, 48,
  abbreviata, Forbes, 48, 49, 50, 377.
  arctica, Sars, 55.
  attenuata, Forb., 52.
  Chinensis, Gr., 52.
  costellata, 49, 52, 377.
  cuspidata, 52, 53, 377.

NEÆRA (continued).
  renovata, Tib., 52.
  rostrata, 51, 53, 54, 55, 377.
  rostrato-costellata, Act., 51.
  sulcata, Lov., 51.
  vitrea, Lov., 49.
Neara, Gr., 47.
Nerita, 343, 361.
  littoralis, L., 360.
  littorea, Fabr., 368.
  pallidulus, Da Costa, 351.
  reticulatus, &c., List., 367.
  rufa, Mont., 351.
Neritina fluviatilis, 66, 120.
Nucleobranchiata, De Bl., 200.
Nucula, 300.
  sulcata, 117.
Nudibranchiata, Cuv., 200.
Nudibranchs, 1.

Odoncincta, Da Costa, 34.
Odostomia, 337.
  truncatula, 216.
Opisthobranches, Milne-Edw., 200.
Osteodesma, Desh., 29.
  elongata, Gr., 30.

Paludina, 337.
PANDORA, Hwass, 23, 27, 29.
  flexuosa, Ph., 28.
  glacialis, Leach, 28.
  oblonga, Ph., 28.
  obtusa, Leach, 24, 25, 26.
  inæquivalvis, 24, 25, 26, 27, 377.
  margaritacea, Lam., 27.
  rostrata, Lam., 24, 26, 28.
  Pandora? æquivalvis, Ph., 32.
PANDORIDÆ, Gr., 22, 23, 28.
Pandorina, Sc., 29.
Panopæa, Goldf., 1, 75.
  Bivonæ, Ph., 81.
  Middendorffii, A. Adams, 81.
  Norvegica, 74, 78.
  Spengleri, Valenc., 81.
PANOPEA, Mén., 74, 75.
  Aldrovandi, 74.
  australis, 74.
  glycimeris, 74, 77, 79.
  glycymeris, 75.
  plicata, 75, 377.
  Norvagica, 60.
Panopia, Sw., 75.
Panopæa, Nyst, 75.
Parapholas, 115.

Parapholas (continued).
  ovoideus, Gould, 115.
PATELLA, List., 95, 97, 189, 190, 204, 209, 229, 230, 233, 234, 235, 242, 245, 246, 253, 266, 277.
  æqualis, J. Sow., 249.
  albida, Don., 275.
  amæna, Say, 248.
  anomala, L., 235.
  apertura, Mont., 267.
  Asmi, Midd., 249.
  aspera, Ph., 238.
  athletica, Bean, 237.
  bimaculata, Mont., 245.
  Bonnardi, Payr., 237.
  cæca, Müll., 252.
  cærulea, L., 245.
  cærulea, Mont., 245.
  candida, Couth., 253.
  cerea, Müll., 253.
  Chinensis, L., 273.
  Clealandi, Couch, 248.
  Clealandi, Sow., 248.
  Clealandiana, Leach, 248.
  clypeus, Brown, 248.
  cornea, Mich., 245.
  crepidula, L., 276.
  depressa, Gm., 237.
  depressa, Penn., 237.
  equestris, L., 273.
  extinctorium, Turt., 235.
  fissura, L., 259, 261.
  fissurella, Müll., 258.
  Forbesii, Sm., 251.
  fulva, Müll., 250.
  græca, L., 266, 267.
  intorta, Penn., 245.
  lævis, Penn., 243.
  larva, reticulata, Da Costa, 267.
  minima, Gm., 249.
  minor, Wall., 245.
  muricata, Brocchi, 275.
  Noachina, L., 257.
  orbiculata, Walk., 235.
  parva, Da Costa, 249.
  pectinata, L., 245.
  pellucida, L., 242.
  pellucida, Ph., 250.
  pileolus, Midd., 249.
  Pileus Morionis major, Da Costa, 271.
  pulchella, Forb., 250.
  squamulata, Ren., 275.
  Tarentina, Lam., 237.

PATELLA (*continued*).
tessellata, Müll., 248.
tessulata, Müll., 248.
testudinalis, Müll., 246.
testudinaria, L., 248.
testudinaria, Müll., 248.
tricornis, Turt., 235.
ungarica, L., 269.
unguis, L., 235.
virginea, Müll., 248.
vulgaris, Bel., 241.
vulgata, L., 231, 236, 371, 378.
Zetlandica, Fl., 258.
Patella? *Ancyloides*, Forb., 254.
exigua, Forb., 255.
Patelladæ, Guild., 229.
PATELLÆ, 354.
PATELLIDÆ, Guild, 199, 229, 255, 268.
Patelloidea, Quoy & Gaim., 246.
Patelloides vitrea, Cantr., 250.
Patellus, De Montf., 235.
Patina, Leach, 242.
Pecten, 98.
maximus, 90.
PECTINIBRANCHIATA, Cuv., 199, 229.
Pellibranchiata, Ald. & Hanc., 200.
Peloris, 114.
Periploma, Sch., 36.
myalis, Coll., 36.
Petricola lithophaga, 58.
Pharella, Gr., 9.
Pharus, Leach, 9.
PHASIANELLA, Lam., 337, 339, 341.
bifasciata, Brown, 348.
cornea, Brown, 348.
fasciata, Brown, 348.
pulchella, Récl., 341.
pulla, 338, 339, 379.
pullus, F. & H., 338.
striata, Brown, 348.
Pholadacea, Tryon, 90.
Pholadaires, Lam., 90.
PHOLADES, 26, 94, 98, 99, 100, 101, 134.
PHOLADIDÆ, Gray, 89, 90, 93, 94, 95, 100, 119, 122, 123, 156.
PHOLADIDEA, Good., 93, 95, 100, 114, 115, 116, 119.
Goodallii, De Bl., 118.
Loscombiana, Good., 118.
papyracea, Turt., 116, 378, 380.
Pholadidoides Anglicanus, 118.

PHOLAS, L., 61, 63, 94, 95, 96, 97, 98, 99, 100, 101, 102, 103, 104, 106, 111, 114, 115, 118, 119, 130, 136, 138, 149, 151, 152, 166.
bifrons, Da Costa, 114.
callosa, Lam., 112.
candida, L., 94, 99, 107, 110, 111, 378.
candidus, L., 107.
clavata, Lam., 114.
conoides, Pars., 114.
crenulatus, Sol., 112.
crispata, L., 89, 94, 98, 109, 111, 112, 117, 118, 378.
cylindrica, J. Sow., 109.
dactyloides, Delle Ch., 109.
dactyloides, Lam., 112.
dactylus, L., 84, 103, 104, 106, 107, 108, 109, 112, 378.
faba, Pult., 93.
hians, Brocchi, 93.
hians, Ch., 93.
hians, Pult., 107.
lamellata, Turt., 118.
ligamentina, Desh., 112.
lignorum, Rumph., 114.
muricatus, Da Costa, 107.
nanus, Sol., 114.
papyracea, Turt., 116, 117.
papyraceus, Sol., 109, 118.
parva, 108, 109, 111, 112, 378.
parvus, Penn., 109.
pusilla, L., 93, 114.
pusilla, Poli, 93.
pypyraceus, Sol., 118.
sulcata, Brown, 114.
teredo, Müll. & Fabr., 166, 180.
Teredula, Pall., 181.
vibonensis, Ph., 113.
xylophaga, Desh., 122.
Pholeobia, Leach, 77.
Phyllidia, 235.
Physa fontinalis, 66.
Pileopsis, Lam., 269.
Hungaricus, F. & H., 269.
Pilidium, F. & H., 246, 253.
fulvum, F. & H., 250.
Piliscus commodus, Midd., 272, 379.
Pinna, 2.
rudis, 250.
Pleurobranchiata, Gr., 200.
Pleurotoma, 283, 329.
Pleurotomaria, 282.

Pleurotomatidæ, 201.
Polyplaxiphora, De Bl., 203.
POROMYA, Forb., 44, 46.
anatinoides, Forb., 46.
granulata, 45, 46, 377.
Proboscidifera, Gr., 201.
PROPILIDIUM, F. & H., 230, 252, 253.
Ancyloide, F. & H., 254.
ancyloides, 254, 379.
PROSOBRANCHES, Milne-Edw., 200.
Psammobia, 1, 4, 98.
scopula, Turt., 5.
tæniata, Turt., 7.
Pulmonobranchiata, Sow., 200, 355.
PUNCTURELLA, Lowe, 254, 256, 257, 265, 266, 282.
Noachina, 257, 258, 260, 379.
Pupa, 356.

Rhomboides, De Bl., 77.
Rhombus, De Bl., 77.
Rimula, Defr., 256.
Flemingii, Macg., 258.
Rissoa, 287, 342.
Zetlandica, 287.
Rissoæ, 342.
Rocellaria, Fl., 90.
Rostrifera, Gr., 201.
Rupicola, Fl., 43.
concentrica, Récl., 43.

Salpæ, 1.
SAXICAVA, Fleur., 60, 72, 74, 77, 81, 84, 134, 138.
arctica, F. & H., 82, 85.
fragilis?, Nyst, 76.
Norvegica, 78, 80, 85, 377.
rugosa, 26, 33, 71, 77, 81, 84, 87, 89, 377.
rugosa, young?, F. & H., 75.
SAXICAVÆ, 69, 72, 80, 84, 86.
SAXICAVIDÆ, Sw., 72, 73, 89.
Scalaria, 337.
Schismope, Jeffr., 257, 282, 283.
SCISSURELLA, D'Orb., 257, 259, 282, 283.
angulata, Lov., 285.
aspera, Ph., 285.
crispa, Sow., 285.
crispata, Fl., 282, 283, 285, 379.
SCISSURELLÆ, 283.
SCISSURELLIDÆ, Gr., 282.
Serpula Teredo, Da Costa, 174.

Siliquaria, 283.
bidens, Ch., 7.
Sipho, Brown, 257.
radiata, Brown, 267.
striata, Brown, 258.
Siphonium, Browne, 167.
Siphonobranchiata, Goldf., 199, 256, 311.
Siphonodentalium vitreum, Sars, 190.
Skenea, Fl., 287, 337.
Cutleriana, Cl., 287.
divisa, Fl., 291.
Serpuloides, Macg., 309.
Skenea? Cutleriana, F. & H., 287.
divisa, F. & H., 290.
lævis, F. & H., 289.
Solarium, 317.
turbinoides, Nyst, 303.
SOLECURTUS, De Bl., 2, 3, 8.
antiquatus, 3, 6. 377.
candidus, 3, 6, 7, 377.
coarctatus, F. & H., 6.
strigilatus, L., 3, 5, 6, 7.
SOLEN, 1, 2, 8, 11, 77, 94.
albicans, Chier., 5.
antiquatus, Pult., 6.
candidus, Ren., 3, 5.
caribbæus, Lam., 8.
centralis, Say, 8.
coarctatus, Gm., 7.
crispus, Gm., 114.
cultellus, L., 7.
cultellus, Penn., 7.
curvus, List., 18.
declivis, Turt., 8.
divisus, Sp., 7.
ensiformis, S. Wood, 18.
ensis, L., 2, 16, 18, 19, 377.
fragilis, Pult., 7.
gibbus, Sp., 8.
gladiolus, Gr., 20.
gladius, Bolt., 20.
Guineensis, Ch., 8.
legumen, L., 9, 10.
ligula, Turt., 19.
marginatus, Pult., 20, 22.
minutus, L., 82.
multistriatus, Sc., 4.
novacula, Mont., 19.
pellucidus, Penn., 2, 8, 14, 18, 19, 376.
pellucidus, Sp., 16, 377.
pinna, Mont., 28.
pygmæus, Lam., 16.

SOLEN (*continued*).
siliqua, L., 18, 20, 21, 22, 377.
tenuis, Ph., 15.
truncata, W. Wood, 22.
vagina, L., 20, 21, 22, 377.
SOLENIDÆ, Latr., 1.
Solenoconches, Lacaze-Duth., 185.
SOLENOCONCHIA, Lacaze-Duth., 185.
Solenocurtis, Sw., 3.
Solenocurtus, Sow., 3.
Sphænia Binghami, F. & H., 70.
Sphenia, Turt., 60.
  Binghami, Turt., 70, 84.
  cylindrica, S. Wood, 76.
  Swainsoni, Turt., 70.
  Swainsonii, Lov., 70.
Spirula australis, 167.
Stomatia, 281.

Tapes, 86.
  decussatus, 88.
  pullastra var. *perforans*, 72, 85.
TECTURA, Cuv., 230, 245, 246, 252.
  fulva, 250, 251, 252, 253, 379.
  testudinalis, 246, 248, 251, 378.
  virginea. Müll., 248, 251, 378.
Tecture, Cuv., 245, 246.
Tellina, 156.
  balthica, 66.
  Cornubiensis, Penn., 88.
  coruscans, Sc., 32.
  cuspidata, Ol., 53.
  fragilis, L., 38.
  fragilis, Penn., 38.
  fragilissima, Chier., 36.
  gibba, Ol., 56.
  inæquivalvis, L., 23, 24.
  naticuta, Chier., 51.
  papyracea, Poli, 36.
  parthenopæa, Della Ch., 58.
Temina, Leach. 344.
  rufa, Leach, 351.
  Turtoniana, Leach, 351.
  variabilis, Leach, 351.
Terebratula, 93, 209.
  caput-serpentis, 304.
Teredarius, Dum., 167.
TEREDINES, 73, 132, 133, 134, 135, 139, 144, 145, 147, 158, 160, 167.
TEREDINIDÆ, Flem., 90, 100, 119, 122, 123.
TEREDO, Sell., 90, 94, 95, 96, 100, 101, 115, 116, 119, 122, 123-138,

TEREDO (*continued*).
  140-142, 144-146, 148-152, 154, 156-163, 165, 166, 170, 178, 181, 183, 192.
  batavus, Sp., 174.
  bipalmata, Delle Ch., 184.
  bipalmulata, Delle Ch., 184.
  bipartita, 183.
  bipennata, Turt., 182.
  bipinnata, Flem., 183.
  bipinnata, Turt., 167, 181, 182.
  Bruguierii, Delle Ch., 171.
  communis, Osl., 171.
  cucullata, 167, 182, 183.
  denticulata, Gr., 181.
  Deshaii, Quatr., 171.
  dilatata, St., 179, 180.
  divaricata, Desh., 169.
  excavata, 167, 183.
  fatalis, Quatr., 171.
  fimbriata, 183, 184.
  fusticulus, 183.
  malleolus, Turt., 167, 179, 181, 182.
  marina, Sell., 130, 171.
  megotara, Hanl., 130, 157, 167, 170, 172, 176, 180, 182, 378.
  minima, De Bl., 127, 170, 184.
  navalis, Gm., 171.
  navalis, L., 127, 130, 131, 140, 144, 155, 156, 157, 163, 171, 173, 174, 175, 179, 180, 184, 378.
  navalis, Möll., 180.
  navalis, Sell., 173.
  navis, L., 173.
  navium, Sell., 171.
  navium, Vall., 130.
  nigra, De Bl., 171.
  norvagica, F. & H., 130, 168.
  norvagicus, Sp., 168.
  Norvegica, Sp., 128, 131, 139, 148, 149, 157, 168, 171, 172, 173, 175, 179, 180, 181, 378.
  oceani, Sell., 180.
  oceani, Vall., 130.
  palmulata, F. & H. 183.
  palmulata, Ph., 184.
  palmulatus, Lam., 183.
  pedicellata, Quatr., 132, 139, 145, 148, 174, 175, 378.
  pedicellatus, Quatr., 174.
  pennatifera, De Bl., 183.
  Philippii, Turt., 184.
  Sellii, Van der Hoev., 174.

TEREDO (continued).
Senegalensis, De Bl., 147, 171.
Senegalensis, Laur., 171.
serratus, Desh., 184.
spatha, 183.
Stutchburii, 156, 182, 183.
subericola, 179.
vulgaris, Lam., 174.
Tergipes lacinulatus, 66.
THRACIA, Leach, 29, 32, 33, 34, 42, 45, 46, 52, 61.
brevis, Desh., 43.
convexa, 39, 46, 377.
corbuloides, Desh., 43.
declivis, Macg., 41.
distorta, 33, 34, 38, 40, 41, 72, 377.
elongata, Ph., 43.
fabula, Ph., 43.
Montagui, Leach, 39.
myopsis, Beck, 43.
ovalis, Ph., 43.
ovata, Brown, 38.
papyracea, 36, 38, 39, 68, 377.
phaseolina, F. & H., 36.
prætenuis, 34, 37, 377.
pubescens, 38, 39, 377.
Scheepmakeri, Dunk., 41.
truncata, 43, 380.
ventricosa, Ph., 41.
villosiuscula, F. & H., 37.
THRACIÆ, 41.
Trapezium, Mühlf., 90.
Tricolia, Risso, 337.
TROCHI, 325.
TROCHIDÆ, D'Orb., 199, 282, 286, 337.
Trochita, Sch., 273.
Trochocochlea, Kl., 294, 295, 317.
Trochoidea, D'Orb., 283.
Trochotoma, 282.
TROCHUS, Rond., 285, 286, 292, 293, 294, 312, 319, 322, 325, 343.
Agathensis, Récl., 313.
alabastrum, Beck, 333.
amabilis, Jeffr., 294, 300, 304, 379.
cinerarius, Born, 312.
cinerarius, Fabr., 299.
cinerarius, L., 286, 294, 309, 313, 314, 315, 318, 379.
cinerarius, Ol., 312.
cinerarius, Pult., 315.
cinerarius, var. conica, 325.
cinereus, 304, 379.
cinereus, Da Costa, 312.

TROCHUS (continued).
Clelandi, W. Wood, 327.
Clelandiana, Leach, 327.
conicus, Don., 323.
conuloides, Lam., 332.
conulus, Da Costa, 325.
conulus, L., 325, 332, 333.
Cranchianus, Leach, 332.
crassus, Pult., 317, 320, 371.
crenulatus, Brocchi, 325.
crenulatus, Lam., 325.
Cyrnæus, Req., 322.
depictus, Desh., 323.
discrepans, Brown, 331.
divaricatus, Fabr., 346, 348.
divaricatus, L., 347.
Duminyi, 315, 379.
electissimus, Bean, 311.
elegans, Jeffr., 327.
elegantissimus, 304, 379.
elegantulus, 304.
elegantulus, Jeffr., 304.
erythroleucos, Gm., 323.
exasperatus, Penn., 324, 379.
exiguus, Pult., 324, 325.
exilis, Ph., 288.
formosus, Forb., 333, 336.
fragilis, Gm., 330.
fragilis, Pult., 330.
fuscus, Walk., 309.
granulatus, Born, 327, 329, 330, 379.
granosus, S. Wood, 329.
Grœnlandicus, 298, 299, 379.
grönlandicus, Ch., 298.
helicinus, Fabr., 295, 296, 298, 379.
helicinus, Gm., 297.
inflatus, Brown, 300.
irregularis, Leach, 332.
lævigata, J. Sow., 331.
lineatus, 294, 295, 317, 320, 371, 379.
lineatus, Da Costa, 312.
lineatus, Lam., 320.
lineatus, Pult., 320.
littoralis, Brown, 312.
Lyonsii, Leach, 331.
magus, L., 305, 306, 379.
margaritus, Gr., 297.
Martini, Sm., 327.
Matonii, Payr., 325.
miliaris, Brocchi, 327.
millegranus, Ph., 325, 327, 379.

Trochus (continued).
Montacuti, 320, 321, 323, 379.
Montagui, W. Wood, 320.
Nassaviensis, Ch., 309.
obliquatus, Gm., 315.
occidentalis, Migh., 294, 333, 379.
papillosus, Da Costa, 330.
parvus, Ad., 332.
parvus, Da Costa, 323.
patholatus, Gm., 309.
perforatus, Brown, 312.
Philberti, Récl., 312.
polymorphus, Cantr., 332.
punctulatus, De Bl., 320.
pusillus, F. & H., 289.
pyramidatus, Lam., 325.
quadricinctus, S. Wood, 336.
Racketti, Payr., 309.
Sisyphinus, Macg., 332.
sitis, Récl., 320.
striatus, L., 304, 317, 322, 323, 324, 325, 379.
tenuis, Mont., 330.
tuberculatus, Da Costa, 307.
tumidus, Mont., 294, 307, 379.
umbilicaris, L., 315.
umbilicaris, Penn., 315.
umbilicatus, L., 294, 312, 314, 316, 379.
undulatus, F. & H., 298.
Vahlii, 294.
varius, L., 312.
Zezyphinus, Ch., 332.
ziczak, Ch., 376.
Ziziphinus, Mont., 332.
Zizyphinus, L., 27, 330, 332, 333, 336, 379.
Zyziphinus, Born, 332.
Trutina, Brown, 24.
solenoides, Brown, 28.
Tubulus antalis, Mont., 194.
TURBINIDÆ, Fl., 292, 337.
Turbo, 341, 343.
auricularis, Mont., 349.
Basterotii, Payr., 364.
cærulescens, Lam., 364.
calcar, Mont., 341.
canalis, Mont., 347.
carinata, Gr., 182.
carneus, Lowe, 299.
castanea, Gm., 341.
crassior, Mont., 344.
dispar, Mont., 376.
dorsalis, Turt., 120.

Turbo (continued).
expansus, Brown, 361.
fabalis, Turt., 357.
fuscus, Müll., 299.
incarnatus, Couth., 300.
inflatus, Tott., 297.
jugosus, Mont., 365.
labiatus, Brown, 366.
Lemani, Delle Ch., 364.
lineatus, Da Costa, 312, 317.
littoralis, Baster, 376.
littoreus, L., 368, 371.
mammillatus, Don., 341.
moniliferus, Nyst, 303.
navalis, Sp., 183.
neritoides, L., 361.
neritoides, Pult., 361.
neritoideus, Ol., 297.
obligatus, Say, 366.
obtusatus, L., 356.
palliatus, Say, 361.
pallidus, Don., 346.
patræus, Mont., 364.
petreus, Fl., 364.
petricola, Leach, 364.
pictus, Da Costa, 341.
pullus, L., 338.
puteolus, Turt., 348, 351.
quadrifasciatus, Mont., 347.
retusus, Lam., 361.
rudis, Mat., 364.
rugosus, L., 341.
sanguineus, L., 305.
saxatilis, Ol., 366.
sulcatus, Leach, 365.
tenebrosus, Mont., 361, 365.
ventricosus, Brown, 366.
vestitus, Say, 366.
vinctus, Mont., 348.
Turbonidæ, Fl., 337.
Turritella, 337.

Uperotus, Guett., 90.

Valvata? striata, Ph., 317.
Velutina lævigata, 275.
VENERUPES, 86.
VENERUPIS, Lam., 85, 86, 87.
decussata, Ph., 88.
Irus, 84, 85, 86, 88, 378.
Lajonkairii, Payr., 88.
Venus, 77, 93, 294.
Bottarii, Ren., 88.
cancellata, L., 88.

Venus (continued).
  cancellata, Ol., 88.
  fluctuosa, 59.
  gallina, 27.
  gallina var. *laminosa*, 27.
  gallina var. *striatula*, 27.
  lamellata, 88.
  sinuosa, Penn., 43.
  striata, Humphr., 88.
Vermetus, 201.

Vitrina, 265.
XYLOPHAGA, Turt., 93, 100, 118, 119, 122.
  dorsalis, 120, 378.
Xylophagus, Gron., 167.

Zirphæa, Leach, 112.
Ziziphinus, Leach, 294, 304, 320.
  alabastrites, Gr., 336.
  vulgaris, Gr., 332.

EXPLANATION OF PLATES.

Frontispiece.
Teredo Norvegica.

Plate I.
Fig. 1. *Solecurtus antiquatus.*
2. *Ceratisolen legumen.*
Fig. 3. *Solen siliqua.*
4. *Pandora inæquivalvis.*

Plate II.
Fig. 1. *Lyonsia Norvegica.*
2. *Thracia papyracea.*
3. *Poromya granulata.*
Fig. 4. *Neæra cuspidata.*
5. *Corbula gibba.*

Plate III.
Fig. 1. *Mya truncata.*
2. *Panopea plicata.*
3. *Saxicava rugosa.*
Fig. 4. *Venerupis Irus.*
5. *Gastrochæna dubia.*

Plate IV.
Fig. 1. *Pholas dactylus.*
1a. *P. parva.*
2. *Pholadidea papyracea.*
Fig. 3. *Xylophaga dorsalis.*
4. *Teredo Norvegica.*

Plate V.
Fig. 1. *Dentalium entalis.*
1a. *D. Tarentinum.*
2. *Chiton fascicularis.*
3. *Patella vulgata.*
Fig. 4. *Helcion pellucidum.*
5. *Tectura virginea.*
6. *Lepeta cæca.*

Plate VI.
Fig. 1. *Propilidium ancyloides.*
2. *Puncturella Noachina.*
3. *Emarginula fissura.*
Fig. 4. *Fissurella Græca.*
5. *Capulus Hungaricus.*
6. *Calyptræa Chinensis.*

Plate VII.
Fig. 1. *Haliotis tuberculata.*
2. *Scissurella crispata.*
Fig. 3. *Cyclostrema serpuloides.*
4. *Trochus zizyphinus.*

Plate VIII.
Fig. 1. *Phasianella pulla.*
2. *Lacuna divaricata.*
Fig. 3. *Littorina litorea.*

END OF VOL. III.

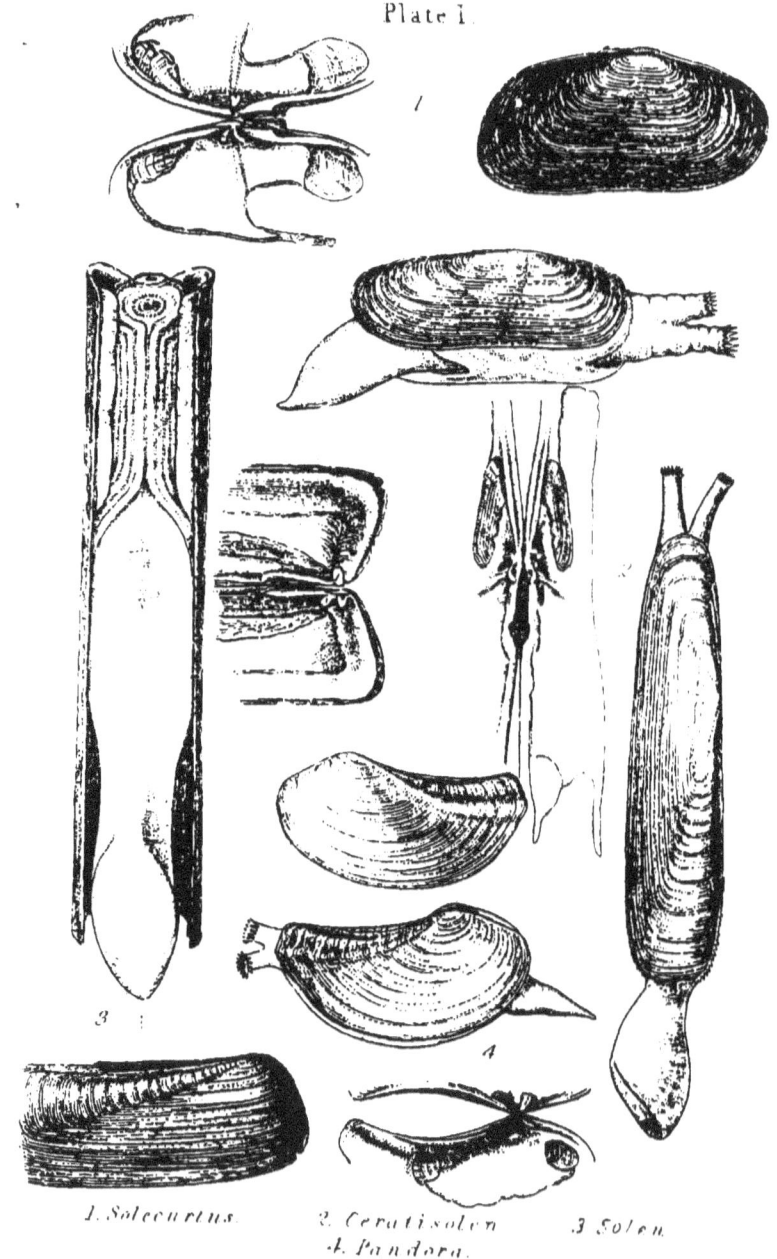

Plate 1.

1. Solecurtus. 2. Ceratisolen. 3. Solen.
4. Pandora.

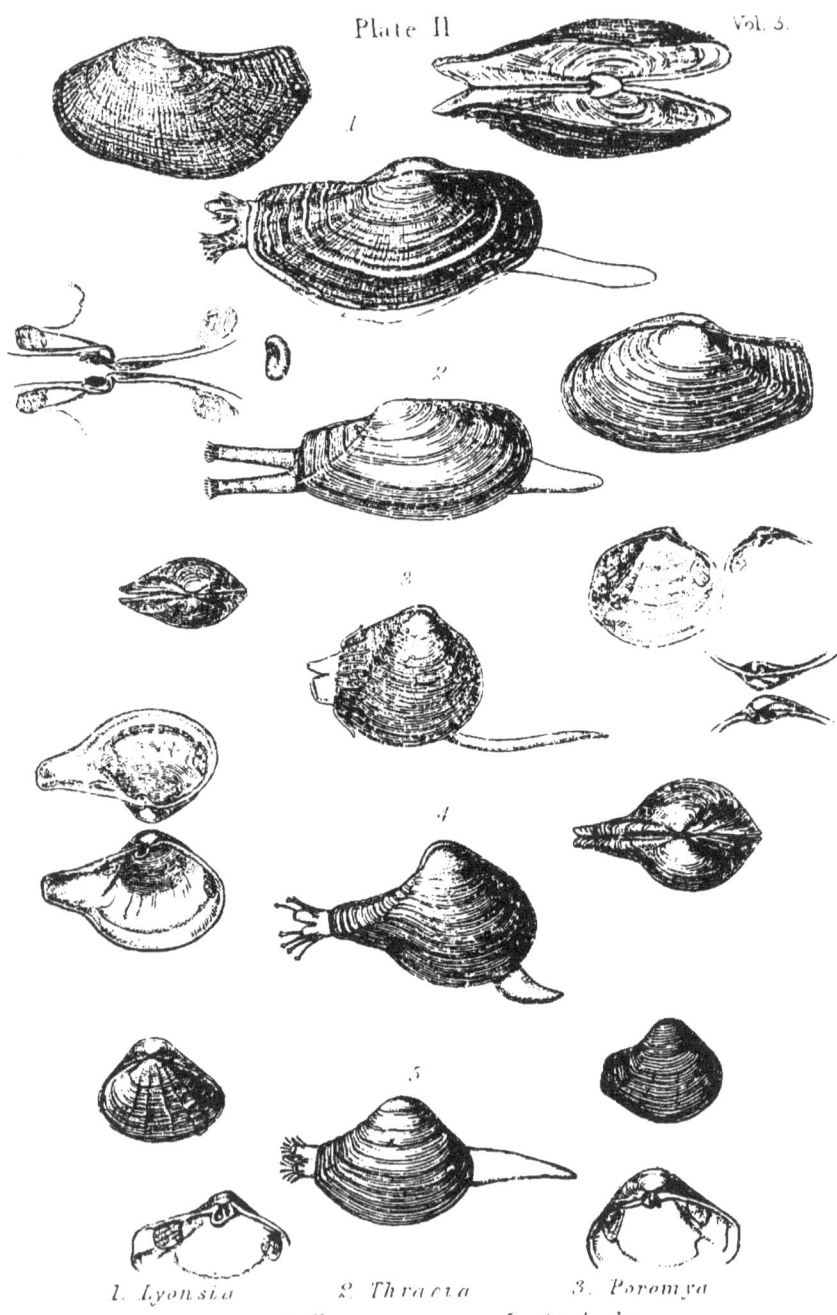

1. Lyonsia 2. Thracia 3. Poromya
4. Neæra 5. Corbula

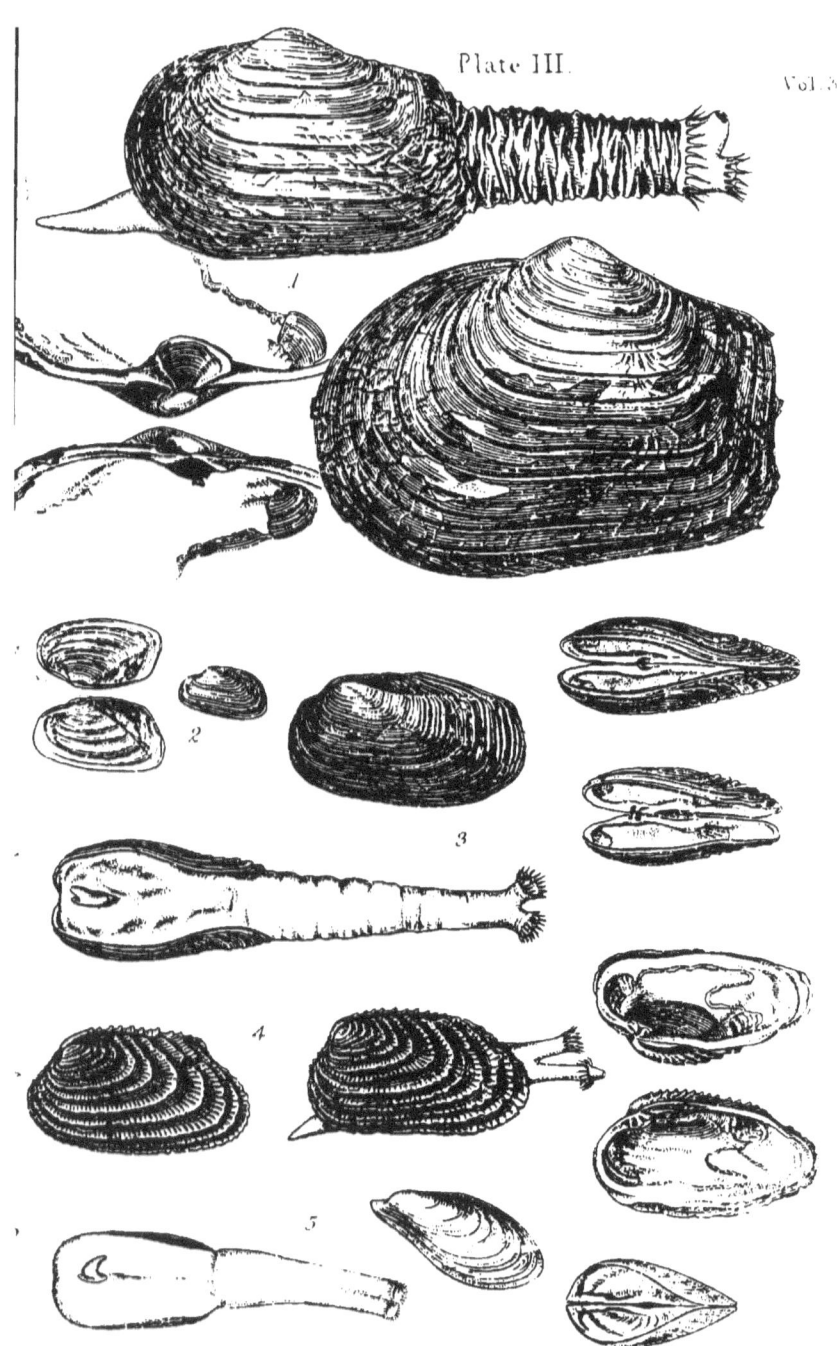

Plate III. Vol. 3

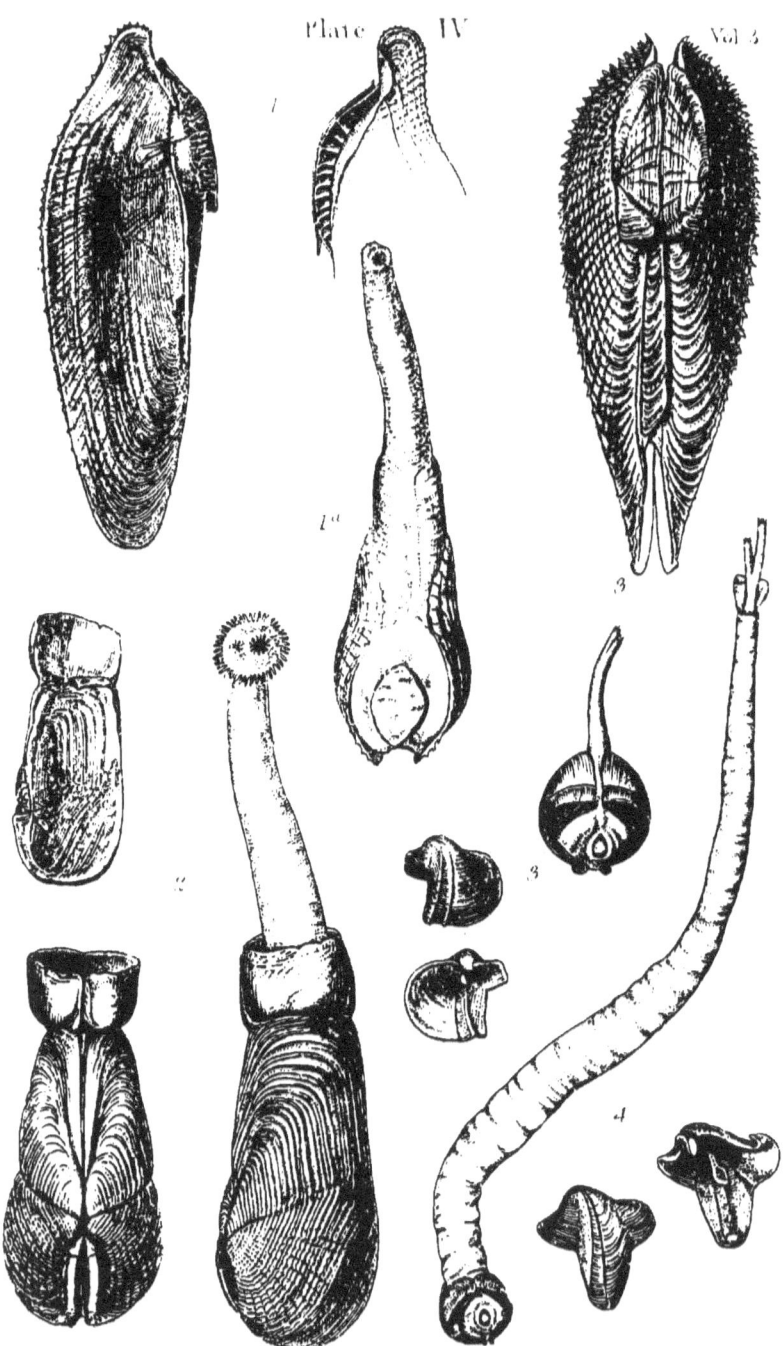

1 Pholas. 2 Pholadidea. 3 Gylophana. 4 Teredo.

Plate V. Vol. 3.

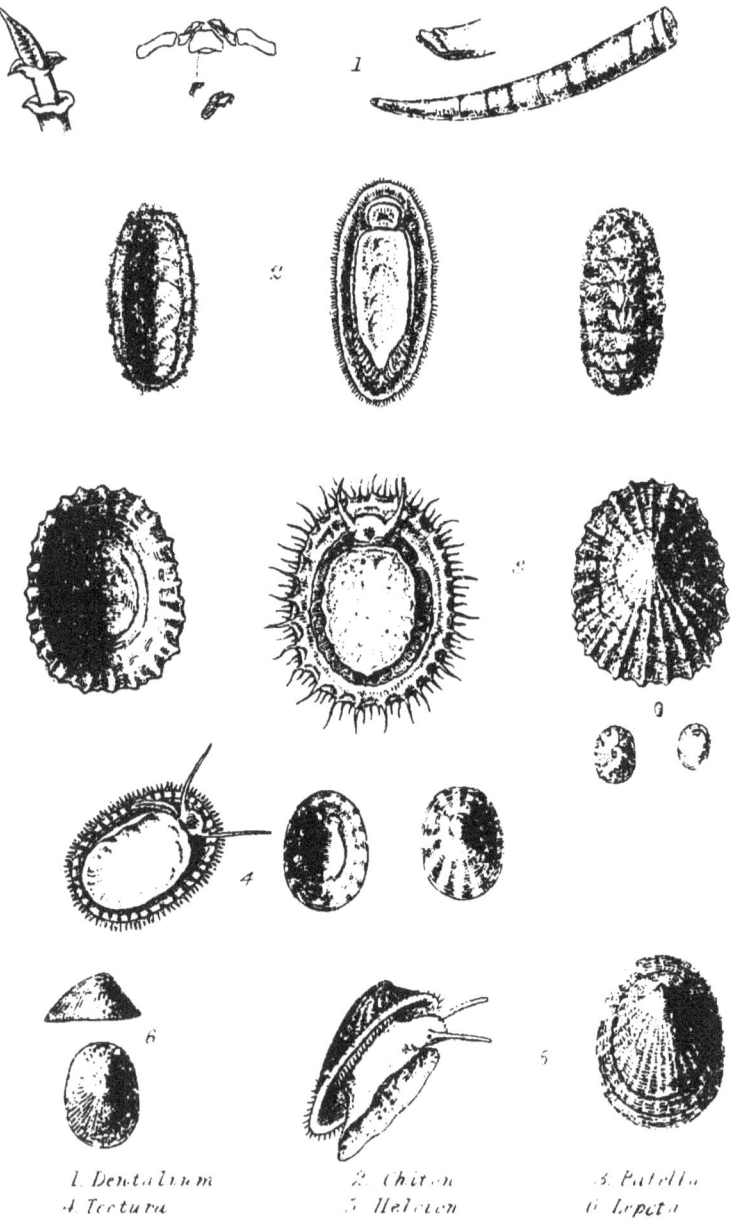

1. Dentalium 2. Chiton 3. Patella
4. Tectura 5. Helcion 6. Lepeta

Plate VI   Vol.3

1. Propilidium   2. Puncturella   3. Emarginula
4. Fissurella   5. Capulus   6. Calyptraea

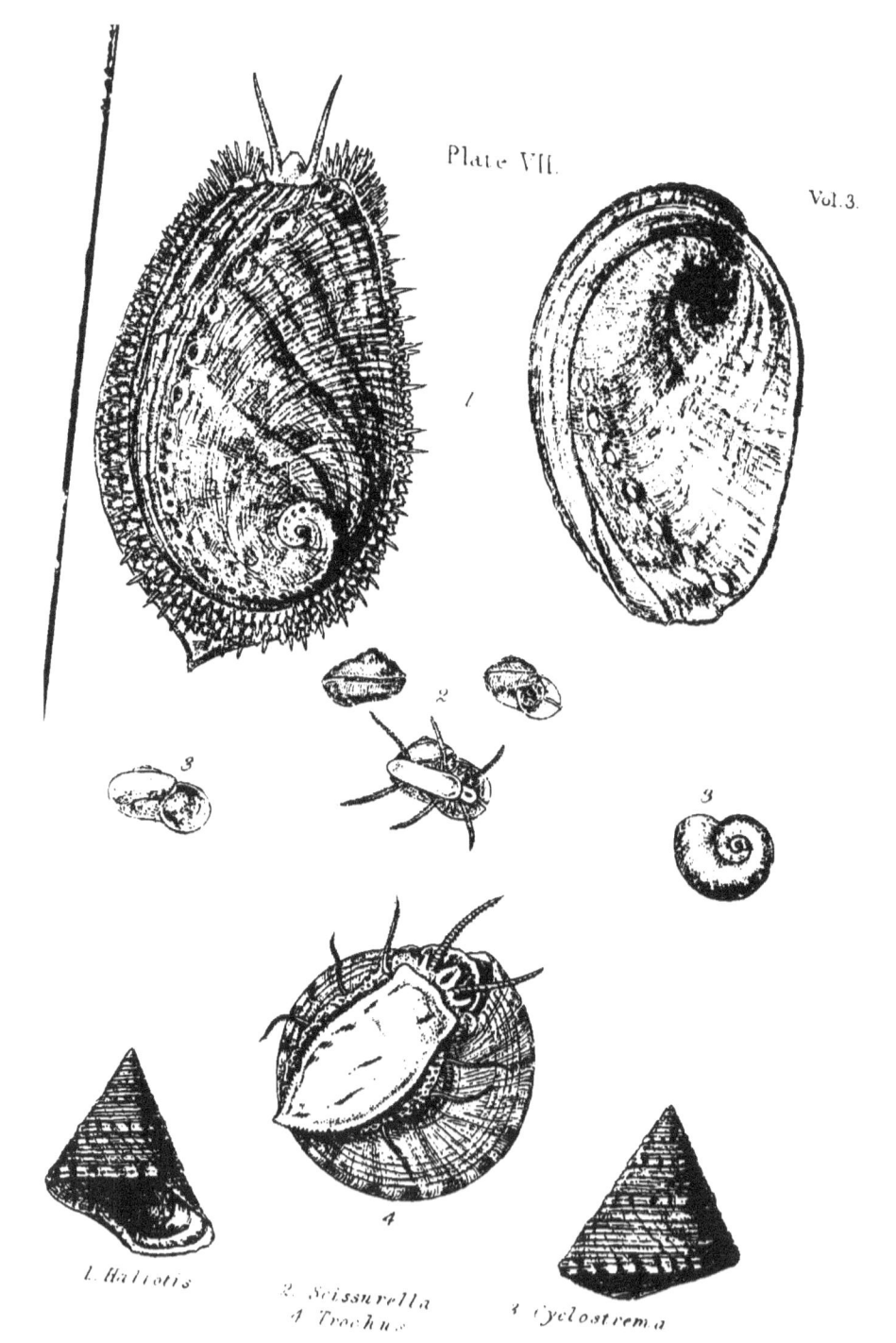

Plate VIII.

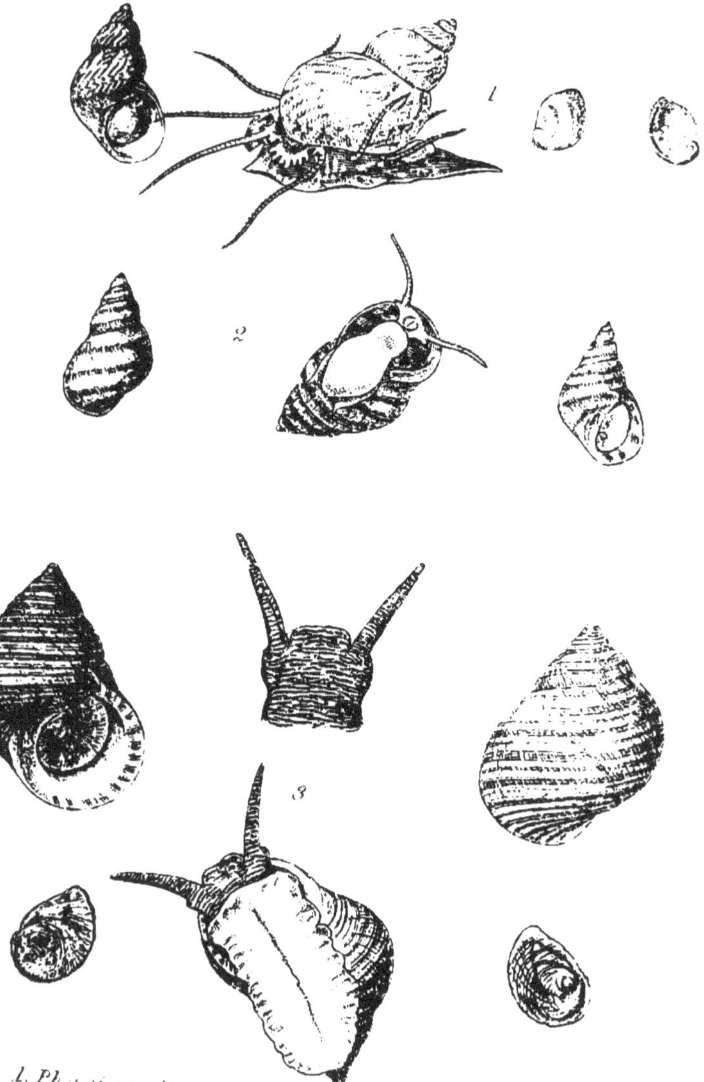

1. Phasianella   2. Lacuna   3. Litorina

www.ingramcontent.com/pod-product-compliance
Lightning Source LLC
Chambersburg PA
CBHW022110290426
44112CB00008B/621